Microbiology

A LABORATORY MANUAL

SEVINTH EDITION

James G. Cappuccino

Natalie Sherman

State University of New York
Rockland Community College

PEARSON
Benjamin Cummings

San Francisco   Boston   New York
Capetown   Hong Kong   London   Madrid   Mexico City
Montreal   Munich   Paris   Singapore   Sydney   Tokyo   Toronto

Publisher: Daryl Fox
Development Manager: Claire Alexander
Sponsoring Editor: Leslie Berriman
Associate Editor: Marie Beaugureau
Managing Editor: Wendy Earl
Production Editor: David Novak

Text Design: Jeanne Calabrese
Cover Design: Yvo Riezebos
Composition: The Left Coast Group, Inc.
Selected Art Rendering: Tara L. Peterson;
    The Left Coast Group; and Shirley Bortoli
Manufacturing Supervisor: Stacey Weinberger
Executive Marketing Manager: Lauren Harp

Color-Plate Photo Credits

Photos 1–5, 7, 13, 16, 17, 22–24, 26–31, 33, 48, 49, 51, 52, 55, 58, 60, 64–68, 70, 74, 79, 80: From *Microbiology: A Photographic Atlas for the Laboratory*, 1e, by Alexander/Strete, © 2001 Benjamin Cummings, an imprint of Addison Wesley Longman, Inc. Reprinted by permission. Photos 8–11, 25, 32, 45–47 © David B. Alexander, University of Portland. Photos 60, 82: Courtesy of the Centers for Disease Control. Photo 53 © Jim Solliday/ Biological Photo Service. Photo 54: © Carolina Biological Supply/Phototake. Photo 56: © 1990, G.W. Willis/Biological Photo Service. Photo 81: © Leon Lebeau/Custom Medical Stock Photography.

The authors and publisher believe that the laboratory experiments described in this publication, when conducted in conformity with the safety precautions described herein and according to the school's laboratory safety procedures, are reasonably safe for the students for whom this manual is directed. Nonetheless, many of the described experiments are accompanied by some degree of risk, including human error, the failure or misuse of laboratory or electrical equipment, mismeasurements, spills of chemicals, and exposure to sharp, objects, heat, bodily fluids, blood, and other biologics. The author and publisher disclaim any liability arising from such risks in the connection with any of the experiments contained in this manual. If students have questions or problems with materials, procedures, or instructions on any experiment, they should always ask their instructor for help before proceeding.

Library of Congress Cataloging-in-Publication Data

Cappuccino, James G.
    Microbiology : a laboratory manual / James G. Cappuccino, Natalie Sherman.—7th ed.
        p. cm.
    Includes index.
    ISBN: 0-0853-2836-X (pbk.)
    1. Microbiology—Laboratory manuals. I. Sherman, Natalie. II. Title.

QR63.C34 2005
579'.078—dc22

2004044548

ISBN 0-8053-2836-X
1 2 3 4 5 6 7 8 9 10—CRS—08 07 06 05 04
www.aw-bc.com

CONTENTS

Appendices

It is with great pride that I dedicate this book
to the memory of Natalie Sherman. She was my
friend, colleague, and coauthor for 32 years.
Her passion for teaching was only exceeded by
her ability to teach her students well. They have
become the beneficiaries of her unique talent.

Microbiology is a dynamic science. It is constantly evolving as more information is added to the continuum of knowledge, and as microbiological techniques are rapidly modified and refined. The seventh edition of *Microbiology: A Laboratory Manual* continues to provide a blend of traditional methodologies with more contemporary procedures to meet the pedagogical needs of all students studying microbiology. As in previous editions, this seventh edition contains a large number of diverse experimental procedures, providing instructors with the flexibility to design a course syllabus that meets their particular instructional approach. For this edition, I have focused on updating the terminology, equipment, and procedural techniques used in the experiments. I also modified and clarified the background information and experimental procedures and revised the color-plate insert. These changes will further facilitate student understanding and performance of microbiological procedures.

The structure of the manual remains essentially the same as in the earlier editions. Comprehensive introductory material is given at the beginning of each major area of study, and specific explanations and detailed directions precede each experiment. This approach augments, enhances, and reinforces course lectures, thereby enabling students to comprehend more readily the concepts and purposes of each experiment. This will be a further asset to those in institutions in which the laboratory and lecture sections are not taught concurrently. Finally, this manual should reduce the time required for explanations at the beginning of each laboratory session and thus make more time available for performing the experiments.

The wide variety of experiments was critically selected and tested to facilitate effective instruction in the basic principles and techniques in a variety of microbiological areas. Thus, this laboratory manual provides a wide spectrum of exercises suitable for use in elementary and advanced general microbiology courses, as well as in allied health programs. Also, the procedures have been carefully designed so that the supplies, equipment, and instrumentation commonly found in undergraduate institutions will suffice for their successful execution.

The manual consists of 79 exercises arranged in 15 parts. The exercises progress from those that are basic and introductory, requiring minimal manipulations, to those that are more complex, requiring more sophisticated skills.

I have created two new experiments for the seventh edition. In Part II, I added an experiment on darkfield microscopy, providing a hands-on introduction to this important means of microbial observation. In Part XIII, I included a bacterial transformation experiment, which addresses the modern biotechnology of DNA transformation and the history of scientific experimentation that has led to our current understanding of it.

> Part I, on **basic laboratory techniques for isolation, cultivation, and cultural characterization of microorganisms,** introduces basic procedures used for isolation and cultivation of microorganisms.

> Part II, on **microscopy,** introduces the use and care of the microscope for the study of microorganisms.

> Part III, on **bacterial staining,** focuses on procedures for bacterial smear preparation, visualization, and differentiation of microorganisms and cell structures.

Part IV focuses on **cultivation of microorganisms, nutritional and physical requirements, and enumeration of microbial populations.**

Part V, on **biochemical activities,** introduces the varied cellular enzymatic activities that may be used for differentiation and identification of specific groups of microorganisms.

Parts VI, VII, and VIII introduce the areas of **protozoology, mycology,** and **virology.**

Part IX, **control of microbial growth,** discusses the antimicrobial activities of various physical and chemical agents.

Parts X and XI are concerned with the sanitary aspects of **food** and **water,** as well as the fermentative role of microorganisms in the production of some beverages and food products.

Part XII, on the **microbiology of soil,** discusses the role of soil microorganisms in the nitrogen cycle and antibiotic production.

Part XIII, on **bacterial genetics,** presents selected experiments to illustrate genetic principles using bacterial systems.

Parts XIV and XV, on **medical microbiology** and **immunology,** highlight both the conventional and the more recent rapid clinical screening methodologies used for the isolation and identification of pathogenic microorganisms. To circumvent the high cost associated with some of the newer experimental methodologies, it is suggested that these procedures be performed as demonstrations.

The format of each exercise is intended to facilitate presentation of the material by the instructor and to maximize the learning experience. To this end, each experiment is designed as follows:

Purpose: Defines the specific principles and/or techniques to be mastered.

Principle: An in-depth discussion of the microbiological concept or technique and the specific experimental procedure.

Materials: To facilitate the preparation of all laboratory sessions, a list of the following materials appears under this heading:

Cultures: These are the selected test organisms that have been chosen to demonstrate effectively the experimental principle or technique under study, as well as their ease of cultivation and maintenance in stock culture. A complete listing of the experimental cultures and prepared slides is presented in Appendix 6.

Media: These are the specific media and their quantities per designated student group. Appendix 3 lists the composition and method of preparation of all the media used in this manual.

Reagents: These include biological stains as well as test reagents. The chemical composition and preparation of the reagents are presented in Appendices 4 and 5.

Equipment: Listed under this heading are the supplies and instrumentation that are needed during the laboratory session. The suggested equipment was selected to minimize expense while reflecting current laboratory technique.

Procedure: Explicit instructions augmented by diagrams aid in the execution and interpretation of the experiments.

Observations and Results: Tear-out sheets located at the end of each exercise facilitate interpretation of data and subsequent review by the instructor.

Review Questions: Questions on tear-out report sheets aid the instructor in determining the student's ability to understand the experimental concepts and techniques. Questions that call for more critical thinking are indicated by the symbol shown to the left.

A caution icon has been placed at the beginning of experiments utilizing procedures that may use **potentially pathogenic materials.** The instructor may wish to perform some of these experiments as demonstrations.

Safety precautions that should be followed during procedures appear throughout the manual. In this edition they have been highlighted as boxes.

I hope that this manual will serve as a vehicle for the development of manipulative skills and techniques essential for understanding the integrated complexity of the biochemical structure and function of the single cell. This will enable an extension of these principles toward a better understanding of the more complex, higher forms of life. Ultimately, I hope that some students might further pursue the study of life at the molecular level or apply these laboratory skills in the vocational fields of applied microbiology and allied health.

Instructor Resources

The *Instructor's Guide* (0-8053-2837-8) has been completely updated for the Seventh Edition, and contains a new Tips section and added tables of media and equipment. The new Instructor's CD-ROM, just added for this edition provides adopters with over 300 photographs of microbiological culture slides, plates, and test tubes.

Acknowledgments

I wish to express my sincere gratitude to the following instructors for their reviews of the sixth edition and/or the two new experiments I have added. Their comments and direction contributed greatly to the seventh edition.

Bernard Arulanandam, University of Texas at San Antonio; Sheila Brady-Root, Nazareth College; Beverly J. Brown, Nazareth College; John Chikwem, Lincoln University; Michael A. Davis, Central Connecticut State University; Ernest M. Hannig, University of Texas at Dallas; Kirkwood M. Land, City College of San Francisco; Sue Looney, University of Alaska Anchorage; S. Jane A. Molinaro, Immaculata University; Tim Mullican, Dakota Wesleyan University; Charles B. Pumpuni, Northern Virginia Community College; Terrill Smith, City College of San Francisco; Curt W. Spanis, University of San Diego; Amy Treonis, Creighton University.

The new edition has also benefited from the contribution of Dr. David B. Alexander at the University of Portland, who provided us with permission to use his fine photos in the color insert.

Likewise, I wish to extend my appreciation to the staff at Benjamin Cummings, whose expertise and technical skills have guided Natalie and me over the many years. David Novak, Production Editor, and Marie Beaugureau, Associate Editor, provided invaluable direction in this new edition; their dedication to the highest standards has left its mark on every page of the manual. Last, but certainly not least, I wish to express my gratitude to the microbiology laboratory technicians at Rockland Community College—Ms. Joan Grace, who early on performed all the experiments to ensure their success when repeated by the students, and Ms. Roz Wehrman, who is presently following in Joan's footsteps.

James G. Cappuccino

General Rules and Regulations

A rewarding laboratory experience demands strict adherence to prescribed rules for personal and environmental safety. The former reflects concern for your personal safety in terms of avoiding laboratory accidents. The latter requires that you maintain a scrupulously clean laboratory setting to prevent contamination of experimental procedures by microorganisms from exogenous sources.

Because most microbiological laboratory procedures require the use of living organisms, an integral part of all laboratory sessions is the use of aseptic techniques. Although the virulence of microorganisms used in the academic laboratory environment has been greatly diminished because of their long-term maintenance on artificial media, *all microorganisms should be treated as potential pathogens* (organisms capable of producing disease). Thus, microbiology students must develop aseptic techniques (free of contaminating organisms) in the preparation of pure cultures that are essential in the industrial and clinical marketplaces.

The following basic steps should be observed at all times to reduce the ever-present microbial flora of the laboratory environment.

1. Upon entering the laboratory, place coats, books, and other paraphernalia in specified locations—never on bench tops.

2. Keep doors and windows closed during the laboratory session to prevent contamination from air currents.

3. At the beginning and termination of each laboratory session, wipe bench tops with a disinfectant solution provided by the instructor.

4. Do not place contaminated instruments, such as inoculating loops, needles, and pipettes, on bench tops. Loops and needles should be sterilized by incineration, and pipettes should be disposed of in designated receptacles.

5. On completion of the laboratory session, place all cultures and materials in the disposal area as designated by the instructor.

6. Rapid and efficient manipulation of fungal cultures is required to prevent the dissemination of their reproductive spores in the laboratory environment.

To prevent accidental injury and infection of yourself and others, observe the following regulations at all times:

1. Wash your hands with liquid detergent, rinse with 95% ethyl alcohol, and dry them with paper towels upon entering and prior to leaving the laboratory.

2. Wear a paper cap or tie back long hair to minimize its exposure to open flames.

3. Wear a laboratory coat or apron while working in the laboratory to protect clothing from contamination or accidental discoloration by staining solutions.

4. Wear closed shoes at all times in the laboratory setting.

5. Never apply cosmetics or insert contact lenses in the laboratory.

6. Do not smoke, eat, or drink in the laboratory. These activities are absolutely prohibited.

7. Carry cultures in a test tube rack when moving around the laboratory. Likewise, keep cultures in a test tube rack on the bench tops when not in use. This serves a dual purpose: to prevent accidents and to avoid contamination of yourself and the environment.

8. Never remove media, equipment, or especially, *bacterial cultures* from the laboratory. Doing so is absolutely prohibited.

9. Immediately cover spilled cultures or broken culture tubes with paper towels and then saturate them with disinfectant solution. After 15 minutes of reaction time,

remove the towels and dispose of them in a manner indicated by the instructor.

10. Report accidental cuts or burns to the instructor immediately.

11. Never pipette by mouth any broth cultures or chemical reagents. Doing so is strictly prohibited. Pipetting is to be carried out with the aid of a mechanical pipetting device only.

12. Do not lick labels. Use only self-stick labels for the identification of experimental cultures.

13. Speak quietly and avoid unnecessary movement around the laboratory to prevent distractions that may cause accidents.

14. Always wear gloves when irradiating cultures.

The following specific precautions must be observed when handling body fluids of unknown origin due to the possible imminent transmission of the HIV and hepatitis B viruses in these test specimens.

1. Wear disposable gloves during the manipulation of test materials such as blood, serum, and other body fluids.

2. Immediately wash hands if contact with any of these fluids occurs and also upon removal of the gloves.

3. Wear masks, safety goggles, and laboratory coats if an aerosol might be formed or splattering of these fluids is likely to occur.

4. Decontaminate spilled body fluids with a 1:10 dilution of household bleach, covered with paper toweling, and allowed to react for 10 minutes before removal.

5. Place test specimens and supplies in contact with these fluids into a container of disinfectant prior to autoclaving.

I have read the above laboratory safety rules and regulations and agree to abide by them.

Name _____   Date _____

Student Preparation for Laboratory Sessions

The efficient performance of laboratory exercises mandates that you attend each session fully prepared to execute the required procedures. Read the assigned experimental protocols to effectively plan and organize the related activities. This will allow you to maximize use of laboratory time.

Preparation of Experimental Materials

Microscope Slides: Meticulously clean slides are essential for microscopic work. Commercially precleaned slides should be used for each microscopic slide preparation. However, wipe these slides with dry lens paper to remove dust and finger marks prior to their use. With a glassware marking pencil, label one end of each slide with the abbreviated name of the organism to be viewed.

Labeling of Culture Vessels: Generally, microbiological experiments require the use of a number of different test organisms and a variety of culture media. To ensure the successful completion of experiments, organize all experimental cultures and sterile media at the start of each experiment. Label culture vessels with non-water-soluble glassware markers and/or self-stick labels prior to their inoculation. The labeling on each of the experimental vessels should include the name of the test organism, the name of the medium, the dilution of sample (if any), your name or initials, and the date. *Place labeling directly below the cap of the culture tube.* When labeling Petri dish cultures, only the name of the organism(s) should be written on the bottom of the plate, close to its periphery, to prevent obscuring observation of the results. The additional information for the identification of the culture should be written on the cover of the Petri dish.

Inoculation Procedures

Aseptic techniques for the transfer or isolation of microorganisms, using the necessary transfer instruments, are described fully in the experiments in Part I of the manual. Technical skill will be acquired through repetitive practice.

Inoculating Loops and Needles: It is imperative that you incinerate the entire wire to ensure absolute sterilization. The shaft should also be briefly passed through the flame to remove any dust or possible contaminants. To avoid killing the cells and splattering the culture, cool the inoculating wire by tapping the inner surface of the culture tube or the Petri dish cover prior to obtaining the inoculum, or touch the edge of the medium in the plate.

When performing an aseptic transfer of microorganisms, a minute amount of inoculum is required. If an agar culture is used, touch only a single area of growth with the inoculating wire to obtain the inoculum. *Never drag the loop or needle over the entire surface, and take care not to dig into the solid medium.* If a broth medium is used, first tap the bottom of the tube against the palm of your hand to suspend the microorganisms. *Caution:* Do not tap the culture vigorously as this may cause spills or excessive foaming of the culture, which may denature the proteins in the medium.

Pipettes: Use only sterile, disposable pipettes or glass pipettes sterilized in a canister. The practice of *pipetting by mouth has been discontinued* to eliminate the possibility of autoinfection by accidentally imbibing the culture or infectious body fluids. Instead, a mechanical pipetting device is to be used to obtain and deliver the material to be inoculated.

Incubation Procedure

Microorganisms exhibit a wide temperature range for growth. However, for most used in this manual, optimum growth occurs at 37°C over a period of 18 to 24 hours. Unless otherwise indicated in specific exercises, incubate all cultures under the conditions cited above. Place culture tubes in a rack for incubation. Petri dishes may be stacked; however, they *must always be incubated in an inverted position (top down)* to prevent water condensation from dropping onto the surface of the culture medium. This resultant excess moisture may then serve as a vehicle for the spread of the microorganisms on the surface of the culture medium, thereby producing confluent rather than discrete microbial growth.

Procedure for Recording Observations and Results

The accurate accumulation of experimental data is essential for the critical interpretation of the observations upon which the final results will be based. To achieve this end, it is imperative that you complete all the preparatory readings that are necessary for your understanding of the basic principles underlying each experiment. Meticulously record all the observed data in the "Observations and Results" section of each experiment.

In the exercises that require drawings to illustrate microbial morphology, it will be advantageous to depict shapes, arrangements, and cellular structures enlarged to 5 to 10 times their actual microscopic size, as indicated by the following illustrations. For this purpose a number 2 pencil is preferable. Stippling may be used to depict different aspects of cell structure (e.g., endospores or differences in staining density).

Review Questions

The review questions are designed to evaluate student's understanding of the principles and the interpretations of observations in each experiment. Completion of these questions will also serve to reinforce many of the concepts that are discussed in the lectures. At times, this will require the use of ancillary sources such as textbooks, microbiological reviews, or abstracts. The designated critical-thinking questions are designed to stimulate further refinement of cognitive skills.

Procedure for Termination of Laboratory Sessions

1. Return all equipment, supplies, and chemical reagents to their original locations.

2. Neatly place all capped test tube cultures and closed Petri dishes in a designated collection area in the laboratory for subsequent autoclaving.

3. Place contaminated materials, such as swabs, disposable pipettes, and paper towels, in a biohazard receptacle prior to autoclaving.

4. Carefully place hazardous biochemicals, such as potential carcinogens, into a sealed container and store in a fume hood prior to their disposal according to the institutional policy.

5. Wipe down table top with recommended disinfectant.

6. Wash hands before leaving the laboratory.

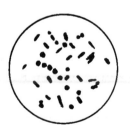
Poor drawing

Good drawing

Basic Laboratory Techniques for Isolation, Cultivation, and Cultural Characterization of Microorganisms

LEARNING OBJECTIVES

Once you have completed the experiments in this section, you should be familiar with

1. The types of laboratory equipment and culture media needed to develop and maintain pure cultures.

2. The concept of sterility and the procedures necessary for successful subculturing of microorganisms.

3. Streak-plate and spread-plate inoculation of microorganisms in a mixed microbial population for subsequent pure culture isolation.

4. Cultural and morphological characteristics of microorganisms grown in pure culture.

INTRODUCTION

Microorganisms are ubiquitous. They are found in soil, air, water, food, sewage, and on body surfaces. In short, every area of our environment is replete with them. The microbiologist separates these mixed populations into individual species for study. A culture containing a single unadulterated species of cells is called a **pure culture.** To isolate and study microorganisms in pure culture, the microbiologist requires basic laboratory apparatus and the application of specific techniques, as illustrated in Figure I.1.

Media

The survival and continued growth of microorganisms depend on an adequate supply of nutrients and a favorable growth environment. For the former, most microbes must use soluble low-molecular-weight substances that are frequently derived from the enzymatic degradation of complex nutrients. A solution containing these nutrients is a **culture medium.** Basically, all culture media are liquid, semi-solid, or solid. A liquid medium lacks a solidifying agent and is called a **broth medium.** A broth medium supplemented with a solidifying agent called **agar** results in a solid or semi-solid medium. Agar is an extract of seaweed, a complex carbohydrate composed mainly of galactose, and is without nutritional value. Agar serves as an excellent solidifying agent because it liquefies at 100°C and solidifies at 40°C. Because of these properties, organisms, especially pathogens, can be cultivated at temperatures of 37.5°C or slightly higher without fear of the medium liquefying. A completely solid medium requires an agar concentration of about 1.5 to 1.8%. A concentration of less than 1% agar results in a **semisolid medium.** A solid medium has the advantage that it presents a hardened surface on which microorganisms can be grown using specialized techniques for the isolation of discrete colonies. Each colony is a cluster of cells that originates from the multiplication of a single cell and represents the growth of a single species of microorganism. Such a defined and well-isolated colony is a **pure culture.** Also, while

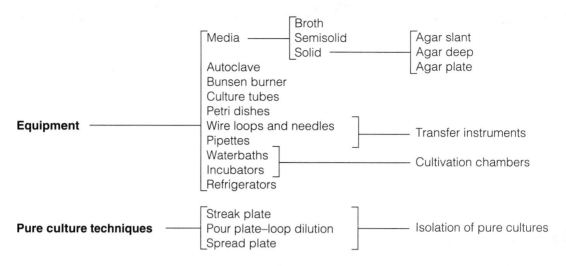

FIGURE I.1 Laboratory apparatus and culture techniques

in the liquefied state, solid media can be placed in test tubes, which are then allowed to cool and harden in a slanted position, producing **agar slants.** These are useful for maintaining pure cultures. Similar tubes that, following preparation, are allowed to harden in the upright position are designated as **agar deep tubes.** Agar deep tubes are used primarily for the study of the gaseous requirements of microorganisms. However, they may be liquefied in a boiling water bath and poured into Petri dishes, producing **agar plates,** which provide large surface areas for the isolation and study of microorganisms. The various forms of solid media are illustrated in Figure I.2.

In addition to nutritional needs, the environmental factors must also be regulated, including proper pH, temperature, gaseous requirements, and osmotic pressure. A more detailed explanation is presented in Part IV, which deals with cultivation of microorganisms; for now, you should simply bear in mind that numerous types of media are available.

Sterilization

Sterility is the hallmark of successful work in the microbiology laboratory. To achieve sterility, it is mandatory that you use sterile equipment and sterile techniques. **Sterilization** is the process of rendering a medium or material free of all forms of life. Although a more detailed discussion is presented in Part IX, which describes the control of microorganisms, Figure I.3 is a brief outline of the routine techniques used in the microbiology laboratory.

Culture Tubes and Petri Dishes

Glass **test tubes** and glass or plastic **Petri dishes** are used to cultivate microorganisms. A suitable nutrient medium in the form of broth or agar may be added to the tubes, while only a solid medium is used in Petri dishes. A sterile environment is maintained in culture tubes by various types of closures. Historically, the first type, a cotton plug, was developed by Schröeder and von Dusch in the nineteenth century. Today most laboratories use sleevelike caps (Morton closures) made of metal, such as stainless steel, or heat-resistant plastics. The advantage of these closures over the cotton plug is that they are labor-saving and, most of all, slip on and off the test tubes easily.

Petri dishes provide a larger surface area for growth and cultivation. They consist of a bottom dish portion that contains the medium and a larger top portion that serves as a loose cover. Petri dishes are manufactured in various sizes to meet different experimental requirements. For routine purposes, dishes approximately 15 cm in diameter are used. The sterile agar medium is dispensed to previously sterilized dishes from molten agar deep tubes containing 15 to 20 ml of medium, or from a molten sterile medium prepared in bulk and contained in 250-, 500-, and 1000-ml flasks, depending on the volume of medium required. When cooled to 40°C, the medium will solidify. Remember that *after inoculation, Petri dishes are incubated in an inverted position* (top down) to prevent condensation that forms on the cover during

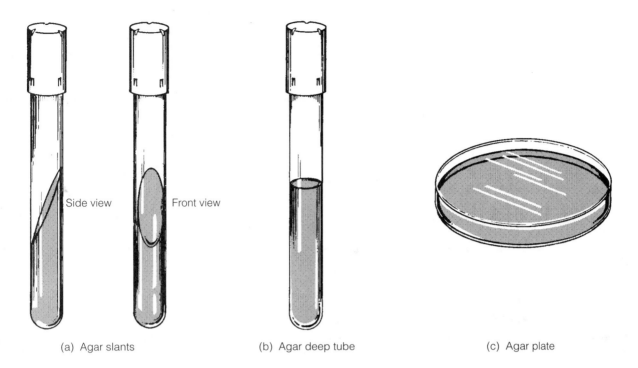

(a) Agar slants   (b) Agar deep tube   (c) Agar plate

FIGURE I.2 Forms of solid (agar) media

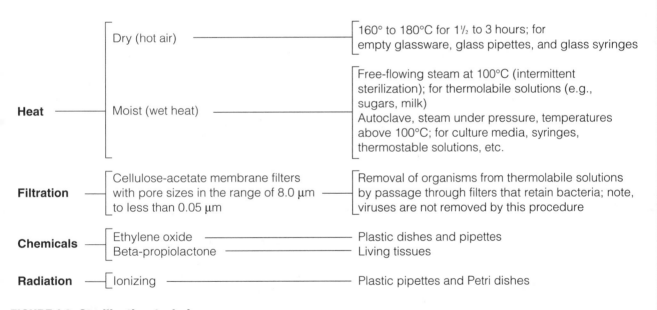

Heat	Dry (hot air)		160° to 180°C for 1½ to 3 hours; for empty glassware, glass pipettes, and glass syringes	
	Moist (wet heat)		Free-flowing steam at 100°C (intermittent sterilization); for thermolabile solutions (e.g., sugars, milk) Autoclave, steam under pressure, temperatures above 100°C; for culture media, syringes, thermostable solutions, etc.	
Filtration	Cellulose-acetate membrane filters with pore sizes in the range of 8.0 µm to less than 0.05 µm		Removal of organisms from thermolabile solutions by passage through filters that retain bacteria; note, viruses are not removed by this procedure	
Chemicals	Ethylene oxide		Plastic dishes and pipettes	
	Beta-propiolactone		Living tissues	
Radiation	Ionizing		Plastic pipettes and Petri dishes	

FIGURE I.3 Sterilization techniques

solidification from dropping down onto the surface of the hardened agar. Figure I.4 illustrates some of the culture vessels used in the laboratory. Built-in ridges on tube closures and Petri dishes provide small gaps necessary for the exchange of air.

Transfer Instruments

Microorganisms must be transferred from one vessel to another or from stock cultures to various media for maintenance and study. Such a transfer is called **subculturing** and must be carried out under sterile conditions to prevent possible contamination.

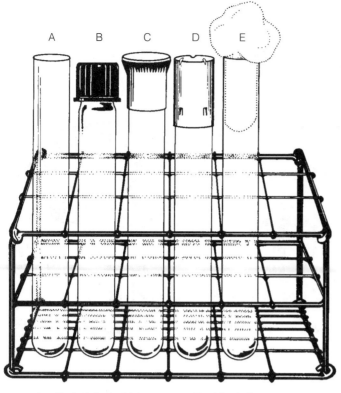

A. Bacteriological tube
B. Screw cap
C. Plastic closure
D. Metal closure
E. Nonabsorbent cotton

(a) Test tube rack with tubes showing various closures

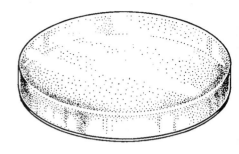

(b) Petri dish

(c) DeLong® shaker flask with closure

FIGURE I.4 Culture vessels

Wire loops and needles are made from inert metals such as nichrome or platinum and are inserted into metal shafts that serve as handles. They are extremely durable instruments and are easily sterilized by incineration in the blue (hottest) portion of the Bunsen burner flame.

A **pipette** is another instrument used for sterile transfers. Pipettes are similar in function to straws; that is, they draw up liquids. They are made of glass or plastic drawn out to a tip at one end and with a mouthpiece forming the other end. They are calibrated to deliver different volumes depending on requirements. Pipettes may be sterilized in bulk inside canisters, or they may be wrapped individually in brown paper and sterilized in an autoclave or dry-heat oven.

Figure I.5 illustrates these transfer instruments. The proper procedure for the use of pipettes will be demonstrated by your instructor.

⚠ **Pipetting by mouth is not permissible! Pipetting is to be performed with the aid of mechanical devices.**

Cultivation Chambers

The specific temperature requirements for growth are discussed in detail in Part IV. However, a prime requirement for the cultivation of microorganisms is that they be grown at their optimum temperature. An incubator is used to maintain optimum temperature during the necessary growth period. It resembles an oven and is thermostatically controlled so that temperature can be varied depending on the requirements of specific microorganisms. Most incubators use dry heat. Moisture is supplied by placing a beaker of water in the incubator during the growth period. A moist environment retards dehydration of the medium and thereby avoids spurious experimental results.

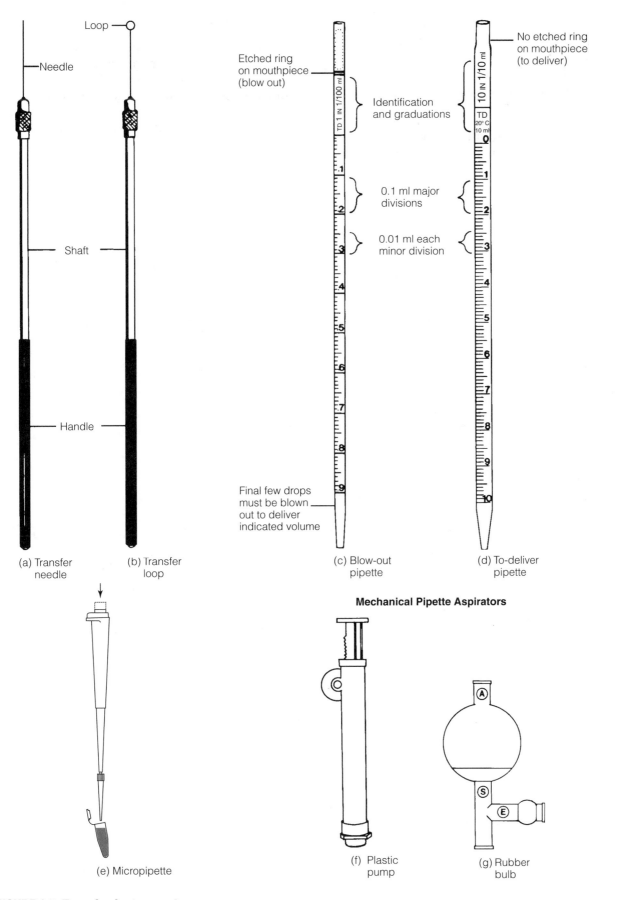

Loop

Needle

Etched ring
on mouthpiece
(blow out)

No etched ring
on mouthpiece
(to deliver)

Identification
and graduations

Shaft

0.1 ml major
divisions

0.01 ml each
minor division

Handle

Final few drops
must be blown
out to deliver
indicated volume

(a) Transfer
needle

(b) Transfer
loop

(c) Blow-out
pipette

(d) To-deliver
pipette

Mechanical Pipette Aspirators

(e) Micropipette

(f) Plastic
pump

(g) Rubber
bulb

FIGURE I.5 Transfer instruments

A thermostatically controlled **shaking waterbath** is another piece of apparatus used to cultivate microorganisms. Its advantage is that it provides a rapid and uniform transfer of heat to the culture vessel, and its agitation provides increased aeration, resulting in acceleration of growth. The single disadvantage of this instrument is that it can be used only for cultivation of organisms in a broth medium.

Refrigerator

A refrigerator is used for a wide variety of purposes such as maintenance and storage of stock cultures between subculturing periods and storage of sterile media to prevent dehydration. It is also used as a repository for thermolabile solutions, antibiotics, serums, and biochemical reagents.

Culture Transfer Techniques

LEARNING OBJECTIVES

Once you have completed this experiment, you should be able to

1. Carry out the technique for aseptic removal and transfer of microorganisms for subculturing.
2. Correctly sterilize inoculating instruments in the flame of a Bunsen burner.
3. Correctly manipulate your fingers to remove and replace the test tube closure.

PRINCIPLE

Microorganisms are transferred from one medium to another by **subculturing.** This technique is of basic importance and is used routinely in preparing and maintaining stock cultures, as well as in microbiological test procedures.

Microorganisms are always present in the air and on laboratory surfaces, benches, and equipment. They can serve as a source of external contamination and thus interfere with experimental results unless proper techniques are used during subculturing. Described below are essential steps that you must follow for aseptic transfer of microorganisms. The complete procedure is illustrated in Figure 1.1.

1. An inoculating needle or loop must always be sterilized by holding it in the hottest portion of the Bunsen burner flame, the inner blue cone, until the entire wire becomes red hot. Then the upper portion of the handle is rapidly passed through the flame. Once flamed, the loop is never put down but is held in the hand and allowed to cool for 10 to 20 seconds. The stock culture tube and the tube to be inoculated are held in the palm of the other hand and secured with the thumb. The two tubes are then separated to form a V in the hand.

2. The tubes are uncapped by grasping the first cap with the little finger and the second cap with the next finger and lifting the closures upward. *Note: Once removed, these caps must be kept in the hand that holds the sterile inoculating loop or needle, thus the inner aspects of the caps point away from the palm of the hand.* They must never be placed on the laboratory bench because doing so would compromise the sterile procedure. Following removal of the closures, the necks of the tubes are briefly passed through the flame and the sterile transfer instrument is further cooled by touching it to the sterile inside wall of the culture tube before removing a small sample of inoculum.

3. Depending on the culture medium, a loop or needle is used for removal of the inoculum. Loops are commonly used to obtain a sample from a broth culture. Either instrument can be used to obtain the inoculum from an agar slant culture by carefully touching the surface of the solid medium in an area exhibiting growth so as not to gouge into the agar. A straight needle is always used when transferring microorganisms to an agar deep tube from both solid and liquid cultures.

4. The cell-laden loop or needle is inserted into the subculture tube. In the case of a broth medium, the loop or needle is shaken slightly to dislodge the organisms; with an agar slant medium, it is drawn lightly over the hardened surface in a straight or zigzag line. For inoculation of an agar deep tube, a straight needle is inserted to the bottom of the tube in a straight line and rapidly withdrawn along the line of insertion. This is a stab inoculation.

5. Following inoculation, the instrument is removed, the necks of the tubes are

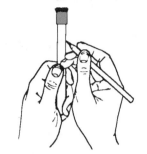

(a) Label the tube to be inoculated with the name of the organism and your initials.

(b) Place the tubes in the palm of your hand, secure with your thumb, and separate to form a V.

(c) Flame the needle or loop until the entire wire is red.

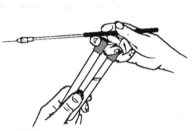

(d) With the sterile loop or needle in hand, uncap the tubes.

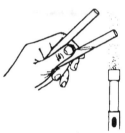

(e) Flame the necks of the tubes by rapidly passing them through the flame once.

(f) **Slant-to-broth transfer:** Dislodge inoculum by slight agitation. **Broth-to-slant transfer:** Following insertion to base of slant, withdraw the loop in a zigzag motion. **Slant-to-agar deep transfer:** Insert the needle to the bottom of the tube and withdraw along the line of insertion.

(g) Flame the necks of the tubes by rapidly passing them through the flame.

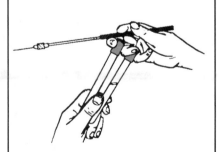

(h) Recap the tubes.

(i) Reflame the loop or needle.

FIGURE 1.1 Subculturing procedure

reflamed, and the caps are replaced on the same tube from which they were removed.

6. The needle or loop is again flamed to destroy remaining organisms.

In this experiment you will master the manipulations required for aseptic transfer of microorganisms in broth-to-slant, slant-to-broth, and slant-to-agar deep transfers. The technique for transfer to and from agar plates is discussed in Experiment 2.

MATERIALS
Cultures

24-hour nutrient broth and nutrient agar slant cultures of *Serratia marcescens*.

Media

Per designated student group: one nutrient broth, one nutrient agar slant, and one nutrient agar deep tube.

Equipment

Bunsen burner, inoculating loop and needle, and glassware marking pencil.

PROCEDURE

1. Label all tubes of sterile media as described in the Laboratory Protocol section on page xv.
2. Following the procedure outlined and illustrated above, perform the following transfers:
   a. *S. marcescens* broth culture to a nutrient agar slant, nutrient agar deep tube, and nutrient broth.
   b. *S. marcescens* agar slant culture to a nutrient broth, nutrient agar slant, and nutrient agar deep tube.
3. Incubate all cultures at 25°C for 24 to 48 hours.

Experiment 1 / Observations and Results

Name ____ Date _______________

1. Examine all cultures for the appearance of growth, which is indicated by turbidity in the broth culture and the appearance of an orange-red growth on the surface of the slant and along the line of inoculation in the agar deep tube.
2. Record your observations in the chart below.

	Nutrient Broth	Nutrient Agar Slant	Nutrient Agar Deep
Growth (+) or (−) Orange-red pigmentation (+) or (−)	________ ________	________ ________	________ ________
Draw the distribution of growth.	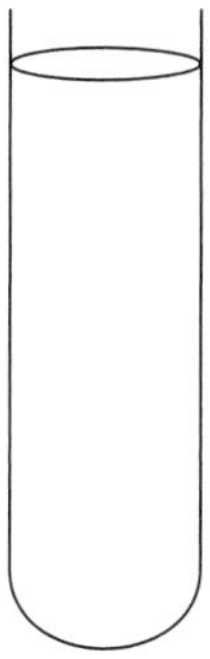	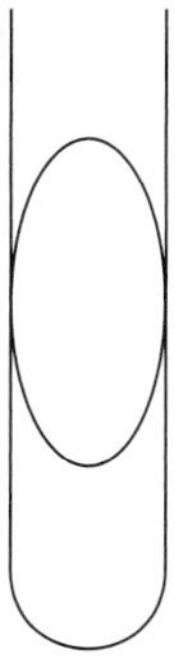	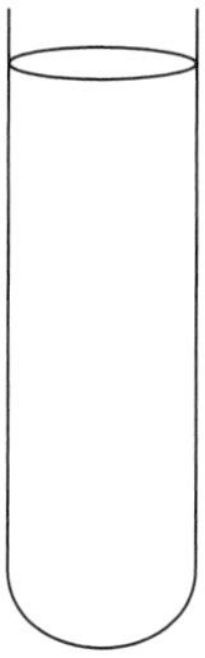

REVIEW QUESTIONS

1. Explain why the following steps are essential during subculturing:
   a. Flaming the inoculating instrument *prior to and after* each inoculation.

   b. Holding the test tube caps in the hand as illustrated in Figure 1.1 on page 8.

   c. Cooling the inoculating instrument prior to obtaining the inoculum.

d. Flaming the neck of the tubes immediately after uncapping and before recapping.

2. Describe the purposes of the subculturing procedure.

3. Explain why a straight inoculating needle is used to inoculate an agar deep tube.

4. There is a lack of orange-red pigmentation in some of the growth on your agar slant labeled *S. marcescens*. Does this necessarily indicate the presence of a contaminant? Explain.

5. Upon observation of the nutrient agar slant culture, you strongly suspect that the culture is contaminated. Outline the method you would follow to ascertain whether your suspicion is justified.

Techniques for Isolation of Pure Cultures

In nature, microbial populations do not segregate themselves by species but exist with a mixture of many other cell types. In the laboratory, these populations can be separated into **pure cultures.** These cultures contain only one type of organism and are suitable for the study of their cultural, morphological, and biochemical properties.

In this experiment, you will first use one of the techniques designed to produce discrete colonies. Colonies are individual, macroscopically visible masses of microbial growth on a solid medium surface, each representing the multiplication of a single organism. Once you have obtained these discrete colonies, you will make an aseptic transfer onto nutrient agar slants for the isolation of pure cultures.

PART A: Isolation of Discrete Colonies from a Mixed Culture

LEARNING OBJECTIVE

Once you have completed this experiment, you should be able to

1. Perform the streak-plate and/or the spread-plate inoculation procedure to separate the cells of a mixed culture so that discrete colonies can be isolated.

PRINCIPLE

The techniques commonly used for isolation of discrete colonies initially require that the number of organisms in the inoculum be reduced. The resulting diminution of the population size ensures that, following inoculation, individual cells will be sufficiently far apart on the surface of the agar medium to effect a separation of the different species present. The

following are techniques that can be used to accomplish this necessary dilution:

1. The **streak-plate** method is a rapid qualitative isolation method. It is essentially a dilution technique that involves spreading a loopful of culture over the surface of an agar plate. Although many types of procedures are performed, the four-way, or quadrant, streak is described. Refer to Figure 2.1, which schematically illustrates this procedure.

   a. Place a loopful of culture on the agar surface in Area 1. Flame the loop, and cool it by touching an unused part of the agar surface close to the periphery of the plate, and then drag it rapidly several times across the surface of Area 1.

   b. Reflame and cool the loop, and turn the Petri dish 90°. Then touch the loop to a

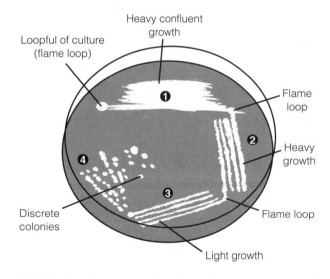

FIGURE 2.1 Four-way streak-plate inoculation

corner of the culture in Area 1 and drag it several times across the agar in Area 2. The loop should never enter Area 1 again.

c. Reflame and cool the loop and again turn the dish 90°. Streak Area 3 in the same manner as Area 2.

d. Without reflaming the loop, again turn the dish 90° and then drag the culture from a corner of Area 3 across Area 4, using a wider streak. Don't let the loop touch any of the previously streaked areas. The flaming of the loop at the points indicated is to effect the dilution of the culture so that fewer organisms are streaked in each area, resulting in the final desired separation.

2. The **spread-plate** technique requires that a previously diluted mixture of microorganisms be used. During inoculation, the cells are spread over the surface of a solid agar medium with a sterile, L-shaped bent rod while the Petri dish is spun on a "lazy-Susan" turntable (Figure 2.2). The

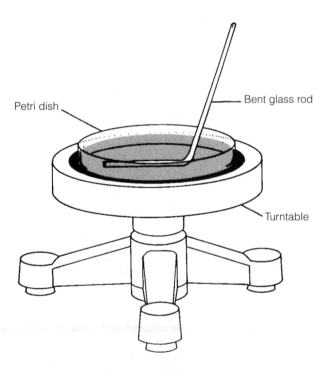

FIGURE 2.2 Petri dish turntable

step-by-step procedure for this technique is as follows:

a. Place the bent glass rod into the beaker and add a sufficient amount of 95% ethyl alcohol to cover the lower, bent portion.

b. Place an appropriately labeled nutrient agar plate on the turntable. With a sterile pipette, place one drop of sterile water on the center of the plate, followed by a sterile loopful of *Micrococcus luteus*. Mix gently with the loop and replace the cover.

c. Remove the glass rod from the beaker, and pass it through the Bunsen burner flame with the bent portion of the rod pointing downward to prevent the burning alcohol from running down your arm. Allow the alcohol to burn off the rod completely. Cool the rod for 10 to 15 seconds.

d. Remove the Petri dish cover and spin the turntable.

e. While the turntable is spinning, lightly touch the sterile bent rod to the surface of the agar and move it back and forth. This will spread the culture over the agar surface.

f. When the turntable comes to a stop, replace the cover. Immerse the rod in alcohol and reflame.

g. In the absence of a turntable, turn the Petri dish manually and spread the culture with the sterile bent glass rod.

3. The **pour-plate** technique requires a serial dilution of the mixed culture by means of a loop or pipette. The diluted inoculum is then added to a molten agar medium in a Petri dish, mixed, and allowed to solidify. The serial dilution and pour-plate procedures are outlined in Experiment 20.

MATERIALS

Cultures

24- to 48-hour nutrient broth cultures of a mixture of one part *Serratia marcescens* and three parts *Micrococcus luteus* and a mixture of one part *Escherichia coli* and ten parts *Micrococcus luteus.* For the spread-plate procedure, adjust the cultures to an optical density (O.D.) of 0.1 at 600 nanometers (nm).

Sources of mixed cultures from the environment could include cultures from a table top, bathroom sink, water fountain, or inside of an incubator. Each student should obtain a mixed culture from one of the environmental sources listed above.

Media

Three trypticase soy agar plates per designated student group for each inoculation technique to be performed.

Equipment

Bunsen burner, inoculating loop, turntable, 95% ethyl alcohol, 500-ml beaker, L-shaped bent glass rod, glassware marking pencil, culture tubes containing 1 ml of sterile water, test tube rack, and sterile cotton swabs.

PROCEDURE

1. Following the procedures previously described, prepare a spread-plate and/or streak-plate inoculation of each test culture on an appropriately labeled plate.
2. Preparation of environmental mixed culture.
    a. Dampen a sterile cotton swab with sterile water. Wring out the excess water by pressing the wet swab against the walls of the tube.
    b. With the moistened cotton swab, obtain your mixed-culture specimen from one of the selected environmental sources listed in the section on Cultures (p. 14).
    c. Place the contaminated swab back into the tube of sterile water. Mix gently and let stand for 5 minutes.
    d. Perform spread-plate and/or streak-plate inoculation on an appropriately labeled plate.
3. Incubate all plates in an inverted position for 48 to 72 hours at 25°C.

PART B: Isolation of Pure Cultures from a Spread-Plate or Streak-Plate Preparation

LEARNING OBJECTIVE

Once you have completed this experiment, you should be able to

1. Prepare a stock culture of an organism using isolates from mixed cultures prepared on an agar streak plate and/or spread plate.

PRINCIPLE

Once discrete, well-separated colonies develop on the surface of a nutrient agar plate culture, each may be picked up with a sterile needle and transferred to separate nutrient agar slants. Each of these new slant cultures represents the growth of a single bacterial species and is designated as a **pure or stock culture.**

MATERIALS
Cultures

Mixed-culture, nutrient agar streak-plate, and/or spread-plate preparations of *S. marcescens* and *M. luteus, M. luteus* and *E. coli,* and the environmental specimen plate from Part A.

Media

Four trypticase soy agar slants per designated student group.

Equipment

Bunsen burner, inoculating needle, and glassware marking pencil.

Procedure

1. Aseptically transfer, from visibly discrete colonies, the yellow *M. luteus,* the white *E.coli,* the red *S. marcescens,* and a discrete colony from the environmental agar plate specimen to the appropriately labeled agar slants as shown in Figure 2.3.
2. Incubate the cultures for 48 to 72 hours at 25°C.

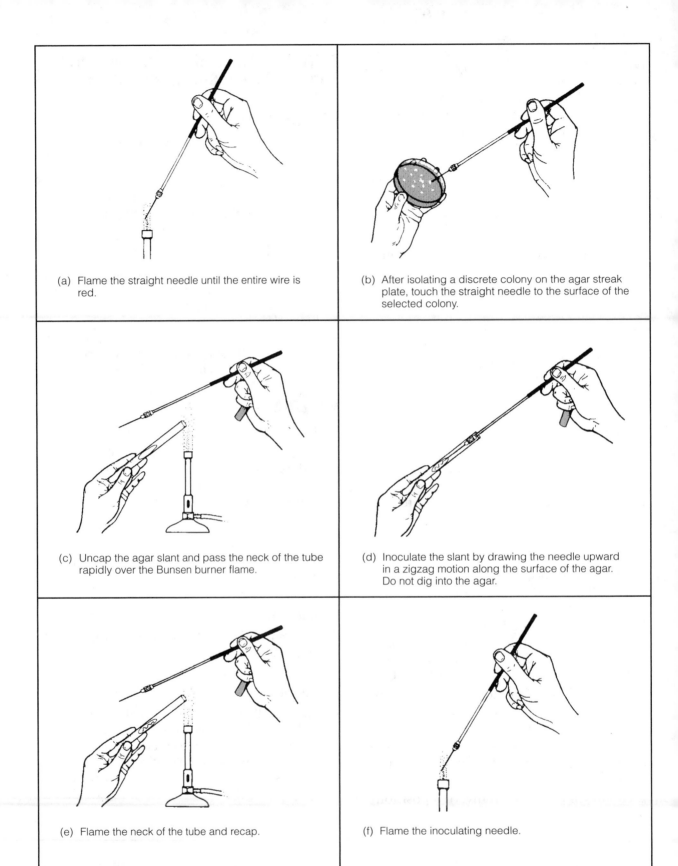

(a) Flame the straight needle until the entire wire is red.

(b) After isolating a discrete colony on the agar streak plate, touch the straight needle to the surface of the selected colony.

(c) Uncap the agar slant and pass the neck of the tube rapidly over the Bunsen burner flame.

(d) Inoculate the slant by drawing the needle upward in a zigzag motion along the surface of the agar. Do not dig into the agar.

(e) Flame the neck of the tube and recap.

(f) Flame the inoculating needle.

FIGURE 2.3 Procedure for the preparation of a pure culture

Name _____   Date _____

PART A: Isolation of Discrete Colonies
from a Mixed Culture

Examine all agar plate cultures to identify the distribution of the colonies. Record your results in the charts below.

1. Make a drawing of the distribution of colonies appearing on each of the agar plate cultures.
2. Select two discrete colonies that differ in appearance on each of the agar plate cultures. Using Figure 3.1 on page 22, describe each colony as to its:

    a. Form: Circular, irregular, or spreading.

    b. Elevation: Flat, slightly raised, or markedly raised.

    c. Pigmentation.

    d. Size: Pinpoint, small, medium, or large.

3. Retain the mixed-culture plates to perform Part B of this experiment.

📷 *Refer to photo number 15 in the color-plate insert for illustration of streak-plate technique.*

Spread-Plate Technique				
	S. marcescens and *M. luteus*		*M. luteus* and *E. coli*	
Draw the colonies that appear on each agar plate.				
Colony description:	Isolate 1	Isolate 2	Isolate 1	Isolate 2
Form	_____	_____	_____	_____
Elevation	_____	_____	_____	_____
Pigmentation	_____	_____	_____	_____
Size	_____	_____	_____	_____

Streak-Plate Technique				
	S. marcescens and *M. luteus*		*M. luteus* and *E. coli*	
Draw the colonies that appear on each agar plate.				
Colony description:	Isolate 1	Isolate 2	Isolate 1	Isolate 2
Form	_____	_____	_____	_____
Elevation	_____	_____	_____	_____
Pigmentation	_____	_____	_____	_____
Size	_____	_____	_____	_____

Environmental Specimen		
	Spread-Plate Technique	Streak-Plate Technique
Draw the colonies that appear on each agar plate.		
Colony description:		
Form	_____	_____
Elevation	_____	_____
Pigmentation	_____	_____
Size	_____	_____

PART B: Isolation of Pure Cultures from a Spread-Plate or Streak-Plate Preparation

1. Using Figure 3.1 on page 22, draw and indicate the type of growth of each pure-culture isolate.
2. Observe the color of the growth and record its pigmentation.
3. Indicate the names of the isolated organisms.

Draw the distribution of growth on the slant surface.				
Pigmentation	_____	_____	_____	_____
Type of growth	_____	_____	_____	_____
Name of organism	_____	_____	_____	_____

REVIEW QUESTIONS

1. Can a pure culture be prepared from a mixed-broth or a mixed–agar-slant culture? Explain.

2. Observation of a streak-plate culture shows more growth in Quadrant 4 than in Quadrant 3. Account for this observation.

3. Why is a needle used to isolate individual colonies from a spread plate or streak plate?

4. How can you determine if the colony that you chose to isolate is a pure culture?

Cultural Characteristics
of Microorganisms

LEARNING OBJECTIVE

Once you have completed this experiment, you should be able to

1. Determine the cultural characteristics of microorganisms as an aid in identifying and classifying organisms into taxonomic groups.

PRINCIPLE

When grown on a variety of media, microorganisms will exhibit differences in the macroscopic appearance of their growth. These differences, called **cultural characteristics,** are used as a basis for separating microorganisms into taxonomic groups. The cultural characteristics for all known microorganisms are contained in *Bergey's Manual of Systematic Bacteriology*. They are determined by culturing the organisms on nutrient agar slants and plates, in nutrient broth, and in nutrient gelatin. The patterns of growth to be considered in each of these media are described below, and some are illustrated in Figure 3.1.

Nutrient Agar Slants

These have a single straight line of inoculation on the surface and are evaluated in the following manner:

1. **Abundance of growth:** The amount of growth is designated as none, slight, moderate, or large.

2. **Pigmentation:** Chromogenic microorganisms may produce intracellular pigments that are responsible for the coloration of the organisms as seen in surface colonies. Other organisms produce extracellular soluble pigments that are excreted into the medium and that also produce a color.

Most organisms, however, are nonchromogenic and will appear white to gray.

3. **Optical characteristics:** Optical characteristics may be evaluated on the basis of the amount of light transmitted through the growth. These characteristics are described as **opaque** (no light transmission), **translucent** (partial transmission), or **transparent** (full transmission).

4. **Form:** The appearance of the single-line streak of growth on the agar surface is designated as:

   a. **Filiform:** Continuous, threadlike growth with smooth edges.

   b. **Echinulate:** Continuous, threadlike growth with irregular edges.

   c. **Beaded:** Nonconfluent to semiconfluent colonies.

   d. **Effuse:** Thin, spreading growth.

   e. **Arborescent:** Treelike growth.

   f. **Rhizoid:** Rootlike growth.

Nutrient Agar Plates

These demonstrate well-isolated colonies and are evaluated in the following manner:

1. **Size:** Pinpoint, small, moderate, or large.

2. **Pigmentation:** Color of colony.

3. **Form:** The shape of the colony is described as follows:

   a. **Circular:** Unbroken, peripheral edge.

   b. **Irregular:** Indented, peripheral edge.

   c. **Rhizoid:** Rootlike, spreading growth.

4. **Margin:** The appearance of the outer edge of the colony is described as follows:

   a. **Entire:** Sharply defined, even.

   b. **Lobate:** Marked indentations.

   c. **Undulate:** Wavy indentations.

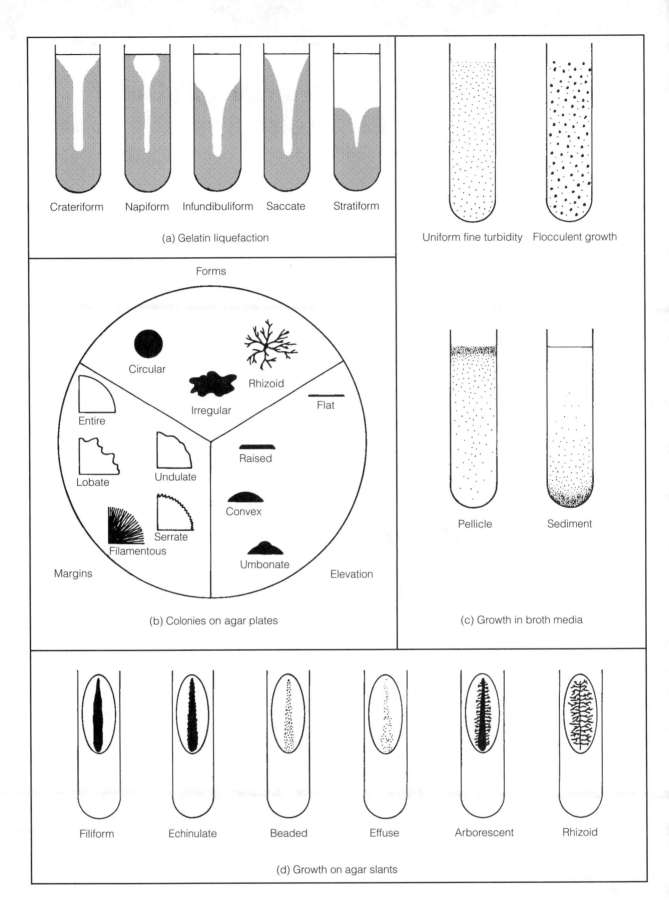

FIGURE 3.1 Cultural characteristics of bacteria

d. **Serrate:** Toothlike appearance.

e. **Filamentous:** Threadlike, spreading edge.

5. **Elevation:** The degree to which colony growth is raised on the agar surface is described as follows:

a. **Flat:** Elevation not discernible.

b. **Raised:** Slightly elevated.

c. **Convex:** Dome-shaped elevation.

d. **Umbonate:** Raised, with elevated convex central region.

Nutrient Broth Cultures

These are evaluated as to the distribution and appearance of the growth as follows:

1. **Uniform fine turbidity:** Finely dispersed growth throughout.

2. **Flocculent:** Flaky aggregates dispersed throughout.

3. **Pellicle:** Thick, padlike growth on surface.

4. **Sediment:** Concentration of growth at the bottom of broth culture may be granular, flaky, or flocculant.

Nutrient Gelatin

This solid medium may be liquefied by the enzymatic action of gelatinase. Liquefaction occurs in a variety of patterns:

1. **Crateriform:** Liquefied surface area is saucer-shaped.

2. **Napiform:** Bulbous-shaped liquefaction at surface.

3. **Infundibuliform:** Funnel-shaped.

4. **Saccate:** Elongated, tubular.

5. **Stratiform:** Complete liquefaction of the upper half of the medium.

MATERIALS

Cultures

24-hour nutrient broth cultures of *Pseudomonas aeruginosa, Bacillus cereus, Micrococcus luteus, Escherichia coli,* and *Staphylococcus aureus.*

Media

Per designated student group: five each of nutrient agar plates, nutrient agar slants, nutrient broth tubes, and nutrient gelatin tubes.

Equipment

Bunsen burner, inoculating loop and needle, and glassware marking pencil.

PROCEDURE

1. Using sterile technique, inoculate each of the appropriately labeled media listed below in the following manner:

a. Nutrient agar plates: With a sterile loop, prepare a streak-plate inoculation of each of the cultures for the isolation of discrete colonies.

b. Nutrient agar slants: With a sterile needle, make a single-line streak of each of the cultures provided, starting at the butt and drawing the needle up the center of the slanted agar surface.

c. Nutrient broth: Using a sterile loop, inoculate each organism into a tube of nutrient broth. Shake the loop a few times to dislodge the inoculum.

d. Nutrient gelatin: Using a sterile needle, prepare a stab inoculation of each of the cultures provided.

2. Incubate all cultures at 37°C for 24 to 48 hours.

Experiment 3 / Observations and Results

Name ______________________________ Date ______________

Refer to Figure 3.1 on page 22 and the descriptions presented in the introductory section of Experiment 3 while making the following observations:

1. Place all gelatin cultures in a refrigerator for 30 minutes, or in a beaker of crushed ice for a few minutes, to determine whether liquefaction of the medium has developed and the organism has produced gelatinase. Record your observations in the chart below according to the presence or absence of liquefaction and its type (if it has occurred).

Nutrient Gelatin Cultures					
	M. luteus	*P. aeruginosa*	*S. aureus*	*E. coli*	*B. cereus*
Draw liquefaction patterns.					
Liquefaction (+) or (−)	______	______	______	______	______
Type of liquefaction	______	______	______	______	______

2. Observe a single, well-isolated colony on each of the nutrient agar plate cultures and identify its size, elevation, margin, form, and pigmentation. Record your observations in the chart below.

Nutrient Agar Plates					
	M. luteus	*P. aeruginosa*	*S. aureus*	*E. coli*	*B. cereus*
Draw distribution of colonies.					
Size	______	______	______	______	______
Elevation	______	______	______	______	______
Margin	______	______	______	______	______
Form	______	______	______	______	______
Pigmentation	______	______	______	______	______

3. Observe each of the nutrient agar slant cultures for the amount, pigmentation, optical characteristics, and form of the growth. Record your observations in the chart below.

		Nutrient Agar Slant Cultures			
	M. luteus	*P. aeruginosa*	*S. aureus*	*E. coli*	*B. cereus*
Draw the distribution of growth on the slant surface.					
Amount of growth	————	————	————	————	————
Pigmentation	————	————	————	————	————
Optical characteristics	————	————	————	————	————
Form	————	————	————	————	————

4. Observe each of the nutrient broth cultures for the appearance of the growth (flocculation, turbidity, sediment, or pellicle). Record your results in the chart below.

		Nutrient Broth Cultures			
	M. luteus	*P. aeruginosa*	*S. aureus*	*E. coli*	*B. cereus*
Draw the distribution of growth.					
Appearance of growth	————	————	————	————	————

Microscopy

LEARNING OBJECTIVES

Once you have completed the experiments in this section, you should be

1. Familiar with the history and diversity of microscopic instruments.
2. Able to understand the components, use, and care of the brightfield microscope.
3. Able to understand the components and use of the darkfield microscope.
4. Able to correctly use the microscope for observation and measurement of microorganisms.

INTRODUCTION

Microbiology, the branch of science that has so vastly extended and expanded our knowledge of the living world, owes its existence to Anton van Leeuwenhoek. In 1673, with the aid of a crude microscope consisting of a biconcave lens enclosed in two metal plates, Leeuwenhoek introduced the world to the existence of microbial forms of life. Over the years, microscopes have evolved from the simple, single-lens instrument of Leeuwenhoek, with a magnification of 300×, to the present-day electron microscopes capable of magnifications greater than 250,000×.

Microscopes are designated as either light microscopes or electron microscopes. The former use visible light or ultraviolet rays to illuminate specimens. They include brightfield, darkfield, phase-contrast, and fluorescent instruments. Fluorescent microscopes use ultraviolet radiations whose wavelengths are shorter than those of visible light and are not directly perceptible to the human eye. Electron microscopes use electron beams (instead of light rays) and magnets (instead of lenses) to observe submicroscopic particles.

Essential Features of Various Microscopes

Brightfield Microscope This instrument contains two-lens systems for magnifying specimens: the ocular lens in the eyepiece and the objective lens located in the nosepiece. The specimen is illuminated by a beam of tungsten light focused on it by a substage lens called a condenser; the result is a specimen that appears dark against a bright background. A major limitation of this system is the absence of contrast between the specimen and the surrounding medium, which makes it difficult to observe living cells. Therefore, most brightfield observations are performed on nonviable, stained preparations.

Darkfield Microscope This is similar to the ordinary light microscope; however, the condenser system is modified so that the specimen is not illuminated directly. The condenser directs the light obliquely so that the light is deflected or scattered from the specimen, which then appears bright against a dark background. Living specimens may be observed more readily with darkfield than with brightfield microscopy.

Phase-Contrast Microscope Observation of microorganisms in an unstained state is possible with this microscope. Its optics include special objectives and a condenser that make visible cellular components that differ only slightly in their refractive indexes. As light is transmitted through a specimen with a refractive index different from that of the surrounding medium, a portion of the light is refracted (bent) due to slight variations in density and thickness of the cellular components. The special optics convert the difference between transmitted light and refracted rays, resulting in a significant variation in the intensity of light and thereby producing a discernible image of the structure under study. The image appears dark against a light background.

Fluorescent Microscope This microscope is used most frequently to visualize specimens that are chemically tagged with a fluorescent dye. The source of illumination is an ultraviolet (UV) light obtained from a high-pressure mercury lamp or hydrogen quartz lamp. The ocular lens is fitted with a filter that permits the longer ultraviolet wavelengths to pass, while the shorter wavelengths are blocked or eliminated. Ultraviolet radiations are absorbed by the fluorescent label, and the energy is reemitted in the form of a different wavelength in the visible light range. The fluorescent dyes absorb at wavelengths between 230 and 350 nanometers (nm) and emit orange, yellow, or greenish light. This microscope is used primarily for the detection of antigen-antibody reactions. Antibodies are conjugated with a fluorescent dye that becomes excited in the presence of ultraviolet light, and the fluorescent portion of the dye becomes visible against a black background.

Electron Microscope This instrument provides a revolutionary method of microscopy, with magnifications up to 1 million×. This permits visualization of submicroscopic cellular particles as well as viral agents. In the electron microscope, the specimen is illuminated by a beam of electrons rather than light, and the focusing is carried out by electromagnets instead of a set of optics. These components are sealed in a tube in which a complete vacuum is established. Transmission electron microscopes require specimens that are prepared as thin filaments, fixed and dehydrated for the electron beam to pass freely through them. As the electrons pass through the specimen, images are formed by directing the electrons onto photographic film, thus making internal cellular structures visible. Scanning electron microscopes are used for visualizing surface characteristics rather than intracellular structures. A narrow beam of electrons scans back and forth, producing a three-dimensional image as the electrons are reflected off the specimen's surface.

While scientists have a variety of optical instruments with which to perform routine laboratory procedures and sophisticated research, the compound brightfield microscope is the "workhorse" and is commonly found in all biological laboratories. Although you should be familiar with the basic principles of microscopy, you probably have not been exposed to this diverse array of complex and expensive equipment. Therefore, only the compound brightfield microscope will be discussed in depth and used to examine specimens.

Microscopic Examination of Stained Cell Preparations

LEARNING OBJECTIVES

Once you have completed this experiment, you should be familiar with the

1. Theoretical principles of brightfield microscopy.
2. Component parts of the compound microscope.
3. Use and care of the compound microscope.
4. Practical use of the compound microscope for visualization of cellular morphology from stained slide preparations.

PRINCIPLE

Microbiology is a science that studies living organisms that are too small to be seen with the naked eye. Needless to say, such a study must involve the use of a good compound microscope. Although there are many types and variations, they all fundamentally consist of a two-lens system, a variable but controllable light source, and mechanical adjustable parts for determining focal length between the lenses and specimen (Figure 4.1).

Components of the Microscope

Stage A fixed platform with an opening in the center allows the passage of light from an illuminating source below to the lens system above the stage. This platform provides a surface for the placement of a slide with its specimen over the central opening. In addition to the fixed stage, most microscopes have a **mechanical stage** that can be moved vertically or horizontally by means of adjustment controls. Less sophisticated microscopes have clips on the fixed stage, and the slide must be positioned manually over the central opening.

Illumination The light source is positioned in the base of the instrument. Some microscopes are equipped with a built-in light source to provide direct illumination. Others are provided with a reversible mirror that has one side flat and the other concave. An external light source, such as a lamp, is placed in front of the mirror to direct the light upward into the lens system. The flat side of the mirror is used for artificial light, and the concave side for sunlight.

Abbé Condenser This component is found directly under the stage and contains two sets of lenses that collect and concentrate light as it passes upward from the light source into the lens systems. The condenser is equipped with an **iris diaphragm,** a shutter controlled by a lever that is used to regulate the amount of light entering the lens system.

Body Tube Above the stage and attached to the arm of the microscope is the body tube. This structure houses the lens system that magnifies the specimen. The upper end of the tube contains the **ocular** or **eyepiece** lens. The lower portion consists of a movable **nosepiece** containing the **objective lenses.** Rotation of the nosepiece positions objectives above the stage opening. The body tube may be raised or lowered with the aid of **coarse-adjustment** and **fine-adjustment knobs** that are located above or below the stage, depending on the type and make of the instrument.

Theoretical Principles of Microscopy

To use the microscope efficiently and with minimal frustration, you should understand the basic principles of microscopy: magnification, resolution, numerical aperture, illumination, and focusing.

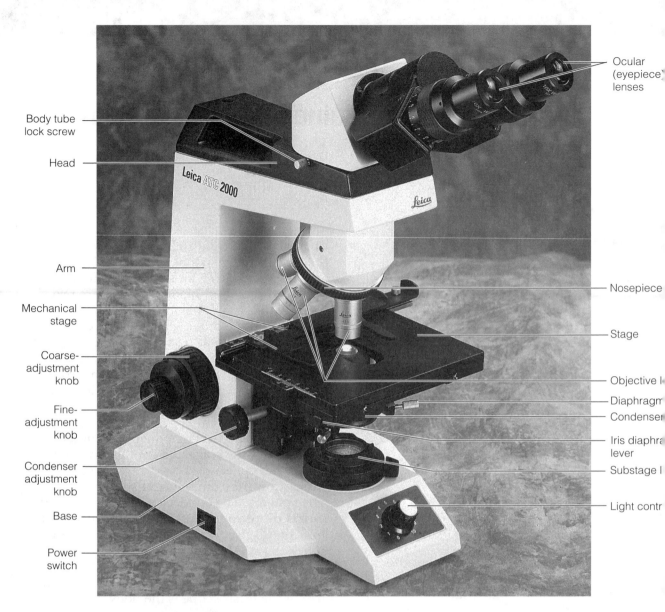

Body tube lock screw

Head

Arm

Mechanical stage

Coarse-adjustment knob

Fine-adjustment knob

Condenser adjustment knob

Base

Power switch

Ocular (eyepiece) lenses

Nosepiece

Stage

Objective l

Diaphragm

Condenser

Iris diaphra lever

Substage l

Light contr

FIGURE 4.1 Leica ATC 2000 compound microscope (Courtesy of Leica Microsystems, Inc.)

Magnification Enlargement, or magnification, of a specimen is the function of a two-lens system; the **ocular lens** is found in the eyepiece, and the **objective lens** is situated in a revolving nosepiece. These lenses are separated by the **body tube.** The objective lens is nearer the specimen and magnifies it, producing the **real image** that is projected up into the focal plane and then magnified by the ocular lens to produce the final image.

The most commonly used microscopes are equipped with a revolving nosepiece containing four objective lenses, each possessing a different degree of magnification. When these are combined with the magnification of the

ocular lens, the total or overall linear magnification of the specimen is obtained. This is shown in Table 4.1.

Resolving Power or Resolution Although magnification is important, you must be aware that unlimited enlargement is not possible by merely increasing the magnifying power of the lenses or by using additional lenses, because lenses are limited by a property called **resolving power.** By definition, resolving power is the ability of a lens to show two adjacent objects as discrete entities. When a lens cannot discriminate, that is, when the two objects appear as one, it has lost resolution. Increased

TABLE 4.1 Overall Linear Magnification

| Magnification | | Total Magnification |
Objective Lenses	Ocular Lens	Objective Multiplied by Ocular
Scanning 4×	10×	40×
Low-power 10×	10×	100×
High-power 45×	10×	450×
Oil-immersion 97×	10×	970×

magnification will not rectify the loss and will, in fact, blur the object. The resolving power of a lens is dependent on the wavelength of light used and the **numerical aperture,** which is a characteristic of each lens and imprinted on each objective. The numerical aperture is defined as a function of the diameter of the objective lens in relation to its focal length. It is doubled by use of the substage condenser, which illuminates the object with rays of light that pass through the specimen obliquely as well as directly. Thus, resolving power is expressed mathematically, as follows:

$$\text{Resolving power} = \frac{\text{Wavelength of light}}{2 \times \text{numerical aperture}}$$

Based on this formula, the shorter the wavelength, the greater the resolving power of the lens. Thus, for the same numerical aperture, short wavelengths of the electromagnetic spectrum are better suited for higher resolution than are longer wavelengths.

However, as with magnification, resolving power also has limits. You might rationalize that merely decreasing the wavelength will automatically increase the resolving power of a lens. Such is not the case, because the visible portion of the electromagnetic spectrum is very narrow and borders on the very short wavelengths found in the ultraviolet portion of the spectrum.

The relationship between wavelength and numerical aperture is valid only for increased resolving power when light rays are parallel. Therefore, the resolving power is also dependent on another factor, the **refractive index.** This is the bending power of light passing through air from the glass slide to the objective lens. The refractive index of air is lower than that of glass; as light rays pass from the glass slide into the air, they are bent or refracted so that they do not pass into the

objective lens. This would cause a loss of light, which would reduce the numerical aperture and diminish the resolving power of the objective lens. Loss of refracted light can be compensated for by interposing mineral oil, which has the same refractive index as glass, between the slide and the objective lens. In this way, decreased light refraction occurs and more light rays enter directly into the objective lens, producing a vivid image with high resolution (Figure 4.2).

Illumination Effective illumination is required for efficient magnification and resolving

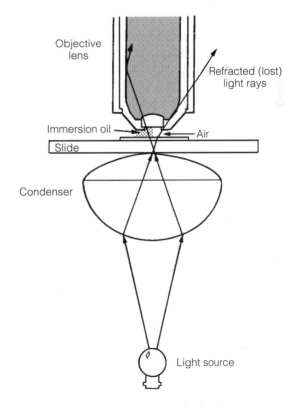

FIGURE 4.2 Refractive index in air and mineral oil

power. Since the intensity of daylight is an uncontrolled variable, artificial light from a tungsten lamp is the most commonly used light source in microscopy. The light is passed through the condenser located beneath the stage. The condenser contains two lenses that are necessary to produce a maximum numerical aperture. The height of the condenser can be adjusted with the **condenser knob.** Always keep the condenser close to the stage, especially when using the oil-immersion objective.

Between the light source and the condenser is the iris diaphragm, which can be opened and closed by means of a lever, thereby regulating the amount of light entering the condenser. Excessive illumination may actually obscure the specimen because of lack of contrast. The amount of light entering the microscope differs with each objective lens used. A rule of thumb is that *as the magnification of the lens increases, the distance between the objective lens and slide, called working distance, decreases, whereas the numerical aperture of the objective lens increases* (Figure 4.3).

Use and Care of the Microscope

You will be responsible for the proper care and use of microscopes. Since microscopes are expensive, you must observe the following regulations and procedures.

The instruments are housed in special cabinets and must be moved by users to their laboratory benches. The correct and only acceptable way to do this is to grip the microscope arm firmly with the right hand and the base with the left hand, and lift the instrument from the cabinet shelf. Carry it close to the body and gently place it on the laboratory bench. This will prevent collision with furniture or coworkers and will protect the instrument against damage.

Once the microscope is placed on the laboratory bench, observe the following rules:

1. Remove all unnecessary materials (such as books, papers, purses, and hats) from the laboratory bench.
2. Uncoil the microscope's electric cord and plug it into an electrical outlet.
3. Clean all lens systems; the smallest bit of dust, oil, lint, or eyelash will decrease the efficiency of the microscope. The ocular, scanning, low-power, and high-power lenses may be cleaned by wiping several times with acceptable lens tissue. Never use paper toweling or cloth on a lens surface. If the oil-immersion lens is gummy or tacky, a piece of lens paper moistened with xylol is used to wipe it clean. The xylol is immediately removed with a tissue moistened with 95% alcohol, and the lens is wiped dry with lens paper. *Note: This xylol cleansing procedure should be performed only by the instructor and only if necessary; consistent use of xylol may loosen the lens.*

The following routine procedures must be followed to ensure correct and efficient use of the microscope.

1. Place the microscope slide with the specimen within the stage clips on the fixed stage. Move the slide to center the specimen over the opening in the stage directly over the light source.
2. Raise the microscope stage up as far as it will go. Rotate the scanning lens or low-power lens into position. Lower the body tube with the coarse-adjustment knob to its lowest position. *Note: Never lower the body tube while looking through the ocular lens.*
3. While looking through the ocular lens, use the fine-adjustment knob, rotating it back and forth slightly, to bring the specimen into sharp focus.
4. Adjust the substage condenser to achieve optimal focus.
5. Routinely adjust the light source by means of the light-source transformer setting, and/or the iris diaphragm, for optimum illumination for each new slide and for each change in magnification.
6. Most microscopes are **par focal,** which means that when one lens is in focus, other lenses will also have the same focal length and can be rotated into position without further major adjustment. In practice, however, usually a half-turn of the fine-adjustment knob in either direction is necessary for sharp focus.
7. Once you have brought the specimen into sharp focus with a low-powered lens, preparation may be made for visualizing the specimen under oil immersion. Place a drop of oil on the slide directly over the area to be viewed. Rotate the nosepiece until the oil-immersion objective locks into position. *Note: Care should be taken not to allow the high-power objective to touch the drop of oil.* The slide is observed from the side as the objective is rotated slowly into

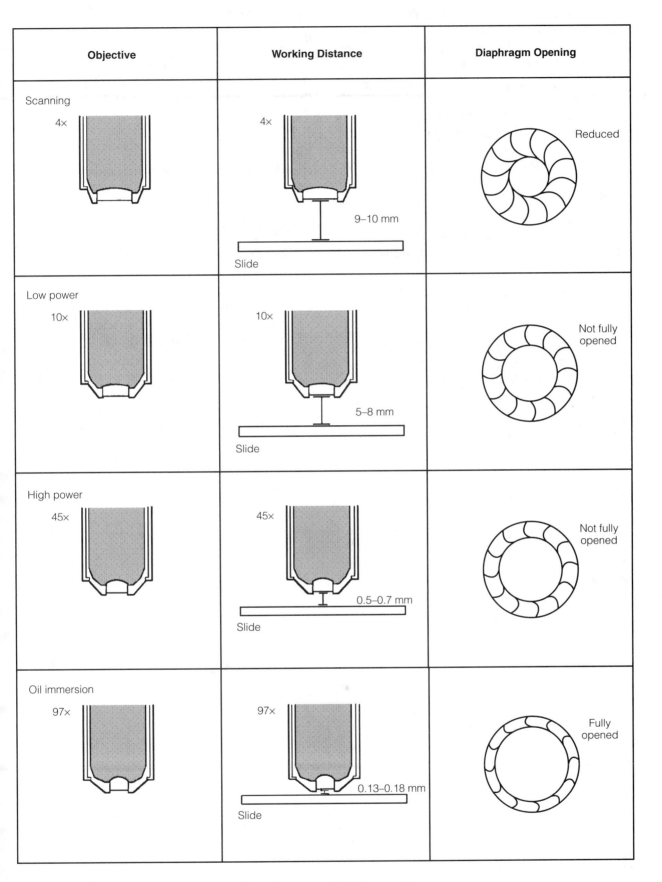

Objective	Working Distance	Diaphragm Opening
Scanning 4×	4× 9–10 mm Slide	Reduced
Low power 10×	10× 5–8 mm Slide	Not fully opened
High power 45×	45× 0.5–0.7 mm Slide	Not fully opened
Oil immersion 97×	97× 0.13–0.18 mm Slide	Fully opened

FIGURE 4.3 Relationship between working distance, objective, and diaphragm opening

position. This will ensure that the objective will be properly immersed in the oil. The fine-adjustment knob is readjusted to bring the image into sharp focus.

8. During microscopic examination of microbial organisms, it is always necessary to observe several areas of the preparation. This is accomplished by scanning the slide without the application of additional immersion oil. *Note: This will require continuous, very fine adjustments by the slow, back-and-forth rotation of the fine-adjustment knob only.*

On completion of the laboratory exercise, return the microscope to its cabinet in its original condition. The following steps are recommended:

1. Clean all lenses with dry, clean lens paper. *Note: Use xylol to remove oil from the stage only.*
2. Place the low-power objective in position and lower the body tube completely.
3. Center the mechanical stage.
4. Coil the electric cord around the body tube and the stage.
5. Carry the microscope to its position in its cabinet in the manner previously described.

MATERIALS

Slides

Commercially prepared slides of *Staphylococcus aureus, Bacillus subtilis, Aquaspirillum itersonii, Saccharomyces cerevisiae,* and a human blood smear.

Equipment

Compound microscope, lens paper, and immersion oil.

PROCEDURE

1. Review the parts of the microscope, making sure you know the names and understand the function of each of these components.
2. Review instructions for the use of the microscope, giving special attention to the use of the oil-immersion objective.
3. Examine the prepared slides, noting the shapes and the relative sizes of the cells under the high-power (also called high-dry, because it is the highest power that does not use oil) and oil-immersion objectives.

Name _____   Date _____

In the chart provided:

1. Draw several cells from a typical microscopic field as viewed under each magnification.
2. Give the total magnification for each objective.

	High Power	Oil Immersion
S. aureus Magnification	◯ ————	◯ ————
B. subtilis Magnification	◯ ————	◯ ————
A. itersonii Magnification	◯ ————	◯ ————
S. cerevisiae Magnification	◯ ————	◯ ————
Blood smear Magnification	◯ ————	◯ ————

REVIEW QUESTIONS

1. Explain why the body tube of the microscope should not be lowered while you are looking through the ocular lens.

2. For what purpose would you adjust each of the following microscope components during a microscopy exercise?
   a. Iris diaphragm:

   b. Coarse-adjustment knob:

   c. Fine-adjustment knob:

   d. Condenser:

   e. Mechanical stage control:

3. As a beginning student in the microbiology laboratory, you experience some difficulties in using the oil-immersion lens. Describe the steps you would take to correct the following problems:
   a. Inability to bring the specimen into sharp focus.

   b. Insufficient light while viewing the specimen.

   c. Artifacts in the microscopic field.

Microscopic Examination of Living Microorganisms Using a Hanging-Drop Preparation or a Wet Mount

LEARNING OBJECTIVES

Once you have completed this experiment, you should know how to

1. Microscopically examine living microorganisms.
2. Make a hanging-drop preparation or wet mount to view living microorganisms.

PRINCIPLE

Bacteria, because of their small size and a refractive index that closely approximates that of water, do not lend themselves readily to microscopic examination in a living, unstained state. Examination of living microorganisms is useful, however, to:

1. Observe cell activities such as motility and binary fission.
2. Observe the natural sizes and shapes of the cells, since **heat fixation** (the rapid passage of the smear over the Bunsen burner flame) and exposure to chemicals during staining cause some degree of distortion.

In this experiment you will use individual cultures of *Pseudomonas aeruginosa, Bacillus cereus, Staphylococcus aureus,* and *Proteus vulgaris* for a hanging-drop preparation (Figure 5.1) or a wet mount. Hay infusion or pond water may be substituted or used in addition to the above organisms. Figure 5.2 illustrates several organisms commonly found in pond water and hay infusions.

You will observe the preparation(s) microscopically for differences in the sizes and shapes of the cells, as well as for motility, a self-directed movement. It is essential to differentiate between actual motility and **Brownian movement,** a vibratory movement of the cells due to their bombardment by water molecules in the suspension. Hanging-drop preparations and wet mounts make the movement of microorganisms easier to see because they slow down the movement of water molecules.

MATERIALS

Cultures

24-hour broth cultures of *P. aeruginosa, B. cereus, S. aureus,* and *P. vulgaris;* and/or hay infusion broth cultures or pond water. (See Appendix 3 for the preparation of hay infusion broth.)

Equipment

Bunsen burner, inoculating loop, depression slides, glass slides, coverslips, microscope, petroleum jelly, and cotton swabs.

PROCEDURE

Hanging-Drop Preparation

Perform the following steps for each culture provided in this experiment.

1. With a cotton swab, apply a ring of petroleum jelly around the concavity of the depression slide.
2. Using sterile technique, place a loopful of the culture in the center of a clean coverslip.
3. Place the depression slide, with the concave surface facing down, over the coverslip so that the depression covers the drop of culture. Press the slide gently to form a seal between the slide and the coverslip.
4. Quickly turn the slide right side up so that the drop continues to adhere to the inner surface of the coverslip.
5. For microscopic examination, first focus on the drop culture under the low-power

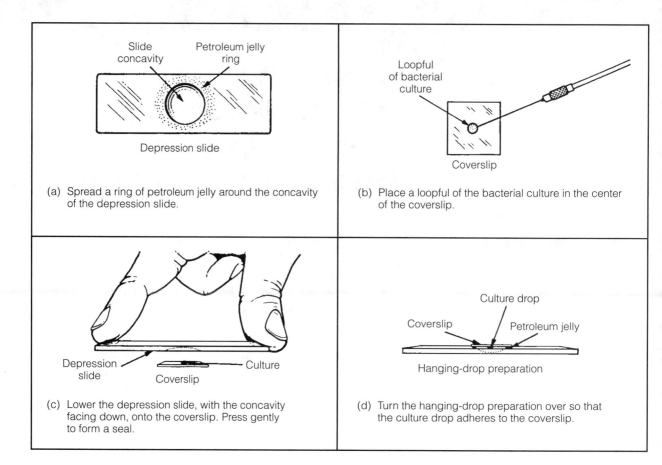

(a) Spread a ring of petroleum jelly around the concavity of the depression slide.

(b) Place a loopful of the bacterial culture in the center of the coverslip.

(c) Lower the depression slide, with the concavity facing down, onto the coverslip. Press gently to form a seal.

(d) Turn the hanging-drop preparation over so that the culture drop adheres to the coverslip.

FIGURE 5.1 Hanging-drop preparation

objective (10×) and reduce the light source by adjusting the Abbé condenser. Repeat using the high-power objective (40×).

Wet Mount

A wet mount may be substituted for the hanging-drop preparation using a similar procedure:

1. With a cotton swab apply a thin layer of petroleum jelly along the edge of the four sides of a coverslip.

2. Using sterile technique, place a loopful of the culture in the center of a clean coverslip.

3. Place a clean glass slide over the coverslip and press the slide gently to form a seal between the slide and the coverslip.

4. Follow Steps 4 and 5 in the hanging-drop procedure.

Bacteria

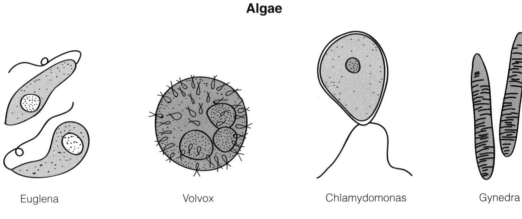

Cocci   Spirals   Rods

Algae

Euglena   Volvox   Chlamydomonas   Gynedra

Protozoa

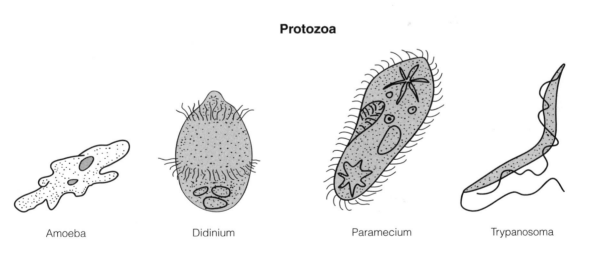

Amoeba   Didinium   Paramecium   Trypanosoma

FIGURE 5.2 Bacteria, algae, and protozoa commonly found in natural infusions and pond water

Name _____ Date _____

1. Examine the hanging-drop or wet-mount preparation to determine shape and motility of the different bacteria present. Record your results in the chart below.

Organisms	Shape	True Motility or Brownian Movement?
S. aureus		
P. aeruginosa		
B. cereus		
P. vulgaris		

2. Draw a representative field of each of the above organisms.

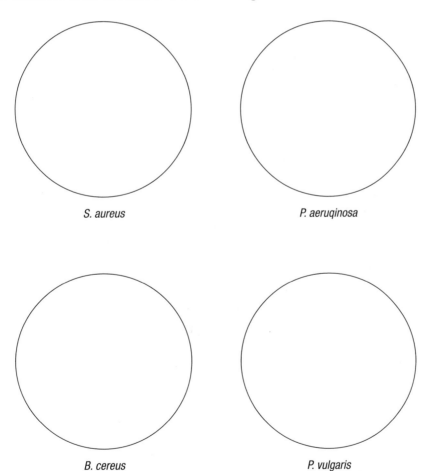

S. aureus

P. aeruqinosa

B. cereus

P. vulgaris

3. Draw representative fields of pond water and hay infusion if you used them. Try to identify some of the organisms that you see by referring to Figure 5.2. Note the shape and type of movement in the following chart.

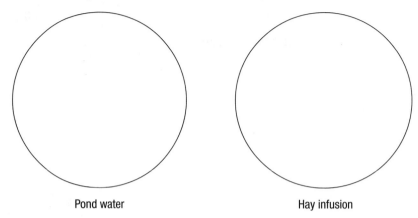

Pond water   Hay infusion

	Pond Water			Hay Infusion		
Shape						
True motility or Brownian movement?						
Organism						

REVIEW QUESTIONS

1. Why are living, unstained bacterial preparations more difficult to observe microscopically than stained preparations?

2. What is the major advantage of using living cell preparations (hanging-drop or wet mount) rather than stained preparations?

3. How do you distinguish between true motility and Brownian movement?

4. During the microscopic observation of a drop of stagnant pond water, what criteria would you use to distinguish viable organisms from non-viable suspended debris?

The Microscopic Measurement of Microorganisms

LEARNING OBJECTIVES

Once you have completed this experiment, you should be able to

1. Calibrate an ocular micrometer.
2. Perform an experimental procedure in the measurement of microorganisms.

PRINCIPLE

Determination of microbial size is not as simple as you might assume. Before an accurate measurement of cells can be made, the diameter of the microscopic field must be established by means of optic devices, namely, an **ocular micrometer** and a **stage micrometer.**

The ocular micrometer (Figure 6.1a), which is placed on a circular shelf inside the eyepiece, is a glass disc with graduations etched on its surface. The distance between these graduations will vary depending on the objective being used, which determines the size of the field. This distance is determined by using a stage micrometer (Figure 6.1b), a special glass slide with etched graduations that are 0.01 mm or 10 micrometers (μm) apart.

The calibration procedure for the ocular micrometer requires that the graduations on both micrometers be superimposed on each other (Figure 6.1c). This is accomplished by rotating the ocular lens. A determination is then made of the number of ocular divisions per known distance on the stage micrometer.

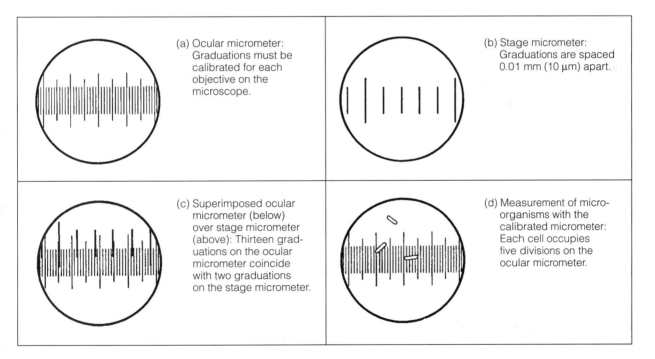

(a) Ocular micrometer: Graduations must be calibrated for each objective on the microscope.

(b) Stage micrometer: Graduations are spaced 0.01 mm (10 μm) apart.

(c) Superimposed ocular micrometer (below) over stage micrometer (above): Thirteen graduations on the ocular micrometer coincide with two graduations on the stage micrometer.

(d) Measurement of microorganisms with the calibrated micrometer: Each cell occupies five divisions on the ocular micrometer.

FIGURE 6.1 Calibration and use of the ocular micrometer

Finally, the calibration factor for one ocular division is calculated as follows:

$$\text{One division on ocular micrometer in mm} = \frac{\text{Known distance between two lines on stage micrometer}}{\text{Number of divisions on ocular micrometer}}$$

Example: If 13 ocular divisions coincide with two stage divisions (2 × 0.01 mm = known distance of 0.02 mm), then:

$$\text{one ocular division} = \frac{0.02 \text{ mm}}{13}$$

$$= 0.00154 \text{ mm or } 1.54 \text{ } \mu m$$

Once the ocular micrometer is calibrated, the size of a microorganism can easily be determined, first by counting the number of spaces occupied by the organism (Figure 6.1d) and second by multiplying this number by the calculated calibration factor for one ocular division.

Example: If an organism occupies five spaces on the ocular micrometer, then:

$$\text{Length of organism} = \frac{\text{Number of ocular divisions occupied}}{} \times \frac{\text{calibration factor for one ocular division}}{}$$

$$= 5 \times 1.54 \text{ } \mu m = 7.70 \text{ } \mu m$$

In this experiment, you will calibrate an ocular micrometer for the oil-immersion objective and determine the sizes of microorganisms such as bacteria, yeast, and protozoa.

MATERIALS

Slides

Prepared slides of yeast cells, a protozoan, and bacterial cocci and bacilli.

Equipment

Ocular micrometer, stage micrometer, microscope, immersion oil, and lens paper.

PROCEDURE

1. With the assistance of a laboratory instructor, carefully place the ocular micrometer into the eyepiece.

2. Place the stage micrometer on the microscope stage and center it over the illumination source.

3. With the stage micrometer in clear focus under the low-power objective, slowly rotate the eyepiece to superimpose the ocular micrometer graduations over those of the stage micrometer.

4. Add a drop of immersion oil to the stage micrometer, bring the oil-immersion objective into position, and focus, if necessary, with the fine-adjustment knob only.

5. Move the mechanical stage so that a line on the stage micrometer coincides with a line on the ocular micrometer at one end. Find another line on the ocular micrometer that coincides with a line on the stage micrometer. Determine the distance on the stage micrometer (number of divisions × 0.01 mm) and the corresponding number of divisions on the ocular micrometer.

6. Determine the value of the calibration factor for the oil-immersion objective.

7. Remove the stage micrometer from the stage.

8. To determine the size of the cocci on the prepared slides under the oil-immersion objective:

   a. Calculate the number of ocular divisions occupied by each of three separate cocci. Record the data in the readings columns of the observation chart.

   b. Determine and record the average of the three measurements.

   c. Determine the size by multiplying the average by your calculated calibration factor, and record this value.

9. Determine the size of the other microorganisms by observing the remaining prepared slides under oil immersion as outlined in Step 8. Since these organisms are not round, both length and width measurements are required.

Experiment 6 / Observations and Results

Name Date

1. Calibration of ocular micrometer for the oil-immersion objective:

   Distance on stage micrometer ______________

   Divisions on ocular micrometer ______________

   Calibration factor for one ocular division ______________

2. Use the chart below to record your observations and measurements of the microorganisms studied.

	Width of Microorganisms in Micrometers (μm)					
	Number of Ocular Divisions					
	Readings					
Organism	1	2	3	Average ×	Calibration Factor =	Size
Cocci						
Bacilli						
Yeast						
Protozoa						

	Length of Microorganisms in Micrometers (μm)					
	Number of Ocular Divisions					
	Readings					
Organism	1	2	3	Average ×	Calibration Factor =	Size
Bacilli						
Yeast						
Protozoa						

REVIEW QUESTIONS

1. Can the same calibration factor be used to determine the size of a microorganism under all objectives? Explain.

2. If one stage micrometer division contains 12 ocular divisions, then the distance between two lines on the ocular micrometer is _____. Show your calculations.

3. If 1 μm is equal to ¹⁄₂₅,₄₀₀ inch, convert the size of a bacterium measuring 3 μm × 1.5 μm, to inches. Show your calculations.

4. A comparative study of microbial size, using a stage micrometer, requires that you measure *Bacillus subtilis* in both the stained and unstained states.
    a. Would you expect the measurements to be comparable in both preparations?

    b. If a variation in size is expected, how would you account for this difference?

Darkfield Microscopy

LEARNING OBJECTIVES

Once you have completed this experiment, you should be able to

1. Visualize living microorganisms using the darkfield microscope.
2. Understand the difference between darkfield and brightfield microscopy.

PRINCIPLE

Darkfield microscopy is often used to observe organisms that do not stain with Gram stain or Giemsa stain, or it is used to view organisms that are very slender. A good example of this type of organism is the highly motile spirochete *Treponema pallidum*, the causative agent of syphilis.

In brightfield microscopy, all of the light from the source shines directly into the objective lens. A darkfield microscope, on the other hand, is fitted with a darkfield condenser containing a darkfield disk. The purpose of the disk is to eliminate the passage of light from the center of the source beam, allowing the darkfield condenser to concentrate light on the slide as a hollow cone (Figure 7.1). The angle of the cone is such that the light will only enter the objective if a specimen is present to change (scatter) its direction. Once the light passes up through the ocular, it will form a bright, magnified image against a black background (dark field). This type of microscopy is limited in that it can give no observations of structures inside the cell. The intensity of the light reflected from the cell surface diminishes and may obscure the less-intense light reflected from intracellular structures or organelles.

MATERIALS

Cultures

Stock cultures will not be used during this experiment. Students will obtain specimens of *Treponema denticola* by gently scraping their gums. An alternative specimen may be obtained from a hay infusion preparation.

Slides

Commercially prepared slides of *Treponema denticola* and *Aquaspirillum itersonii*.

Equipment

Darkfield microscope, flat wooden toothpicks, inoculating loop, glass slides, coverslips, disposable plastic Pasteur pipettes, 250-ml beaker containing disinfectant, and Petri dishes.

PROCEDURE

1. Each student will obtain a specimen of *Treponema denticola* and make a wet-mount preparation.
   a. With a flat-sided toothpick, *gently* scrape your gum line. *Be careful not to injure the gum tissue* (the instructor will demonstrate).
   b. Place a drop (one to two loopfuls) of distilled water on the center of a glass slide and suspend the material from the toothpick in the water by gentle stirring until the specimen is fully suspended.

⚠ **Discard all toothpicks in the beaker of disinfectant. Do the same with all slides and coverslips at the end of the lab session. Do not discard the prepared slides in the disinfectant beaker.**

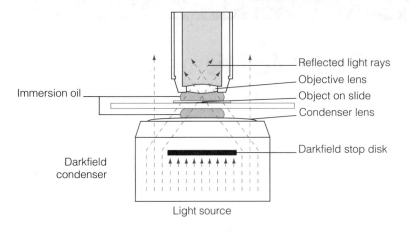

(a) Darkfield microscope

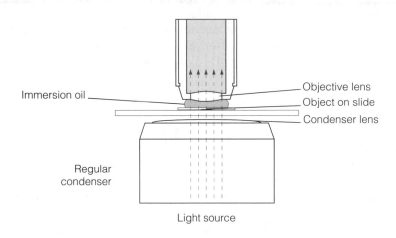

(b) Brightfield microscope

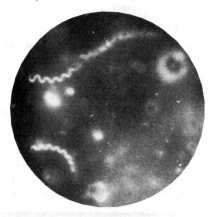

(c) Photomicrograph of spirochetes under a
    darkfield microscope

FIGURE 7.1 Comparison of darkfield and brightfield microscopy. (a) Darkfield microscope. Specimen is illuminated by reflected light and the background is dark. (b) Brightfield microscope. Both the specimen (stained) and the background are illuminated. (c) A photomicrograph of spirochetes under a darkfield microscope (enlarged view).

c. Gently lower the coverslip over the specimen so as not to create air bubbles.

d. A hay infusion preparation may be made by placing one or two loopfuls of hay infusion liquid on the center of a glass slide and gently lowering the coverslip to prevent the formation of bubbles.

2. Place your preparations in a Petri dish and keep it covered to help prevent evaporation. *Note: Remove only one slide at a time for examination.*

3. Place one of the specimen slides on the stage over the central opening.

4. Lower the condenser, clean the lens with lens paper, and place a drop of immersion oil on it. Slowly raise the condenser using the condenser-adjustment knob so that the oil on the lens just touches the bottom of the specimen slide.

5. Rotate the nosepiece so that the 10× (low-power) objective locks into place. Lower the body tube of the microscope and use the coarse-adjustment knob to bring the specimen into focus.

6. Repeat the above by rotating the 40× (high-power) objective into position. It will be necessary to use the fine-adjustment knob to refocus.

7. Consider that most microscopes are par focal (when one lens is in focus, all other lenses will have the same focal length). Place a drop of oil on the coverslip directly over the center of the specimen, and rotate the oil-objective lens so that it is just immersed in the drop of oil without making contact with the glass slide. Use only the fine-adjustment knob to focus. The oil-immersion objective must be used with care; *be sure you know what direction to rotate the nosepiece so that the 100× objective is brought into position.*

8. Slowly scan the slide and observe the shapes and motility of the organisms in the darkfield.

9. Observe both the commercially prepared slides and your preparations.

10. Before returning the microscope to the cabinet, be sure to use lens paper to clean the oil from the condenser lens.

Name _____ Date _____

1. In the spaces provided below, draw several illustrations of the organisms that you see. Indicate the magnification and the shapes that you are able to visualize.

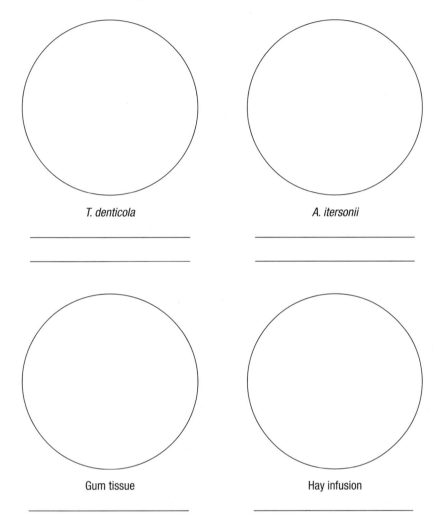

T. denticola

A. itersonii

Magnification: _____   _____

Shape: _____   _____

Gum tissue

Hay infusion

Magnification: _____   _____

Shape: _____   _____

REVIEW QUESTIONS

1. What is the advantage of using darkfield microscopy rather than brightfield microscopy?

2. Why is a specimen brightly illuminated when using a darkfield microscope?

3. Why is immersion oil placed on the center of the darkfield condenser?

4. Aside from syphilis, name a few diseases or infections that can be diagnosed by darkfield microscopy.

Bacterial Staining

LEARNING OBJECTIVES

Once you have completed the experiments in this section, you should be familiar with

1. The chemical and theoretical basis of biological staining.

2. Manipulative techniques of smear preparation.

3. Procedures for simple staining and negative staining.

4. The method for performing differential staining procedures, such as the Gram, acid-fast, capsule, and spore stains.

INTRODUCTION

Visualization of microorganisms in the living state is quite difficult, not only because they are minute, but also because they are transparent and practically colorless when suspended in an aqueous medium. To study their properties and to divide microorganisms into specific groups for diagnostic purposes, biological stains and staining procedures in conjunction with light microscopy have become major tools in microbiology.

Chemically, a stain (dye) may be defined as an organic compound containing a benzene ring plus a chromophore and auxochrome group (Figure III.1).

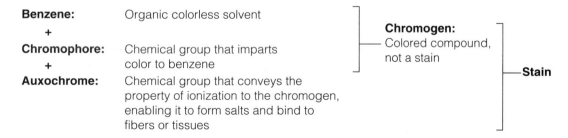

Benzene: Organic colorless solvent

  +

Chromophore: Chemical group that imparts color to benzene

  +

Auxochrome: Chemical group that conveys the property of ionization to the chromogen, enabling it to form salts and bind to fibers or tissues

Chromogen: Colored compound, not a stain

Stain

FIGURE III.1 Chemical composition of a stain

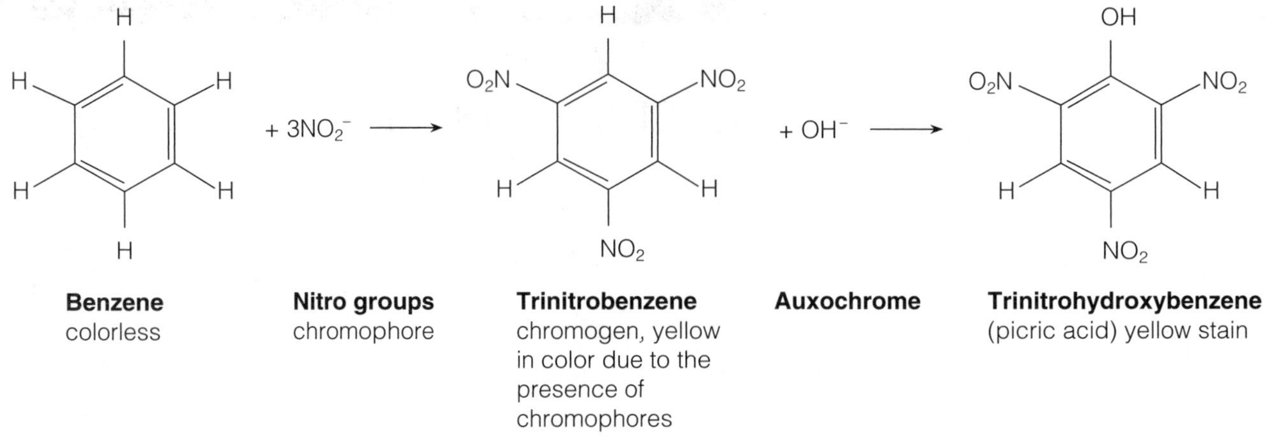

FIGURE III.2 Chemical formation of picric acid

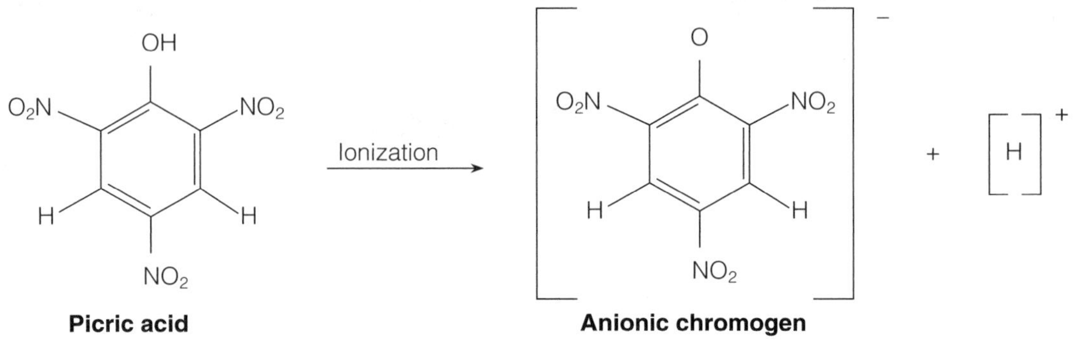

Benzene
colorless

Nitro groups
chromophore

Trinitrobenzene
chromogen, yellow
in color due to the
presence of
chromophores

Auxochrome

Trinitrohydroxybenzene
(picric acid) yellow stain

Picric acid

Anionic chromogen

FIGURE III.3 Picric acid: an acidic stain

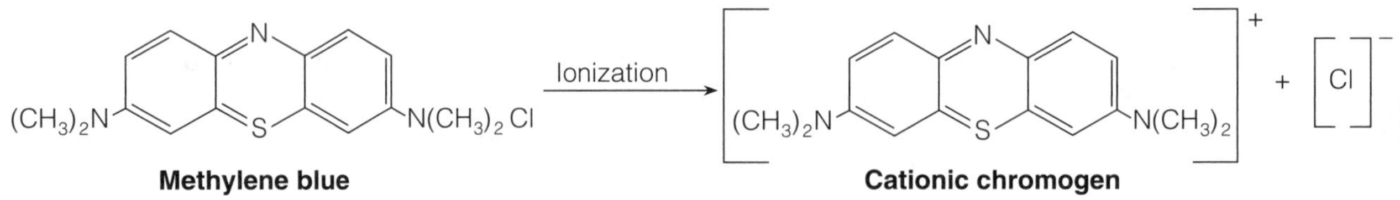

Methylene blue

Cationic chromogen

FIGURE III.4 Methylene blue: a basic stain

The stain picric acid may be used to illustrate this definition (Figure III.2).

The ability of a stain to bind to macromolecular cellular components such as proteins or nucleic acids depends on the electrical charge found on the chromogen portion, as well as on the cellular component to be stained.

Acidic stains are anionic, which means that, on ionization of the stain, the chromogen portion exhibits a negative charge and therefore has a strong affinity for the positive constituents of the cell. Proteins, positively charged cellular components, will readily bind to and accept the color of the negatively charged, an-

ionic chromogen of an acidic stain. Structurally, picric acid is an example of an acidic stain that produces an anionic chromogen as illustrated in Figure III.3

Basic stains are cationic, because on ionization the chromogen portion exhibits a positive charge and therefore has a strong affinity for the negative constituents of the cell. Nucleic acids, negatively charged cellular components, will readily bind to and accept the color of the positively charged, cationic chromogen of a basic stain. Structurally, methylene blue is a basic stain that produces a cationic chromogen as illustrated in Figure III.4.

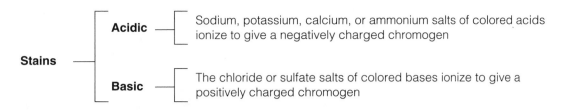

FIGURE III.5 Acidic and basic stains

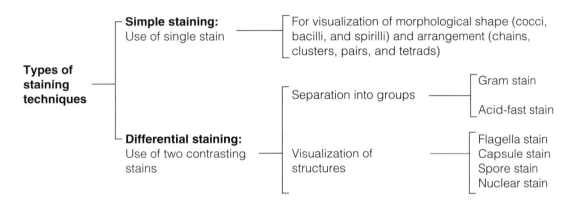

FIGURE III.6 Staining techniques

Figure III.5 is a summary of acidic and basic stains.

Basic stains are more commonly used for bacterial staining. The presence of a negative charge on the bacterial surface acts to repel most acidic stains and thus prevent their penetration into the cell.

Numerous staining techniques are available for visualization, differentiation, and separation of bacteria in terms of morphological characteristics and cellular structures. A summary of commonly used procedures and their purposes is outlined in Figure III.6.

Preparation of Bacterial Smears

LEARNING OBJECTIVE

Once you have completed this experiment, you should be able to

1. Prepare bacterial smears for the microscopic visualization of bacteria.

PRINCIPLE

Bacterial smears must be prepared prior to the execution of any of the staining techniques listed in Figure III.6 on page 55. Although not difficult, the technique requires adequate care. Meticulously follow the rules listed below.

1. **Preparation of the glass microscope slide:** Clean slides are essential for the preparation of microbial smears. Grease or oil from the fingers on slides must be removed by washing the slides with soap and water or scouring powders such as Bon Ami®, followed by a water rinse and a rinse of 95% alcohol. After cleaning, dry the slides and place them on laboratory towels until ready for use. *Note: Remember to hold the clean slides by their edges.*

2. **Labeling of slides:** Proper labeling of the slide is essential. The initials of the organism can be written on either end of the slide with a glassware marking pencil on the surface on which the smear is to be made. Care should be taken that the label does not come into contact with staining reagents.

3. **Preparation of smear:** It is crucial to avoid thick, dense smears. A thick or dense smear occurs when too much of the culture is used in its preparation, which concentrates a large number of cells on the slide. This type of preparation diminishes the amount of light that can pass through and makes it difficult to visualize the morphology of single cells. *Note: Smears require*

only a small amount of the bacterial culture. A good smear is one that, when dried, appears as a thin whitish layer or film. The print of your textbook should be legible through the smear. Those made from broth cultures or cultures from a solid medium require variations in technique.

  a. **Broth cultures:** Resuspend the culture by tapping the tube with your finger. Depending on the size of the loop, one or two loopfuls should be applied to the center of the slide with a sterile inoculating loop and spread evenly over an area about the size of a dime. Set the smears on the laboratory table and allow to air-dry.

  b. **Cultures from solid medium:** Organisms cultured in a solid medium produce thick, dense surface growth and are not amenable to direct transfer to the glass slide. These cultures must be diluted by placing one or two loopfuls of water on the center of the slide in which the cells will be emulsified. Transfer of the cells requires the use of a sterile inoculating loop or a needle, if preferred. Only the tip of the loop or needle should touch the culture to prevent the transfer of too many cells. Suspension is accomplished by spreading the cells in a circular motion in the drop of water with the loop or needle. This helps to avoid cell clumping. The finished smear should occupy an area about the size of a nickel and should appear as a translucent, or semitransparent, confluent whitish film. At this point the smear should be allowed to dry completely. *Note: Do not blow on slide or wave it in the air.*

4. **Heat fixation:** Unless fixed on the glass slide, the bacterial smear will wash away during the staining procedure. This is

avoided by heat fixation, during which the bacterial proteins are coagulated and fixed to the glass surface. Heat fixation is performed by the rapid passage of the air-dried smear two or three times over the flame of the Bunsen burner.

The preparation of a bacterial smear is illustrated in Figure 8.1.

MATERIALS

Cultures

24-hour nutrient agar slant culture of *Bacillus cereus* and a 24-hour nutrient broth culture of *Staphylococcus aureus*.

Equipment

Glass microscope slides, Bunsen burner, inoculating loop and needle, and glassware marking pencil.

PROCEDURE

Broth Cultures

1. Label three clean slides with the initials of the organism, and number them 1, 2, and 3.

2. Resuspend the sedimented cells in the broth culture by tapping the culture tube with your fingers.

3. With a sterile loop, place one loopful of culture on Slide number 1, two loopfuls on Slide 2, and three loopfuls on Slide 3, respectively. Allow all slides to air-dry completely; then heat fix the preparation. *Note: Pass the air-dried slide through the* outer *portion of the Bunsen burner flame to prevent overheating, which can distort the morphology through plasmolysis of the cell wall.*

Solid Medium Cultures

1. Label four clean glass slides with the initials of the organism. Label Slides 1 and 2 with an L for loop and Slides 3 and 4 with an N for needle.

2. Using the loop, add two loopfuls of water to each slide.

3. With a sterile loop, touch the entire loop to the culture and emulsify the cells in the drop of water on Slide 1.

4. With a sterile loop, just touch the tip of the loop to the culture and emulsify the cells in the drop of water on Slide 2.

5. Repeat Steps 3 and 4 using a sterile inoculating needle on Slides 3 and 4.

6. Air-dry completely and heat fix.

From liquid media

From solid media

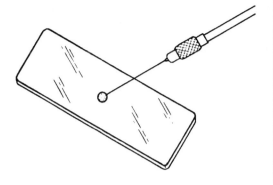

(a) Place one to two loopfuls of the cell suspension on the clean slide.

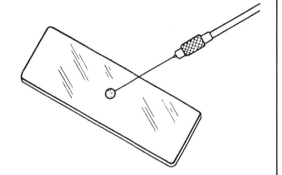

(a) Take one drop of water on the loop and place it on the center of the slide.

(b) With a circular movement of the loop, spread the suspension into a thin area approximately the size of a dime.

(b) Transfer a small amount of the bacterial inoculum from the slant culture into the drop of water. Spread both into a thin area approximately the size of a nickel.

Fixation

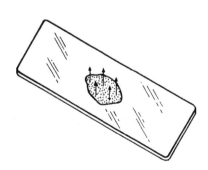

(c) Allow the smear to air-dry.

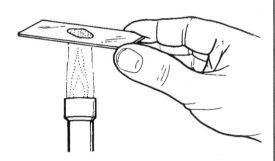

(d) While holding the slide at one end, quickly pass the smear over the flame of the Bunsen burner two to three times.

FIGURE 8.1 Bacterial smear preparation

Name _____ Date _____

1. Examine each slide for the confluent, whitish film or haze.
2. Select the preparation that you think is best and ask your instructor to comment on your choice. Remember that printed material should be legible through a good smear. Indicate by slide number the consistency of smears from both broth and solid cultures that you considered best.

   Broth cultures _____   Solid culture

                                     Loop _____   Needle _____

Refer to photo number 1 in the color-plate insert for an example of a bacterial smear.

REVIEW QUESTIONS

1. Why are thick or dense smears less likely to provide a good smear preparation for microscopic evaluation?

2. Why is it essential that smears be air-dried? Why can't they be gently heated over a flame to speed up the drying process?

3. Why should you be careful not to overheat the smear during the heat-fixing process?

4. Why do you think the presence of grease or dirt on a glass slide will result in a poor smear preparation? Cite two or three reasons.

Simple Staining

LEARNING OBJECTIVES

Once you have completed this experiment, you should be able to

1. Perform a simple staining procedure.
2. Compare the morphological shapes and arrangements of bacterial cells.

PRINCIPLE

In **simple staining,** the bacterial smear is stained with a single reagent, which produces a distinctive contrast between the organism and its background. Basic stains with a positively charged chromogen are preferred because bacterial nucleic acids and certain cell wall components carry a negative charge that strongly attracts and binds to the cationic chromogen. The purpose of simple staining is to elucidate the morphology and arrangement of bacterial cells (Figure 9.1). The most commonly used basic stains are methylene blue, crystal violet, and carbol fuchsin.

MATERIALS

Cultures

24-hour nutrient agar slant cultures of *Escherichia coli* and *Bacillus cereus* and a 24-hour nutrient broth culture of *Staphylococcus aureus.* Alternatively, use the smears prepared in Experiment 8.

Reagents

Methylene blue, crystal violet, and carbol fuchsin.

Equipment

Bunsen burner, inoculating loop, staining tray, microscope, lens paper, bibulous paper, and glass slides.

PROCEDURE

1. Prepare separate bacterial smears of the organisms following the procedure described in Experiment 8. *Note: All smears must be heat fixed prior to staining.*

2. Place a slide on the staining tray and flood the smear with one of the indicated stains, using the appropriate exposure time for each: carbol fuchsin, 15 to 30 seconds; crystal violet, 20 to 60 seconds; methylene blue, 1 to 2 minutes.

3. *Gently* wash the smear with tap water to remove excess stain. During this step, hold the slide parallel to the stream of water; in this way you can reduce the loss of organisms from the preparation.

4. Using bibulous paper, blot dry but *do not* wipe the slide.

5. Repeat this procedure with the remaining two organisms, using a different stain for each.

6. Examine all stained slides under oil immersion.

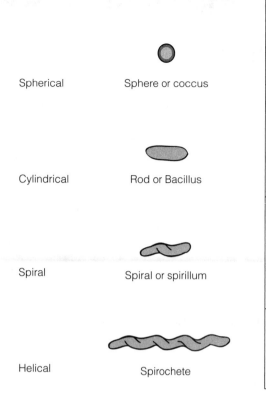

Basic Arrangements	Prefix Meaning
Diplococcus	Diplo = Pair
Streptococcus Streptobacillus	Strepto = chain
Staphylococcus	Staphylo = cluster
Sarcina	Sarcina = package Textrads = packets of 4

Spherical — Sphere or coccus

Cylindrical — Rod or Bacillus

Spiral — Spiral or spirillum

Helical — Spirochete

(a) Basic shapes

(b) Basic arrangements and standard prefixes

FIGURE 9.1 Bacterial shapes and arrangements

Name _____ Date _____

In the space provided:

1. Draw a representative field for each organism (refer to page xvi for proper drawing procedure).

2. Describe the morphology of the organisms with reference to their shapes (bacilli, cocci, spirilli) and arrangements (chains, clusters, pairs).

📷 *Refer to photo numbers 2, 3, and 4 in the color-plate insert for examples of bacilli, cocci, and spirilli.*

	Methylene Blue	Crystal Violet	Carbol Fuchsin
Draw a representative field.	◯	◯	◯
Organism	_____	_____	_____
Cell morphology:			
Shape	_____	_____	_____
Arrangement	_____	_____	_____
Cell color	_____	_____	_____

REVIEW QUESTIONS

1. Why are basic dyes more effective for bacterial staining than acidic dyes?

2. Can simple staining techniques be used to identify more than the morphological characteristics of microorganisms? Explain.

3. During the performance of the simple staining procedure, you failed to heat fix your *E. coli* smear preparation. Upon microscopic examination, how would you expect this slide to differ from the correctly prepared slides?

4. During a coffee break, your friend spills coffee on your lab coat and the fabric is discolored. Is this a true biological stain or simply a compound capable of imparting color? Explain your rationale.

Negative Staining

LEARNING OBJECTIVES

Once you have completed this experiment, you should be able to

1. Perform a negative staining procedure.
2. Understand the benefit obtained from visualizing unstained microorganisms.

PRINCIPLE

Negative staining requires the use of an acidic stain such as India ink or nigrosin (Figure 10.1). The acidic stain, with its negatively charged chromogen, will not penetrate the cells because of the negative charge on the surface of bacteria. Therefore, the unstained cells are easily discernible against the colored background.

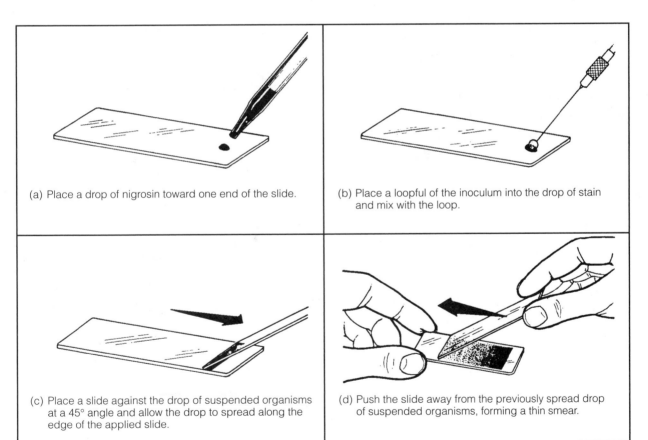

(a) Place a drop of nigrosin toward one end of the slide.

(b) Place a loopful of the inoculum into the drop of stain and mix with the loop.

(c) Place a slide against the drop of suspended organisms at a 45° angle and allow the drop to spread along the edge of the applied slide.

(d) Push the slide away from the previously spread drop of suspended organisms, forming a thin smear.

FIGURE 10.1 Negative staining procedure

The practical application of negative staining is twofold. First, since heat fixation is not required and the cells are not subjected to the distorting effects of chemicals and heat, their natural size and shape can be seen. Second, it is possible to observe bacteria that are difficult to stain, such as some spirilli. Because heat fixation is not done during the staining process, keep in mind that the organisms are not killed and *slides should be handled with care.*

MATERIALS

Cultures

24-hour agar slant cultures of *Micrococcus luteus, Bacillus cereus,* and *Aquaspirillum itersonii.*

Reagent

Nigrosin.

Equipment

Bunsen burner, inoculating loop, staining tray, glass slides, lens paper, and microscope.

PROCEDURE

1. Place a small drop of nigrosin close to one end of a clean slide.

2. Using sterile technique, place a loopful of inoculum from the *M. luteus* culture in the drop of nigrosin and mix.

3. With the edge of a second slide held at a 45° angle and placed in front of the bacterial suspension, push the mixture to form a thin smear.

4. Air-dry. *Note: Do not heat fix the slide.*

5. Repeat Steps 1 to 4 for slide preparations of *Bacillus cereus* and *Aquaspirillum itersonii.*

6. Examine the slides under oil immersion.

Name _____   Date _____

1. Draw representative fields of your microscopic observations.

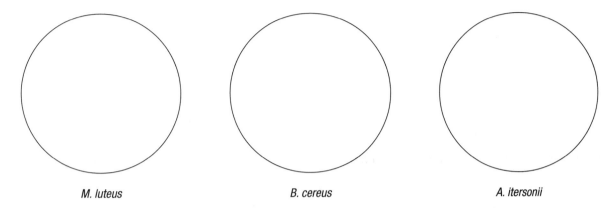

📷 *Refer to photo number 5 in the color-plate insert for illustration of this staining procedure.*

         M. luteus   *B. cereus*   *A. itersonii*

2. Describe the microscopic appearance of the different bacteria using the chart below.

Organism	*M. luteus*	*B. cereus*	*A. itersonii*
Shape			
Arrangement			
Magnification			

REVIEW QUESTIONS

1. Why can't methylene blue be used in place of nigrosin for negative staining? Explain.

2. What are the practical advantages of negative staining?

3. Why doesn't nigrosin penetrate bacterial cells?

Gram Stain

LEARNING OBJECTIVES

Once you have completed this experiment, you should understand

1. The chemical and theoretical basis for differential staining procedures.
2. The chemical basis for the Gram stain.
3. The procedure for differentiating between two principal groups of bacteria: gram-positive and gram-negative.

PRINCIPLE

Differential staining requires the use of at least three chemical reagents that are applied sequentially to a heat-fixed smear. The first reagent is called the **primary stain.** Its function is to impart its color to all cells. In order to establish a color contrast, the second reagent used is the decolorizing agent. Based on the chemical composition of cellular components, the **decolorizing agent** may or may not remove the primary stain from the entire cell or only from certain cell structures. The final reagent, the counterstain, has a contrasting color to that of the primary stain. Following decolorization, if the primary stain is not washed out, the **counterstain** cannot be absorbed and the cell or its components will retain the color of the primary stain. If the primary stain is removed, the decolorized cellular components will accept and assume the contrasting color of the counterstain. In this way, cell types or their structures can be distinguished from each other on the basis of the stain that is retained.

The most important differential stain used in bacteriology is the **Gram stain,** named after Dr. Christian Gram. It divides bacterial cells into two major groups, gram-positive and gram-negative, which makes it an essential tool for classification and differentiation of microorganisms. The Gram stain reaction is based on the difference in the chemical composition of bacterial cell walls. Gram-positive cells have a thick peptidoglycan layer, whereas the peptidoglycan layer in gram-negative cells is much thinner and surrounded by outer lipid-containing layers. Peptidoglycan is mainly a polysaccharide composed of two chemical subunits found only in the bacterial cell wall. These subunits are N-acetylglucosamine and N-acetylmuramic acid. As adjacent layers of peptidoglycan are formed, they are cross-linked by short chains of peptides by means of a transpeptidase enzyme, resulting in the shape and rigidity of the cell wall. Early experiments have shown that if the gram-positive cell is denuded of its cell wall by the action of lysozyme or penicillin, the gram-positive cell will stain gram-negative.

The Gram stain uses four different reagents. Descriptions of these reagents and their mechanisms of action follow. Figure 11.1 shows the microscopic observation of the cell at each step of the Gram staining procedure. See also plates 6–11 in the color-plate insert.

Primary Stain

Crystal Violet (Hucker's) This violet stain is used first and stains all cells purple.

Mordant

Gram's Iodine This reagent is not only a killing agent, but also serves as a mordant, a substance that increases the cells' affinity for a stain. It does this by binding to the primary stain, thus forming an insoluble complex. The resultant crystal-violet-iodine (CV-I) complex serves to intensify the color of the stain. At this point, all cells will appear purple-black.

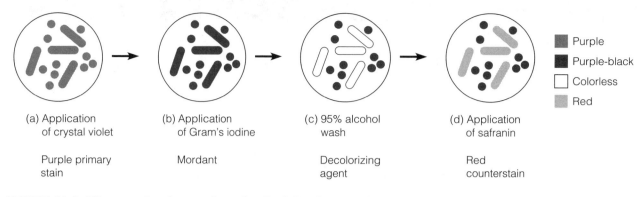

Purple
Purple-black
Colorless
Red

(a) Application
of crystal violet

Purple primary
stain

(b) Application
of Gram's iodine

Mordant

(c) 95% alcohol
wash

Decolorizing
agent

(d) Application
of safranin

Red
counterstain

FIGURE 11.1 Microscopic observation of cells following the Gram staining procedure

Decolorizing Agent

Ethyl Alcohol, 95% This reagent serves a dual function as a protein-dehydrating agent and as a lipid solvent. Its action is determined by two factors, the concentration of lipids and the thickness of the peptidoglycan layer in bacterial cell walls. In gram-negative cells, the alcohol increases the porosity of the cell wall by dissolving the lipids in the outer layers. Thus, the CV-I complex can be more easily removed from the thinner and less highly cross-linked peptidoglycan layer. Therefore, the washing-out effect of the alcohol facilitates the release of the unbound CV-I complex, leaving the cells colorless or unstained. The much thicker peptidoglycan layer in gram-positive cells is responsible for the more stringent retention of the CV-I complex, as the pores are made smaller due to the dehydrating effect of the alcohol. Thus the tightly bound primary stain complex is difficult to remove, and the cells remain purple.

Counterstain

Safranin This is the final reagent, used to stain red those cells that have been previously decolorized. Since only gram-negative cells undergo decolorization, they may now absorb the counterstain. Gram-positive cells retain the purple color of the primary stain.

The preparation of adequately stained smears requires that you bear in mind the following precautions:

1. The most critical phase of the procedure is the decolorization step, which is based on the ease with which the CV-I complex is released from the cell. Remember that over-decolorization will result in loss of the primary stain, causing gram-positive organisms to appear gram-negative. Under-decolorization, however, will not completely remove the CV-I complex, causing gram-negative organisms to appear gram-positive. Strict adherence to all instructions will help remedy part of the difficulty, but individual experience and practice are the keys to correct decolorization.

2. It is imperative that, between applications of the reagents, slides be thoroughly washed under running water or water applied with an eyedropper. This removes excess reagent and prepares the slide for application of the subsequent reagent.

3. The best Gram stained preparations are made with fresh cultures, that is, not older than 24 hours. As cultures age, especially in the case of gram-positive cells, the organisms tend to lose their ability to retain the primary stain and may appear to be **gram-variable;** that is, some cells will appear purple, while others will appear red.

MATERIALS
Cultures

24-hour nutrient agar slant cultures of *Escherichia coli*, *Staphylococcus aureus*, and *Bacillus cereus*.

Reagents

Crystal violet, Gram's iodine, 95% ethyl alcohol, and safranin.

Equipment

Bunsen burner, inoculating loop or needle, staining tray, glass slides, bibulous paper, lens paper, and microscope.

PROCEDURE

The steps are pictured in Figure 11.2.

1. Obtain four clean glass slides.
2. Using sterile technique, prepare a smear of each of the three organisms and on the remaining slide prepare a smear consisting of a mixture of *S. aureus* and *E. coli*. Do this by placing a drop of water on the slide, and then transferring each organism separately to the drop of water with a sterile, cooled loop. Mix and spread both organisms by means of a circular motion of the inoculating loop.
3. Allow smears to air-dry and then heat fix in the usual manner.
4. *Gently* flood smears with crystal violet and let stand for 1 minute.
5. *Gently* wash with tap water.
6. *Gently* flood smears with the Gram's iodine mordant and let stand for 1 minute.
7. *Gently* wash with tap water.
8. Decolorize with 95% ethyl alcohol. *Note: Do not over-decolorize.* Add reagent drop by drop until alcohol runs almost clear, showing only a blue tinge.
9. *Gently* wash with tap water.
10. Counterstain with safranin for 45 seconds.
11. *Gently* wash with tap water.
12. Blot dry with bibulous paper and examine under oil immersion.

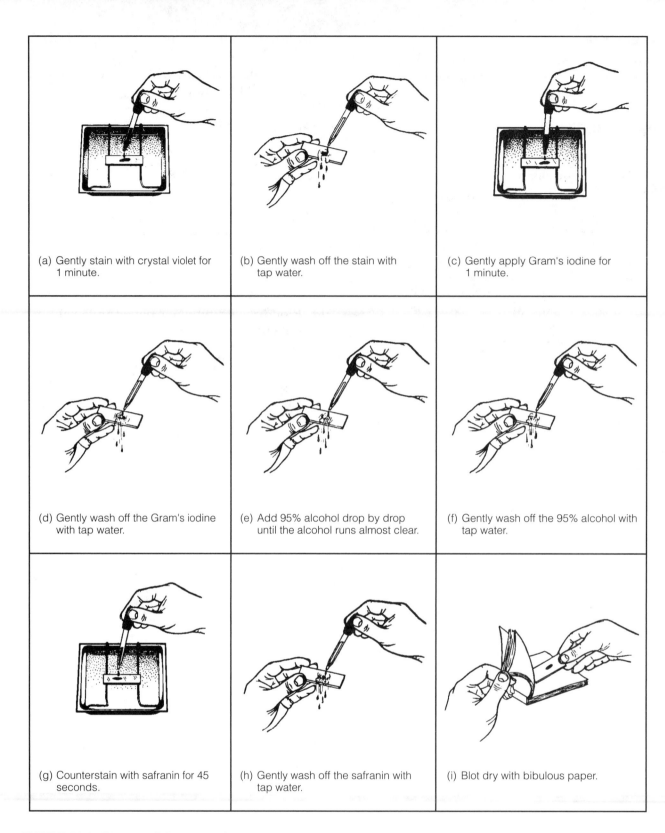

(a) Gently stain with crystal violet for 1 minute.

(b) Gently wash off the stain with tap water.

(c) Gently apply Gram's iodine for 1 minute.

(d) Gently wash off the Gram's iodine with tap water.

(e) Add 95% alcohol drop by drop until the alcohol runs almost clear.

(f) Gently wash off the 95% alcohol with tap water.

(g) Counterstain with safranin for 45 seconds.

(h) Gently wash off the safranin with tap water.

(i) Blot dry with bibulous paper.

FIGURE 11.2 Gram staining procedure

Name _____   Date _____

Following your observation of all slides under oil immersion, record your results in the chart.

1. Make a drawing of a representative microscopic field.
2. Describe the cells according to their morphology and arrangement.
3. Describe the color of the stained cells.
4. Classify the organism as to the Gram reaction: gram-positive or gram-negative.

Refer to photo numbers 6–11 in the color-plate insert for illustration of this staining procedure.

	E. coli	B. cereus	S. aureus	Mixture
Draw a representative field.				
Cell morphology:				
Shape	_____	_____	_____	_____
Arrangement	_____	_____	_____	_____
Cell color	_____	_____	_____	_____
Gram reaction	_____	_____	_____	_____

REVIEW QUESTIONS

1. What are the advantages of differential staining procedures over the simple staining technique?

2. Cite the purpose of each of the following reagents in a differential staining procedure.
   a. Primary stain:

   b. Counterstain:

c. Decolorizing agent:

d. Mordant:

3. Why is it essential that the primary stain and the counterstain be of contrasting colors?

4. Which is the most crucial step in the performance of the Gram staining procedures? Explain.

5. Because of a snowstorm, your regular laboratory session was cancelled and the Gram staining procedure was performed on cultures incubated for a longer period of time. Examination of the stained *Bacillus cereus* slides revealed a great deal of color variability, ranging from an intense blue to shades of pink. Account for this result.

Acid-Fast Stain
(Ziehl-Neelsen Method)

LEARNING OBJECTIVES

Once you have completed this experiment, you should understand

1. The chemical basis of the acid-fast stain.
2. The procedure for differentiating bacteria into acid-fast and non–acid-fast groups.

PRINCIPLE

While the majority of bacterial organisms are stainable by either simple or Gram staining procedures, a few genera, particularly the members of the genus *Mycobacterium*, are resistant and can only be visualized by the **acid-fast** method. Since *M. tuberculosis* and *M. leprae* represent bacteria that are pathogenic to humans, the stain is of diagnostic value in identifying these organisms.

The characteristic difference between mycobacteria and other microorganisms is the presence of a thick, waxy (lipoidal) wall that makes penetration by stains extremely difficult. Once the stain has penetrated, however, it cannot be readily removed even with the vigorous use of acid alcohol as a decolorizing agent. Because of this property, these organisms are called acid-fast, while all other microorganisms, which are easily decolorized by acid-alcohol, are non–acid-fast.

The acid-fast stain uses three different reagents.

Primary Stain

Carbol Fuchsin Unlike cells that are easily stained by ordinary aqueous stains, most species of mycobacteria are not stainable with common dyes such as methylene blue and crystal violet. Carbol fuchsin, a dark red stain in 5% phenol that is soluble in the lipoidal materials that constitute the major portion of the mycobacterial cell wall, does penetrate these bacteria and

is retained. Penetration is further enhanced by the application of heat, which drives the carbol fuchsin through the lipoidal wall and into the cytoplasm. A modification of the Ziehl-Neelsen method circumvents the use of heat by addition of a wetting agent (Turgitol™) to this stain, which reduces surface tension between the cell wall of the mycobacteria and the stain. Following application of the primary stain, all cells will appear red.

Decolorizing Agent

Acid-Alcohol (3% HCl + 95% Ethanol) Prior to decolorization, the smear is cooled, which allows the waxy cell substances to harden. On application of acid-alcohol, acid-fast cells will be resistant to decolorization since the primary stain is more soluble in the cellular waxes than in the decolorizing agent. In this event, the primary stain is retained and the mycobacteria will stay red. This is not the case with non–acid-fast organisms which lack cellular waxes. The primary stain is more easily removed during decolorization, leaving these cells colorless or unstained.

Counterstain

Methylene Blue This is used as the final reagent to stain previously decolorized cells. As only non–acid-fast cells undergo decolorization, they may now absorb the counterstain and take on its blue color, while acid-fast cells retain the red of the primary stain.

MATERIALS
Cultures

72- to 96-hour trypticase soy broth culture of *Mycobacterium smegmatis* and 18- to 24-hour culture of *Staphylococcus aureus*.

Reagents

Carbol fuchsin, acid-alcohol, and methylene blue.

Equipment

Bunsen burner, hot plate, inoculating loop, glass slides, bibulous paper, lens paper, staining tray, and microscope.

PROCEDURE

The steps are pictured in Figure 12.1.

1. Obtain three clean glass slides.
2. Using sterile technique, prepare a bacterial smear of each organism plus a third mixed smear of *M. smegmatis* and *S. aureus*.
3. Allow smears to air-dry and then heat fix in the usual manner.
4. Flood smears with carbol fuchsin and place over a beaker of water on a warm hot plate, allowing the preparation to steam for 5 minutes. *Note: Do not allow stain to evaporate; replenish stain as needed. Also, prevent stain from boiling by adjusting the hot plate temperature.* For heatless method, flood smear with carbol fuchsin containing Turgitol for 3 to 5 minutes.
5. Wash with tap water. Heated slides must be cooled prior to washing.
6. Decolorize with acid-alcohol, adding the reagent drop by drop until the alcohol runs almost clear with a slight red tinge.
7. Wash with tap water.
8. Counterstain with methylene blue for 2 minutes.
9. Wash smear with tap water.
10. Blot dry with bibulous paper and examine under oil immersion.

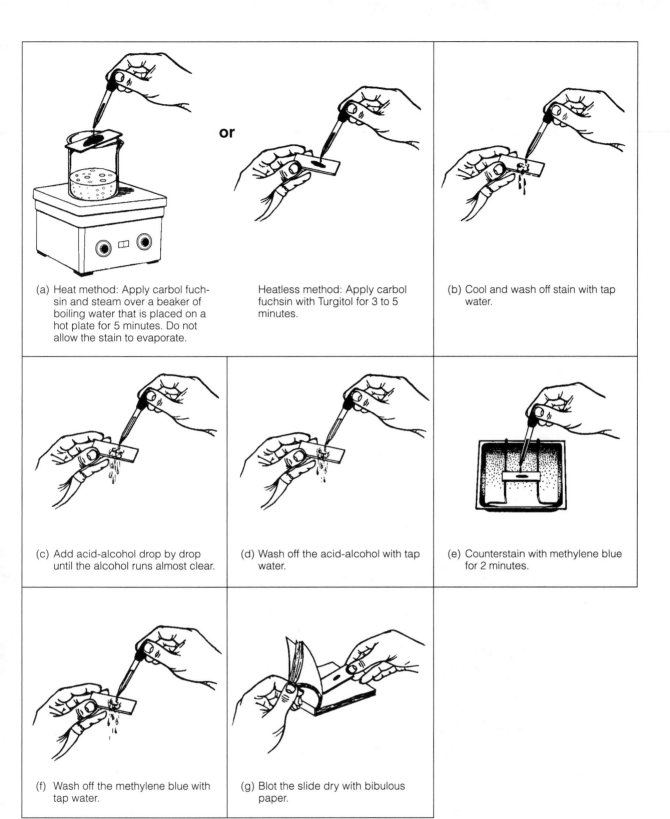

(a) Heat method: Apply carbol fuchsin and steam over a beaker of boiling water that is placed on a hot plate for 5 minutes. Do not allow the stain to evaporate.

Heatless method: Apply carbol fuchsin with Turgitol for 3 to 5 minutes.

(b) Cool and wash off stain with tap water.

(c) Add acid-alcohol drop by drop until the alcohol runs almost clear.

(d) Wash off the acid-alcohol with tap water.

(e) Counterstain with methylene blue for 2 minutes.

(f) Wash off the methylene blue with tap water.

(g) Blot the slide dry with bibulous paper.

FIGURE 12.1 Acid-fast staining procedure

Name _____ Date _____

Following your observation of all slides under oil immersion, record your results in the chart.

1. Make drawings of a representative microscopic field of each preparation.
2. Describe the cells according to their shapes and arrangements.
3. Describe the color of the stained cells.
4. Classify the organisms as to reaction: acid-fast or non–acid-fast.

Refer to photo number 12 in the color-plate insert for illustration of this staining procedure.

	M. smegmatic	*S. aureus*	**Mixture**
Draw a representative field.			

Cell morphology:
   Shape
   Arrangement
Cell color
Acid-fast reaction
	_____
	_____
	_____
	_____
		_____
		_____
		_____
		_____
			_____
			_____
			_____
			_____

REVIEW QUESTIONS

1. Why must heat or a surface-active agent be used with application of the primary stain during acid-fast staining?

2. Why is acid-alcohol rather than ethyl alcohol used as a decolorizing agent?

3. What is the specific diagnostic value of this staining procedure?

4. Why is the application of heat or a surface-active agent not required during the application of the counterstain in acid-fast staining?

5. A child presents symptoms suggestive of tuberculosis, namely a respiratory infection with a productive cough. Microscopic examination of the child's sputum reveals no acid-fast rods. However, examination of gastric washings reveals the presence of both acid-fast and non–acid-fast bacilli. Do you think the child has active tuberculosis? Explain.

Differential Staining
for Visualization
of Bacterial Cell Structures

LEARNING OBJECTIVES

Once you have completed this experiment, you should understand

1. The chemical basis for the spore and capsule stains.
2. The procedure for differentiation between the bacterial spore and vegetative cell forms.
3. The procedure to distinguish capsular material from the bacterial cell.

PART A: Spore Stain (Schaeffer-Fulton Method)

PRINCIPLE

Members of the anaerobic genera *Clostridium* and *Desulfotomaculum* and the aerobic genus *Bacillus* are examples of organisms that have the capacity to exist either as metabolically active **vegetative cells** or as highly resistant, metabolically inactive cell types called **spores.** When environmental conditions become unfavorable for continuing vegetative cellular activities, particularly with the exhaustion of a nutritional carbon source, these cells have the capacity to undergo **sporogenesis** and give rise to a new intracellular structure called the **endospore,** which is surrounded by impervious layers called spore coats. As conditions continue to worsen, the endospore is released from the degenerating vegetative cell and becomes an independent cell called a **spore.** Because of the chemical composition of spore layers, the spore is resistant to the deleterious effects of excessive heat, freezing, radiation, desiccation, and chemical agents, as well as to the commonly employed microbiological stains. With

the return of favorable environmental conditions, the free spore may revert to a metabolically active and less resistant vegetative cell through **germination** (see Figure 13.1). It should be emphasized that sporogenesis and germination are not means of reproduction but merely mechanisms that ensure cell survival under all environmental conditions.

In practice, the spore stain uses two different reagents.

Primary Stain

Malachite Green Unlike most vegetative cell types that stain by common procedures, the spore, because of its impervious coat, will not accept the primary stain easily. For further penetration, the application of heat is required. After the primary stain is applied and the smear is heated, both the vegetative cell and spore will appear green.

Decolorizing Agent

Water Once the spore accepts the malachite green, it cannot be decolorized by tap water, which removes only the excess primary stain. The spore remains green. On the other hand, the stain does not demonstrate a strong affinity for vegetative cell components; the water removes it, and these cells will be colorless.

Counterstain

Safranin This contrasting red stain is used as the second reagent to color the decolorized vegetative cells, which will absorb the counterstain and appear red. The spores retain the green of the primary stain.

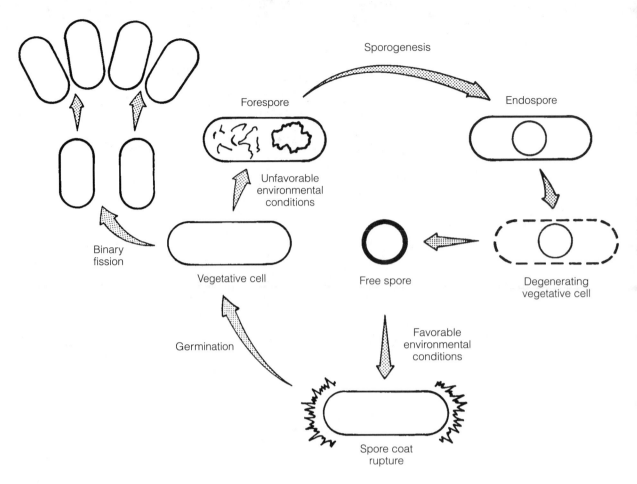

FIGURE 13.1 Life cycle of a spore-forming bacterium

MATERIALS

Cultures

48- to 72-hour nutrient agar slant culture of *Bacillus cereus* and thioglycollate culture of *Clostridium butyricum*.

Reagents

Malachite green and safranin.

Equipment

Bunsen burner, hot plate, staining tray, inoculating loop, glass slides, bibulous paper, lens paper, and microscope.

PROCEDURE

The steps are pictured in Figure 13.2.

1. Obtain two clean glass slides.

2. Make individual smears in the usual manner using sterile technique.

3. Allow smear to air-dry, and heat fix in the usual manner.

4. Flood smears with malachite green and place on top of a beaker of water sitting on a warm hot plate, allowing the preparation to steam for 2 to 3 minutes. *Note: Do not allow stain to evaporate; replenish stain as needed.* Prevent the stain from boiling by adjusting the hot plate temperature.

5. Remove slides from hot plate, cool, and wash under running tap water.

6. Counterstain with safranin for 30 seconds.

7. Wash with tap water.

8. Blot dry with bibulous paper and examine under oil immersion.

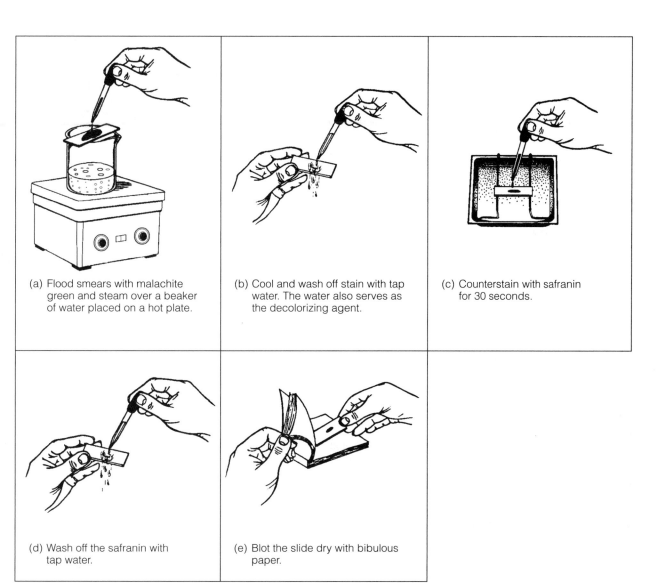

FIGURE 13.2 Spore staining procedure

(a) Flood smears with malachite green and steam over a beaker of water placed on a hot plate.

(b) Cool and wash off stain with tap water. The water also serves as the decolorizing agent.

(c) Counterstain with safranin for 30 seconds.

(d) Wash off the safranin with tap water.

(e) Blot the slide dry with bibulous paper.

PART B: Capsule Stain (Anthony Method)

PRINCIPLE

A **capsule** is a gelatinous outer layer that is secreted by the cell and that surrounds and adheres to the cell wall. It is not common to all organisms. Cells that have a heavy capsule are generally virulent and capable of producing disease, since the structure protects bacteria against the normal phagocytic activities of host cells. Chemically, the capsular material is a polysaccharide, a glycoprotein, or a polypeptide.

Capsule staining is more difficult than other types of differential staining procedures because the capsular materials are water-soluble and may be dislodged and removed with vigorous washing. Smears should not be heated because the resultant cell shrinkage may create a clear zone around the organism that is an artifact that can be mistaken for the capsule.

The capsule stain uses two reagents.

Primary Stain

Crystal Violet (1% aqueous)
A violet stain is applied to a non–heat-fixed smear. At this point, the cell and the capsular material will take on the dark color.

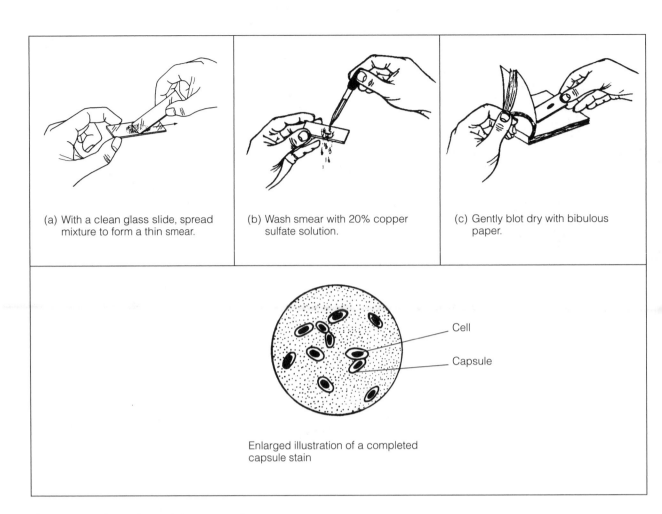

(a) With a clean glass slide, spread mixture to form a thin smear.

(b) Wash smear with 20% copper sulfate solution.

(c) Gently blot dry with bibulous paper.

Cell

Capsule

Enlarged illustration of a completed capsule stain

FIGURE 13.3 Capsule staining procedure

Decolorizing Agent

Copper Sulfate (20%) Because the capsule is non-ionic, unlike the bacterial cell, the primary stain adheres to the capsule but does not bind to it. In the capsule staining method, copper sulfate is used as a decolorizing agent rather than water. The copper sulfate washes the purple primary stain out of the capsular material without removing the stain bound to the cell wall. At the same time, the decolorized capsule absorbs the copper sulfate, and the capsule will now appear blue in contrast to the deep purple color of the cell.

MATERIALS

Cultures

48-hour-old skimmed milk cultures of *Alcaligenes viscolactis*, *Leuconostoc mesenteroides*, and *Enterobacter aerogenes*.

Reagents

1% crystal violet and 20% copper sulfate ($CuSO_4 \bullet 5H_2O$).

Equipment

Bunsen burner, inoculating loop or needle, staining tray, bibulous paper, lens paper, glass slides, and microscope.

PROCEDURE

The steps are pictured in Figure 13.3.

1. Obtain one clean glass slide.

2. Place several drops of crystal violet stain on a clean glass slide. Using sterile technique, add three loopfuls of a culture to the stain and *gently* mix with the inoculating loop.

3. With a clean glass slide spread the mixture over the entire surface of the slide to create a very thin smear. Let stand for 5 to 7 minutes.

4. Allow smears to air-dry. *Note: Do not heat fix.*

5. Wash smears with 20% copper sulfate solution.

6. *Gently* blot dry and examine under oil immersion.

7. Repeat Steps 1 to 6 for each of the remaining test cultures.

Name _____ Date _____

PART A: Spore Staining Procedure

Following your observation of all slides under oil immersion, record your results in the chart.

1. Make drawings of a representative microscopic field of each preparation.
2. Describe the location of the endospore within the vegetative cell as being central, subterminal, or terminal on each preparation.
3. Indicate color of the spore and vegetative cell on each preparation.

📷 *Refer to photo number 13 in the color-plate insert for illustration of this staining procedure.*

	C. butyricum	*B. cereus*
Draw a representative field.		
Color of spores	_____	_____
Color of vegetative cells	_____	_____
Location of endospore	_____	_____

PART B: Capsule Staining Procedure

Following your observation of all slides under oil immersion, record your results in the chart on the following page.

1. Make drawings of a representative microscopic field of each preparation.
2. Record the comparative size of the capsule, that is, small, moderate, or large.
3. Indicate the color of the capsule and the cell on each preparation.

📷 *Refer to photo number 14 in the color-plate insert for illustration of this staining procedure.*

	A. viscolactis	L. mesenteroides	E. aerogenes
Draw a representative field.			
Capsule size			
Color of capsule			
Color of cell			

REVIEW QUESTIONS

1. Why is heat necessary in spore staining?

2. Explain the function of water in spore staining.

3. Assume that during the performance of this exercise you made several errors in your spore-staining procedure. In each of the following cases, indicate how your microscopic observations would differ from those observed when the slides were prepared correctly.

a. You used acid-alcohol as the decolorizing agent.

b. You used safranin as the primary stain and malachite green as the counterstain.

c. You did not apply heat during the application of the primary stain.

4. Explain the medical significance of a capsule.

5. Explain the function of copper sulfate in this procedure.

Cultivation of Microorganisms: Nutritional and Physical Requirements, and Enumeration of Microbial Populations

LEARNING OBJECTIVES

Once you have completed the experiments in this section, you should be familiar with

1. The basic nutritional and environmental requirements for the cellular activities of all forms of life.

2. The principles associated with the use of routine and special-purpose media for microbial cultivation.

3. The diversified physical factors essential for microbial cultivation.

4. Specialized techniques for the cultivation of anaerobic microorganisms.

5. The serial dilution–agar plate technique for enumeration of viable microorganisms.

6. The growth dynamics of bacterial populations.

INTRODUCTION

As do all other living organisms, microorganisms require certain basic nutrients and physical factors for the sustenance of life. However, their particular requirements vary greatly. Understanding these needs is necessary for successful cultivation of microorganisms in the laboratory.

Nutritional Needs

Nutritional needs of microbial cells are supplied in the laboratory through a variety of media. The following list illustrates the nutritional diversity that exists among microbes.

1. **Carbon:** This is the most essential and central atom common to all cellular structures and functions. Among microbial cells, two carbon-dependent types are noted:
   a. **Autotrophs:** These organisms can be cultivated in a medium consisting solely of inorganic compounds; specifically, they use inorganic carbon in the form of carbon dioxide.
   b. **Heterotrophs:** These organisms cannot be cultivated in a medium consisting solely of inorganic compounds; they must be supplied with organic nutrients, primarily glucose.

2. **Nitrogen:** This is also an essential atom in many cellular macromolecules, particularly proteins and nucleic acids. Proteins serve as the structural molecules forming the so-called fabric of the cell and as functional molecules, enzymes, that are responsible for the metabolic activities of the cell. Nucleic acids include DNA, the genetic basis of cell life, and RNA, which plays an active role in protein synthesis within the cell. Some microbes use atmospheric nitrogen, others rely on inorganic compounds such as ammonium or nitrate salts, and still others require nitrogen-containing organic compounds such as amino acids.

3. **Nonmetallic elements:** The major non-metallic ions used for cellular nutrition are:
    a. **Sulfur:** This is integral to some amino acids and is therefore a component of proteins. Sources include organic compounds such as sulfur-containing amino acids, inorganic compounds such as sulfates, and elementary sulfur.
    b. **Phosphorus:** This is necessary for the formation of the nucleic acids DNA and RNA and also for synthesis of the high-energy organic compound adenosine triphosphate (ATP). Phosphorus is supplied in the form of phosphate salts for use by all microbial cells.
4. **Metallic elements: Ca^{++}, Zn^{++}, Na^+, K^+, Cu^{++}, Mn^{++}, Mg^{++}, and $Fe^{+2,+3}$** are some of the metallic ions necessary for continued efficient performance of varied cellular activities. Some of these activities are osmoregulation, regulation of enzyme activity, and electron transport during biooxidation. Remember that these ions are micronutrients and are required in trace concentrations only. Inorganic salts supply these materials.
5. **Vitamins:** These organic substances contribute to cellular growth and are essential in minute concentrations for cell activities. They are also sources of coenzymes, which are required for the formation of active enzyme systems. Some microbes require vitamins to be supplied in a preformed state for normal metabolic activities. Some possess extensive vitamin-synthesizing pathways, whereas others can synthesize only a limited number from other compounds present in the medium.
6. **Water:** All cells require water in the medium so that the low-molecular-weight nutrients can cross the cell membrane.
7. **Energy:** Active transport, biosynthesis, and biodegradation of macromolecules are the metabolic activities of cellular life. These activities can be sustained only if there is a constant availability of energy within the cell. Two bioenergetic types of microorganisms exist:
    a. **Phototrophs:** These use radiant energy as their sole energy source.
    b. **Chemotrophs:** These depend on oxidation of chemical compounds as their energy source. Some microbes use organic molecules such as glucose; others utilize inorganic compounds such as H_2S or $NaNO_2$.

Physical Factors

Three of the most important physical factors that influence the growth and survival of cells are temperature, pH, and the gaseous environment. An understanding of the roles they play in cell metabolism is essential.

1. **Temperature** influences the rate of chemical reactions through its action on cellular enzymes. Bacteria, as a group of organisms, exist over a wide range of temperatures. However, individual species can exist only within a narrower spectrum of temperatures. Low temperatures slow down or inhibit enzyme activity, thereby slowing down or inhibiting cell metabolism and, consequently, cell growth. High temperatures cause coagulation and thus irreversibly denature thermolabile enzymes. Although enzymes differ in their degree of heat sensitivity, generally temperatures in the range of 70°C will destroy most essential enzymes and cause cell death.
2. **The pH of the extracellular environment** greatly affects cells' enzymatic activities. Most commonly, the optimum pH for cell metabolism is in the neutral range of 7. An increase in the hydrogen ion concentration resulting in an acidic pH (below 7) or a decrease in the hydrogen ion concentration resulting in an alkaline pH (above 7) is often detrimental. Either increase or decrease will slow down the rate of chemical reactions because of the destruction of cellular enzymes, thereby affecting the rate of growth and, ultimately, survival.
3. **The gaseous requirement** in most cells is atmospheric oxygen, which is necessary for the biooxidative process of respiration. Atmospheric oxygen plays a vital role in ATP formation and the availability of energy in a utilizable form for cell activities. Other cell types, however, lack the enzyme systems for respiration in the presence of oxygen and therefore must use an anaerobic form of respiration or fermentation.

The following exercises will demonstrate the diversity of nutritional and environmental requirements among microorganisms.

Nutritional Requirements: Media for the Routine Cultivation of Bacteria

LEARNING OBJECTIVES

Once you have completed this experiment, you should know how to evaluate

1. The abilities of several types of media to support the growth of different bacterial species.
2. The nutritional needs of the bacteria under study.

PRINCIPLE

To satisfy the diverse nutritional needs of bacteria, bacteriologists employ two major categories of media for routine cultivation.

Chemically Defined Media

These are composed of known quantities of chemically pure, specific organic and/or inorganic compounds. Their use requires knowledge of the organism's specific nutritional needs. The following two chemically defined media are used in this exercise:

1. **Inorganic synthetic broth:** This completely inorganic medium is prepared by incorporating the following salts per 1000 ml of water:

Sodium chloride (NaCl)	5.0 g
Magnesium sulfate ($MgSO_4$)	0.2 g
Ammonium dihydrogen phosphate ($NH_4H_2PO_4$)	1.0 g
Dipotassium hydrogen phosphate (K_2HPO_4)	1.0 g
Atmospheric carbon dioxide (CO_2)	

2. **Glucose salts broth:** This medium is composed of salts incorporated into the inorganic synthetic broth medium plus **glucose,** 5 g per liter, which serves as the sole organic carbon source.

Complex Media

The exact chemical composition of these media is not known. They are made of extracts of plant and animal tissue and are variable in their chemical composition. Most contain abundant amino acids, sugars, vitamins, and minerals; however, the quantities of these constituents are not known. They are capable of supporting the growth of most heterotrophs. The following two complex media are used in this exercise.

1. **Nutrient broth:** This basic complex medium is prepared by incorporating the following ingredients per 1000 ml of distilled water:

Peptone	5.0 g
Beef extract	3.0 g

   Peptone, a semidigested protein, is primarily a nitrogen source. The **beef extract,** a beef derivative, is a source of organic carbon, nitrogen, vitamins, and inorganic salts.

2. **Yeast extract broth:** This is composed of the basic artificial medium ingredients used in the nutrient broth plus **yeast extract,** 5 g per liter, which is a rich source of vitamin B and provides additional organic nitrogen and carbon compounds.

   The yeast extract broth is an example of an **enriched medium** and is used for the cultivation of **fastidious** microorganisms, organisms that have highly elaborate and specific nutritional needs. These bacteria do not grow or grow poorly on a basic artificial medium and require the addition of one or more growth-supporting substances, enrichments such as additional plant or animal extracts, vitamins, or blood.

MATERIALS

Cultures

Saline suspension of 24-hour trypticase soy broth cultures, adjusted to 0.05 optical density at a wavelength of 600 nm, of *Escherichia coli*, *Alcaligenes faecalis*, and *Streptococcus mitis*.

Media

Per designated student group: three test tubes (13 × 100 mm) of each type of broth: inorganic synthetic broth, glucose salts broth, nutrient broth, and yeast extract broth.

Equipment

Bunsen burner, sterile 1-ml serological pipettes, mechanical pipetting device, glassware marking pencil, test tube rack, and Bausch & Lomb Spectronic 20 spectrophotometer.

PROCEDURE

1. Using a sterile 1-ml pipette, add 0.1 ml of the *E. coli* culture to one test tube of each of the appropriately labeled media.

2. Repeat step 1 for inoculation with *A. faecalis* and *S. mitis*.

3. Incubate the test cultures for 24 to 48 hours at 37°C.

Name   Date

In this experiment you will evaluate (1) the abilities of media to support the growth of different species of bacteria, and (2) the nutritional needs of the bacteria. You will observe the amount of growth, measured by turbidity, present in each culture following incubation. To evaluate more accurately the amount of growth, a Bausch & Lomb Spectronic 20 spectrophotometer will be used.

This instrument measures the amount of light transmitted (T) or absorbed (A). It transmits a beam of light at a single wavelength (monochromatic light) through a liquid culture. The cells suspended in the culture interrupt the passage of light, and the amount of light energy transmitted through the suspension is measured on a photoelectric cell and converted into electrical energy. The electrical energy is then recorded on a galvanometer as 0% to 100% T. A schematic representation of a spectrophotometer is shown in Figure 14.1.

In practice, the density of a cell suspension is expressed as absorbance or optical density (O.D.) rather than percent T, since O.D. is directly proportional to the concentration of cells, whereas percent T is inversely proportional to the concentration of suspended cells. Therefore, as the turbidity of a culture increases, the O.D. increases and percent T decreases, indicating growth of the cell population in the culture. For example, in comparing three cultures with O.D. readings of 0.10 (percent T = 78), 0.30 (percent T = 49), and 0.50 (percent T = 30), the O.D. reading of 0.50 would be indicative of the most abundant growth, and the 0.10 reading would be indicative of the least amount of growth.

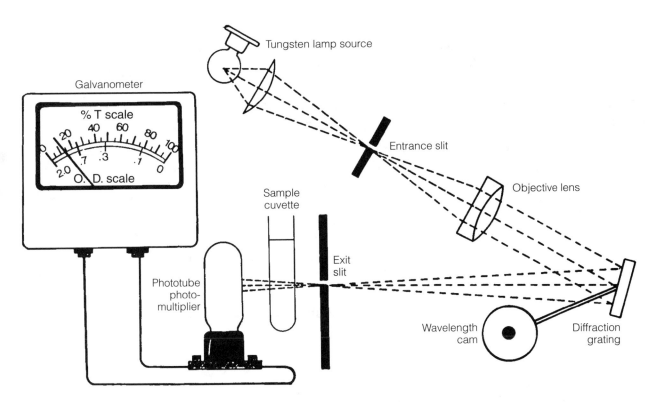

FIGURE 14.1 Schematic diagram of a spectrophotometer

a. Power switch/zero control d. Wavelength

b. Sample holder e. 100% control

c. Pilot lamp

FIGURE 14.2 The Bausch & Lomb Spectronic 20 spectrophotometer (Courtesy of Milton Roy Co., Rochester, New York)

Follow the instructions below and refer to Figure 14.2 for the use of the spectrophotometer to obtain the optical density readings of all your cultures.

1. Turn instrument on 10 to 15 minutes prior to use.
2. Set wavelength at 600 nm.
3. Set percent transmittance to 0% (O.D. to 2) by turning the knob on the left.
4. Read the four yeast broth cultures as follows:
   a. Wipe clean the provided test tube of sterile yeast broth that will serve as the blank for the yeast broth culture readings. Fingerprints on the test tube will obscure the light path of the spectrophotometer.
   b. Insert the yeast broth blank into the tube holder, close the cover, and set the O.D. to 0 (percent T = 100) by turning the knob on the right.
   c. Shake one of the tubes of yeast broth culture to resuspend the bacteria, wipe the test tube clean, and allow it to sit for several seconds for the equilibration of the bacterial suspension.
   d. Remove the yeast broth blank from the tube holder.
   e. Insert a yeast broth culture into the tube holder, close the cover, and read and record the optical density reading in the chart provided.
   f. Remove the yeast broth culture from the tube holder.
   g. Reset the spectrophotometer to an O.D. of 2 with the tube holder empty and to an O.D. of 0 with the yeast broth blank.
   h. Repeat Steps c through g to read and record the optical density of the remaining yeast broth cultures.
5. Repeat Step 4 (a–h) to read and record the optical density of the nutrient broth cultures. Use the provided nutrient broth blank to set the spectrophotometer to an O.D. of 0.

6. Repeat Step 4 (a–h) to read and record the optical density of the glucose salts broth cultures. Use the provided glucose salts broth blank to set the spectrophotometer to an O.D. of 0.

7. Repeat Step 4 (a–h) to read and record the optical density of the inorganic synthetic broth cultures. Use the provided inorganic synthetic broth blank to set the spectrophotometer to an O.D. of 0.

8. At the end of the experiment, return all cultures to the area designated for their disposal.

Optical Density Readings

	Yeast Broth	Nutrient Broth	Glucose Broth	Inorganic Synthetic Broth
E. coli				
A. faecalis				
S. mitis				

9. On the basis of the above data, list the media in order (from best to worst) according to their ability to support the growth of bacteria.

10. List the three bacterial species in order of their increasing fastidiousness.

11. Why did the most fastidious organism grow poorly in the chemically defined medium?

REVIEW QUESTIONS

1. Explain the advantages of using O.D. readings rather than percent T as a means of estimating microbial growth.

2. Explain the reason for the use of different medium blanks in adjusting the spectrophotometer prior to obtaining O.D. readings.

3. Why are complex media preferable to chemically defined media for routine cultivation of microorganisms?

4. Would you expect a heterotrophic organism to grow in an inorganic synthetic medium? Explain.

5. A soil isolate is found to grow poorly in a basic artificial medium. You suspect that a vitamin supplement is required.

   a. What supplement would you use to enrich the medium to support and maintain the growth of the organisms? Explain.

   b. Outline the procedure you would follow to determine the specific vitamins required by the organism to produce a more abundant growth.

Use of Differential, Selective, and Enriched Media

LEARNING OBJECTIVES

Once you have completed this experiment, you should be familiar with

1. The use and function of specialized media for the selection and differentiation of microorganisms.
2. How an enriched medium like blood agar can also function as both a selective and differential medium.

PRINCIPLE

Numerous special-purpose media are available for functions such as:

1. Isolation of bacterial types from a mixed population of organisms.
2. Differentiation among closely related groups of bacteria on the basis of macroscopic appearance of the colonies and biochemical reactions within the medium.
3. Enumeration of bacteria in sanitary microbiology, such as in water and sewage, and also in food and dairy products.
4. Assay of naturally occurring substances such as antibiotics, vitamins, and products of industrial fermentation.
5. Characterization and identification of bacteria by their abilities to produce chemical changes in different media.

In addition to nutrients necessary for the growth of all bacteria, special-purpose media contain one or more chemical compounds that are essential for their functional specificity. In this exercise, three types of media will be studied and evaluated.

Selective Media

These media are used to select (isolate) specific groups of bacteria. They incorporate chemical substances that inhibit the growth of one type of bacteria while permitting growth of another, thus facilitating bacterial isolation.

1. **Phenylethyl alcohol agar:** This medium is used for the isolation of most gram-positive cocci. The phenylethyl alcohol is partially inhibitory to gram-negative organisms, which may form visible colonies whose size and number are much smaller than on other media.
2. **Crystal violet agar:** This medium is selective for most gram-negative microorganisms. Crystal violet dye exerts an inhibitory effect on most gram-positive organisms.
3. **7.5% sodium chloride agar:** This medium is inhibitory to most organisms other than halophilic (salt-loving) microorganisms. It is most useful in the detection of members of the genus *Staphylococcus*.

Differential Media

These can distinguish among morphologically and biochemically related groups of organisms. They incorporate chemical compounds that, following inoculation and incubation, produce a characteristic change in the appearance of bacterial growth and/or the medium surrounding the colonies, which permits differentiation.

1. **Mannitol salt agar:** This medium contains a high salt concentration, 7.5% NaCl, which is inhibitory to the growth of most bacteria other than the staphylococci.

The medium also performs a differential function: it contains the carbohydrate mannitol, which some staphylococci are capable of fermenting, and phenol red, a pH indicator for detecting acid produced by mannitol-fermenting staphylococci. These staphylococci exhibit a yellow zone surrounding their growth; staphylococci that do not ferment mannitol will not produce a change in coloration.

2. **MacConkey agar:** The inhibitory action of crystal violet on the growth of gram-positive organisms allows the isolation of gram-negative bacteria. Incorporation of the carbohydrate lactose, bile salts, and the pH indicator neutral red permits differentiation of enteric bacteria on the basis of their ability to ferment lactose. On this basis, enteric bacteria are separated into two groups:

   a. **Coliform bacilli** produce acid as a result of lactose fermentation. The bacteria exhibit a red coloration on their surface. *Escherichia coli* produce greater quantities of acid from lactose than other coliform species. When this occurs, the medium surrounding the growth also becomes red because of the action of the acid that precipitates the bile salts, followed by absorption of the neutral red.

   b. **Dysentery, typhoid, and paratyphoid bacilli** are not lactose fermenters and therefore do not produce acid. The colonies appear uncolored and frequently transparent.

3. **Eosin-methylene blue agar (Levine):** Lactose and the dyes eosin and methylene blue permit differentiation between enteric lactose fermenters and nonfermenters as well as identification of the colon bacillus, *E. coli*. The *E. coli* colonies are blue-black with a metallic green sheen caused by the large quantity of acid that is produced and that precipitates the dyes onto the growth's surface. Other coliform bacteria, such as *Enterobacter aerogenes*, produce thick, mucoid, pink colonies on this medium. Enteric bacteria that do not ferment lactose produce colorless colonies, which, because of their transparency, appear to take on the purple color of the medium. This medium is also partially inhibitory to the growth of gram-positive organisms, and thus gram-negative growth is more abundant.

Enriched Media

Enriched media are media that have been supplemented with highly nutritious materials, such as blood, serum, or yeast extract, for the purpose of cultivating fastidious organisms.

1. **Blood agar:** The blood that is incorporated into this medium is an enrichment ingredient for the cultivation of fastidious organisms such as the *Streptococcus* spp. The blood also permits demonstration of the hemolytic properties of some microorganisms, particularly the streptococci, whose hemolytic activities are classified as follows:

   a. **Gamma hemolysis:** No lysis of red blood cells results in no significant change in the appearance of the medium surrounding the colonies.

   b. **Alpha hemolysis:** Incomplete lysis of red blood cells, with reduction of hemoglobin to methemoglobin, results in a greenish halo around the bacterial growth.

   c. **Beta hemolysis:** Lysis of red blood cells with complete destruction and use of hemoglobin by the organism results in a clear zone surrounding the colonies. This hemolysis is produced by two types of beta hemolysins, namely **streptolysin O,** an antigenic, oxygen-labile enzyme, and **streptolysin S,** a nonantigenic, oxygen-stable lysin. The hemolytic reaction is enhanced when blood agar plates are streaked and simultaneously stabbed to show subsurface hemolysis by streptolysin O in an environment with reduced oxygen tension.

MATERIALS

Cultures

24- to 48-hour trypticase soy broth cultures of *Enterobacter aerogenes, Escherichia coli, Streptococcus* var. Lancefield Group E, *Streptococcus mitis, Enterococcus faecalis, Staphylococcus aureus, Staphylococcus epidermidis,* and *Salmonella typhimurium.*

Media

Per designated student group: one each of phenylethyl alcohol agar, crystal violet agar, 7.5% sodium chloride agar, mannitol salt agar, MacConkey agar, eosin-methylene blue agar, and blood agar.

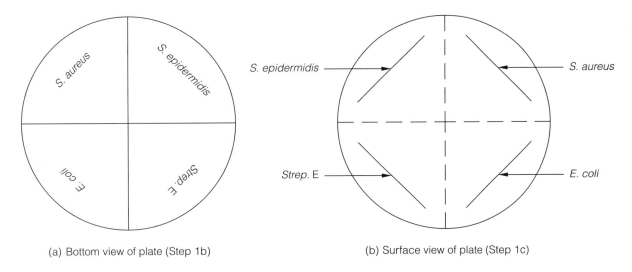

(a) Bottom view of plate (Step 1b)

(b) Surface view of plate (Step 1c)

FIGURE 15.1 Mannitol salt agar plate preparation and inoculation procedure

Equipment

Bunsen burner, inoculating loop, and glassware marking pencil.

PROCEDURE

1. Using the bacterial organisms listed in Step 2, prepare and inoculate each of the plates in the following manner:

   a. Appropriately label the cover of each plate as indicated in the section entitled "Laboratory Protocol" on page xv.

   b. Divide each of the Petri dishes into the required number of sections (one section for each different organism) by marking the *bottom of the dish*. Label each section with the name of the organism to be inoculated as illustrated in Figure 15.1.

   c. Using sterile technique, inoculate all plates, except the blood agar plate, with the designated organisms by making a single line of inoculation of each organism in its appropriate section. Be sure to close the Petri dish and flame the inoculating needle between inoculations of the different organisms. Refer to Figure 15.1 for an illustration of this procedure.

   d. Using sterile technique, inoculate the blood agar plate as described in Step 1c. Upon completion of each single line of inoculation, use the inoculating loop and make three or four stabs at a 45° angle across the streak.

2. Inoculate each of the different media with the following:

   a. Phenylethyl alcohol agar: *E. coli, S. aureus,* and *E. faecalis.*

   b. Crystal violet agar: *E. coli, S. aureus,* and *E. faecalis.*

   c. 7.5% sodium chloride agar: *S. aureus, S. epidermidis,* and *E. coli.*

   d. Mannitol salt agar: *S. aureus, S. epidermidis, Streptococcus* var. Lancefield Group E, and *E. coli.*

   e. MacConkey agar and eosin-methylene blue agar: *E. coli, E. aerogenes, S. typhimurium,* and *S. aureus.*

   f. Blood agar: *E. faecalis, S. mitis,* and *Streptococcus* var. Lancefield Group E.

3. Incubate the phenylethyl alcohol agar plate in an inverted position for 48 to 72 hours at 37°C. Incubate the remaining plates in an inverted position for 24 to 48 hours at 37°C.

Experiment 15 / Observations and Results

Name Date

1. Carefully examine each of the plates. Note and record the following on the chart below:
   a. Amount of growth along line of inoculation as follows: 0 = none; 1+ = scant; and 2+ = moderate to abundant.
   b. Appearance of the growth: coloration, transparency.
   c. Change in the appearance of the medium surrounding the growth: coloration, transparency indicative of hemolysis.

Refer to photo numbers 16–24 in the color-plate insert for illustration of these reactions.

Type of Medium	Medium	Bacterial Species	Amount of Growth	Appearance of Growth	Appearance of Medium
Selective	Phenylethyl alcohol agar	*E. coli* *S. aureus* *E. faecalis*			
	Crystal violet agar	*E. coli* *S. aureus* *E. faecalis*			
	7.5% sodium chloride agar	*E. coli* *S. aureus* *S. epidermidis*			
Differential	Mannitol salt agar	*E. coli* *Streptococcus* var. Lancefield Group E *S. aureus* *S. epidermidis*			
	MacConkey agar	*E. coli* *E. aerogenes* *S. typhimurium* *S. aureus*			
	Eosin-methylene blue agar	*E. coli* *E. aerogenes* *S. typhimurium* *S. aureus*			
Enriched	Blood agar	*S. mitis* *E. faecalis* *Streptococcus* var. Lancefield Group E			

2. Indicate the specific selective and/or differential purpose of each of the following media:
   a. Phenylethyl alcohol agar:

   b. Crystal violet agar:

c. 7.5% sodium chloride agar:

d. Mannitol salt agar:

e. MacConkey agar:

f. Eosin-methylene blue agar (Levine):

g. Blood agar:

REVIEW QUESTIONS

1. Explain the purpose of:
   a. Crystal violet in the MacConkey agar medium:

   b. Blood in the blood agar medium:

   c. Eosin and methylene blue dyes in the eosin-methylene blue agar medium:

   d. High salt concentration in the mannitol salt agar medium:

   e. Lactose in the MacConkey agar medium:

   f. Phenylethyl alcohol in the phenylethyl alcohol agar medium:

2. Why are crystal violet agar and 7.5% sodium chloride agar considered selective media?

3. A patient exhibits a boil on his neck. You, as a microbiology technician, are asked to identify and determine whether the causative organism is pathogenic. Describe the procedure that you would follow to make this determination.

Physical Factors: Temperature

LEARNING OBJECTIVES

Once you have completed this experiment, you should know

1. The diverse growth temperature requirements of bacteria.

2. How to determine whether the optimum growth temperature is also the ideal temperature for enzyme-regulated cell activities such as pigment production and carbohydrate fermentation.

PRINCIPLE

Microbial growth is directly dependent on how temperature affects cellular enzymes. With increasing temperatures, enzyme activity increases until the three-dimensional configuration of these molecules is lost because of denaturation of their protein structure. On the other hand, as the temperature is lowered toward the freezing point, enzyme inactivation occurs and cellular metabolism gradually diminishes. At 0°C, biochemical reactions cease in most cells.

Bacteria, as a group of living organisms, are capable of growth within an overall temperature range of minus 5°C to 80°C. Each species, however, requires a narrower range that is determined by the heat sensitivity of its enzyme systems. Specific temperature ranges consist of the following **cardinal (significant) temperature points:**

1. **Minimum growth temperature:** The lowest temperature at which growth will occur. Below this temperature, enzyme activity is inhibited and the cells are metabolically inactive so that growth is negligible or absent.

2. **Maximum growth temperature:** The highest temperature at which growth will occur.

Above this temperature, most cell enzymes are destroyed and the organism dies.

3. **Optimum growth temperature:** The temperature at which the rate of reproduction is most rapid; however, it is not necessarily optimum or ideal for all enzymatic activities of the cell.

All bacteria can be classified into one of three major groups, depending on their temperature requirements:

1. **Psychrophiles:** Bacterial species that will grow within a temperature range of minus 5°C to 20°C. The distinguishing characteristic of all psychrophiles is that they will grow between 0° and 5°C.

2. **Mesophiles:** Bacterial species that will grow within a temperature range of 20°C to 45°C. The distinguishing characteristics of all mesophiles are their ability to grow at human body temperature (37°C) and their inability to grow at temperatures above 45°C. Included among the mesophiles are two distinct groups:

   a. Those whose optimum growth temperature is in the range of 20°C to 30°C are plant saprophytes.

   b. Those whose optimum growth temperature is in the range of 35°C to 40°C are organisms that prefer to grow in the bodies of warm-blooded hosts.

3. **Thermophiles:** Bacterial species that will grow at 35°C and above. Two groups of thermophiles exist:

   a. **Facultative thermophiles:** Organisms that will grow at 37°C, with an optimum growth temperature of 45°C to 60°C.

   b. **Obligate thermophiles:** Organisms that will grow only at temperatures above 50°C, with optimum growth temperatures above 60°C.

The ideal temperature for specific enzymatic activities may not coincide with the optimum growth temperature for a given organism. To understand this concept, you will investigate pigment production and carbohydrate fermentation by selected organisms at a variety of incubation temperatures.

1. The production of an endogenous red or magenta pigment by *Serratia marcescens* is determined by the presence of an orange to deep red coloration on the surface of the colonial growth.

2. Carbohydrate fermentation by *Saccharomyces cerevisiae* is indicated by the presence of gas, one of the end products of this fermentative process. Detection of this accumulated gas may be noted as an air pocket, of varying size, in an inverted inner vial (Durham tube) within the culture tube. Refer to Experiment 23 for a more extensive discussion of carbohydrate fermentation.

MATERIALS

Cultures

24- to 48-hour nutrient broth cultures of *Escherichia coli*, *Bacillus stearothermophilus*, *Pseudomonas savastanoi*, *Serratia marcescens*, and Sabouraud broth culture of *Saccharomyces cerevisiae*.

Media

Per designated student group: four trypticase soy agar plates and four Sabouraud broth tubes containing inverted Durham tubes.

Equipment

Bunsen burner, inoculating loop, refrigerator set at 4°C, two incubators set at 37°C and 60°C, sterile Pasteur pipette, test tube rack, and glassware marking pencil.

PROCEDURE

1. Score the underside of all plates into four quadrants with a glassware marker. Label each section with the name of the test organism to be inoculated. When labeling the cover of each plate, include the temperature of incubation (4°C, 20°C, 37°C, or 60°C).

2. Aseptically inoculate each of the plates with *E. coli*, *B. stearothermophilus*, *P. savastanoi*, and *S. marcescens* by means of a single line of inoculation of each organism in its appropriately labeled section.

3. Appropriately label the four Sabouraud broth tubes, including the temperatures of incubation as indicated above.

4. Gently shake the *S. cerevisiae* culture to suspend the organisms. Using a sterile Pasteur pipette, aseptically add one drop of the culture into each of the four tubes of broth media.

5. Incubate all plates in an inverted position and the broth cultures at each of the four experimental temperatures (4°C, 20°C, 37°C, or 60°C) for 24 to 48 hours.

Experiment 16 / Observations and Results

Name ____ Date _______________

1. Observe all the cultures for the presence of growth. Record your observations in the chart below: (1+) for scant growth; (2+) for moderate growth; (3+) for abundant growth; and (−) for the absence of growth. Evaluate the amount of growth in the *S. cerevisiae* cultures by noting the degree of developed turbidity.
2. Observe the *S. marcescens* growth on all the plate cultures for the presence or absence of orange to deep red pigmentation. In the chart below, record the presence of pigment on a scale of 1+ to 3+, and enter (−) for the absence of pigmentation.
3. Observe the *S. cerevisiae* cultures for the presence of a gas pocket in the Durham tube, which is indicative of carbohydrate fermentation. Record your observations in the chart below using the following designations: (1+) for a minimal amount of gas; (2+) for a moderate amount of gas; (3+) for a large amount of gas; and (−) for the absence of gas.
4. In the chart below, classify the cultures as psychrophiles, mesophiles, facultative thermophiles, or obligate thermophiles.

Refer to photo number 25 in the color-plate insert for illustration of the effect of temperature on bacterial growth.

	Serratia marcescens		***Pseudomonas savastanoi***	***Escherichia coli***	***Bacillus stearothermophilus***	***Saccharomyces cerevisiae***	
Temperature	**Pigment**	**Growth**	**Growth**	**Growth**	**Growth**	**Growth**	**Gas**
4°C (refrig.)							
20°C (room temp.)							
37°C (body temp.)							
60°C							
Classification							

Based on your observations of the *S. marcescens* and *S. cerevisiae* cultures, is the optimum growth temperature the ideal temperature for all cell activities? Explain.

REVIEW QUESTIONS

1. In the following chart, indicate the types of organisms that would grow preferentially in or on various environments, and indicate the optimum temperature for their growth.

Environment	Type of Organism	Optimum Temperature
Ocean bottom near shore		
Ocean bottom near hot vent		
Hot sulfur spring		
Compost pile (middle)		
High mountain lake		
Center of an abscess		
Antarctic ice		

2. Explain the effects of temperatures above the maximum and below the minimum growth temperatures on cellular enzymes.

3. If an organism grew at 20°C, explain how you would determine experimentally whether the organism was a psychrophile or a mesophile.

4. Is it possible for thermophilic organisms to induce infections in warm-blooded animals? Explain.

Physical Factors: pH of the Extracellular Environment

LEARNING OBJECTIVE

Once you have completed this experiment, you should be familiar with

1. The pH requirements of microorganisms.

PRINCIPLE

Growth and survival of microorganisms are greatly influenced by the pH of the environment, and all bacteria and other microorganisms differ as to their requirements. Based on their pH optima, microorganisms may be classified as acidophilic, neutralophilic, or alkalophilic (Figure 17.1). Each species has the ability to grow within a specific pH range, which may be broad or limited, with the most rapid growth occurring within a narrow optimum range. These specific pH needs reflect the organisms' adaptations to their natural environment. For example, enteric bacteria are capable of survival within a broad pH range, which is characteristic of their natural habitat, the digestive system. Bacterial blood parasites, on the other hand, can tolerate only a narrow range; the pH of the circulatory system remains fairly constant at approximately 7.4.

Despite this diversity and the fact that certain organisms can grow at extremes of the pH scale, generalities can be made. The specific range for bacteria is between 4 and 9, with the optimum being 6.5 to 7.5. Fungi, molds, and yeasts prefer an acidic environment, with optimum activities at a pH of 4 to 6.

Because a neutral or nearly neutral environment is generally advantageous to the growth of microorganisms, the pH of the laboratory medium is frequently adjusted to approximately 7. Metabolic activities of the microorganism will result in the production of wastes, such as acids from carbohydrate degradation and alkali from protein breakdown, and these will cause shifts in pH that can be detrimental to growth.

To retard this shift, chemical substances that act as **buffers** are frequently incorporated when the medium is prepared. A commonly used **buffering system** involves the addition of equimolar concentrations of K_2HPO_4, a salt of a weak base, and KH_2PO_4, a salt of a weak acid. In a medium that has become acidic, the K_2HPO_4 absorbs excess H^+ to form a weakly acidic salt and a potassium salt with the anion of the strong acid.

$$K_2HPO_4 \quad + \quad HCl \quad \rightarrow \quad KH_2PO_4 \quad + \quad KCl$$

| **Salt of a weak base** | **Strong acid** | **Salt of a weak acid** | **Potassium chloride salt** |

In a medium that has become alkaline, KH_2PO_4 releases H^+ to form water by combining with the excess OH^-, and the remaining anionic portion of the weakly acidic salt combines with the cation of the alkali.

$$KH_2PO_4 \quad + \quad KOH \quad \rightarrow \quad K_2HPO_4 \quad + \quad H_2O$$

| **Salt of a weak acid** | **Strong base** | **Salt of a weak base** | **Water** |

Most media contain amino acids, peptones, and proteins, which because of their amphoteric nature, can act as natural buffers. For example, amino acids are zwitterions, molecules in which the amino group and the carboxyl group ionize to form dipolar ions. These behave in the following manner:

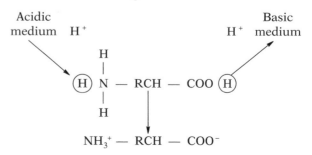

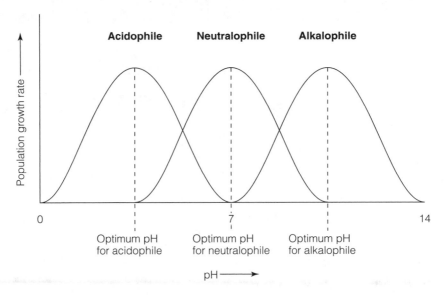

FIGURE 17.1 The effect of pH on the growth of microorganisms

MATERIALS

Cultures

Saline suspensions of 24-hour nutrient broth cultures, adjusted to an O.D. of 0.05 at a wavelength of 600 nm, of *Alcaligenes faecalis*, *Escherichia coli*, and *Saccharomyces cerevisiae*.

Media

Per designated student group: 12 trypticase soy broth (TSB) tubes, three at each of the following pH designations: 3, 6, 7, and 9. The pH has been adjusted with 1N sodium hydroxide or 1N hydrochloric acid.

Equipment

Bunsen burner, sterile 1-ml pipettes, mechanical pipetting device, Bausch & Lomb Spectronic 20 spectrophotometer, test tube rack, and glassware marking pencil.

PROCEDURE

1. Using a sterile pipette, inoculate a series of the appropriately labeled TSB tubes of media, pH values of 3, 6, 7, and 9, with *E. coli* by adding 0.1 ml of the saline culture to each.

2. Repeat Step 1 for the inoculation of *A. faecalis* and *S. cerevisiae*, using a new sterile pipette each time.

3. Incubate the *A. faecalis* and *E. coli* cultures for 24 to 48 hours at 37°C and the *S. cerevisiae* cultures for 48 to 72 hours at 25°C.

Name _____   Date _____

1. Using the spectrophotometer as described in Experiment 14, determine the optical density of all cultures and record the readings in the chart.

Microbial Species	Optical Density Readings			
	pH 3	pH 6	pH 7	pH 9
E. coli				
A. faecalis				
S. cerevisiae				

2. Summarize your results as to the overall range and optimum pH of each organism studied in the chart.

Microbial Species	pH Range	Optimum pH
E. coli		
A. faecalis		
S. cerevisiae		

REVIEW QUESTIONS

1. Explain the mechanism by which buffers prevent radical shifts in pH.

2. Explain why it is necessary to incorporate buffers into media in which microorganisms are grown.

3. Why are proteins and amino acids considered to be natural buffers?

4. Explain why microorganisms differ in their pH requirements.

5. Will all microorganisms grow optimally at a neutral pH? Explain.

6. You are instructed to grow E. coli in a chemically defined medium containing glucose and NH$_4$Cl as the carbon and nitrogen sources and also in nutrient broth that contains beef extract and peptone. Both media are adjusted to a pH of 7. With turbidity as an index for the amount of growth in each of the cultures, the following spectrophotometric readings are obtained following incubation:

| Time (hours) | Optical Density Readings | |
	Chemically Defined Medium	Nutrient Broth Medium
6	0.100	0.100
12	0.300	0.500
18	0.275	0.900
24	0.125	1.500

Based on the above data, explain why E. coli ceased growing in the chemically defined medium but continuted to grow in the nutrient broth.

Physical Factors: Atmospheric Oxygen Requirements

LEARNING OBJECTIVE

Once you have completed this experiment, you should be familiar with

1. The diverse atmospheric oxygen requirements of microorganisms.

PRINCIPLE

Microorganisms exhibit great diversity in their ability to use free oxygen (O_2) for cellular respiration. These variations in O_2 requirements reflect the differences in biooxidative enzyme systems present in the various species. As such, they can be classified into one of five major groups according to their O_2 needs:

1. **Aerobes:** These require the presence of atmospheric oxygen for growth. Their enzyme system necessitates use of O_2 as the final hydrogen (electron) acceptor in the complete oxidative degradation of high-energy molecules such as glucose.

2. **Microaerophiles:** They require limited amounts of atmospheric oxygen for growth. Oxygen in excess of the required amount appears to block the activities of their oxidative enzymes and results in death.

3. **Obligate anaerobes:** They require the absence of free oxygen for growth because their oxidative enzyme system requires the presence of molecules other than O_2 to act as the final hydrogen (electron) acceptor. In these organisms, as in aerobes, the presence of atmospheric oxygen results in the formation of toxic metabolic end products, such as superoxide, O_2^-, a free radical of oxygen. However, these organisms lack the enzymes superoxide dismutase and

catalase, whose function is to degrade the superoxide to water and oxygen as follows:

$$2O_2^- + 2H^+ \xrightarrow{\text{Superoxide Dismutase}} H_2O_2 + O_2$$

$$2H_2O_2 \xrightarrow{\text{Catalase}} 2H_2O + O_2$$

Therefore, in the absence of these enzymes, small amounts of atmospheric oxygen are lethal, and justifiably these organisms are called obligate anaerobes.

4. **Aerotolerant anaerobes:** These are fermentative organisms, and therefore they do not use O_2 as a final electron acceptor. Unlike the obligate anaerobes, they produce catalase and/or superoxide dismutase, and thus they are not killed by the presence of O_2. Hence, these organisms are anaerobes that are termed aerotolerant.

5. **Facultative anaerobes:** These organisms can grow in the presence or absence of free oxygen. They preferentially use oxygen for aerobic respiration. However, in an oxygen-poor environment, cellular respiration may occur anaerobically, utilizing such compounds as nitrates (NO_3^-) or sulfates (SO_4^{2-}) as final hydrogen acceptors, or via a fermentative pathway (refer to Experiment 23).

The oxygen needs of microorganisms can be determined by noting their growth distributions following a **shake-tube inoculation.** This procedure requires introduction of the inoculum into a melted agar medium, shaking of the test tube to disperse the microorganisms throughout the agar, and rapid solidification of the medium to ensure that the cells remain dispersed. Following incubation, the growth distribution indicates the organisms' oxygen requirements. Aerobes exhibit surface growth, whereas anaerobic growth is limited to the bottom of the deep tube. Facultative anaerobes, because of their indifference to the presence or absence of oxygen, exhibit growth throughout the medium. Microaerophiles grow in a zone slightly below the surface. Figure 18.1 illustrates the shake-tube inoculation procedure and the distribution of growth following an appropriate incubation period.

MATERIALS
Cultures

24- to 48-hour nutrient broth cultures of *Staphylococcus aureus, Corynebacterium xerosis,* and *Enterococcus faecalis;* 48- to 72-hour Sabouraud broth cultures of *Saccharomyces cerevisiae and Aspergillus niger;* and a 48-hour thioglycollate broth culture of *Clostridium sporogenes.*

Media

Six brain–heart infusion agar deep tubes per designated student group.

Equipment

Bunsen burner, water bath, ice-water bath, thermometer, sterile Pasteur pipettes, test tube rack, and glassware marking pencil.

PROCEDURE

1. Liquefy the sterile infusion agar by boiling in a water bath at 100°C.

2. Cool molten agar to 45°C; check temperature with a thermometer inserted into the water bath.

3. Using sterile technique, inoculate each experimental organism by introducing two drops of the culture from a sterile Pasteur pipette into the appropriately labeled tubes of molten agar.

4. Vigorously rotate the freshly inoculated molten infusion agar between the palms of the hands to distribute the organisms.

5. Place inoculated test tubes in an upright position in the ice-water bath to solidify the medium rapidly.

6. Incubate the *S. aureus, C. xerosis, E. faecalis,* and *C. sporogenes* cultures for 24 to 48 hours at 37°C and the *A. niger* and *S. cerevisiae* cultures for 48 to 72 hours at 25°C.

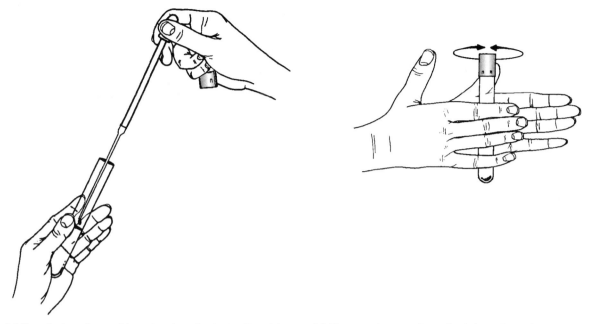

(a) Transfer two drops of inoculum from the test culture into a melted agar deep tube.

(b) Disperse the organisms throughout the molten agar medium by rapidly rotating the tube between the palms of your hands.

(c) Cool rapidly by immersion in an ice-water bath.

(d) Incubate at 37°C.

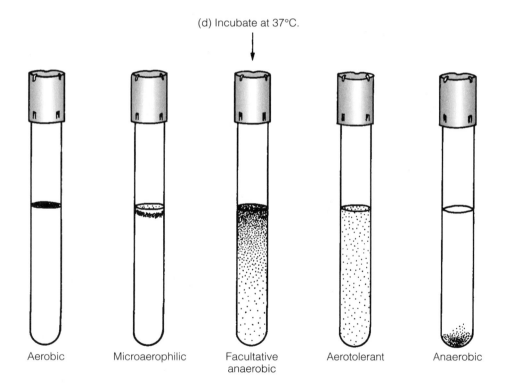

Aerobic   Microaerophilic   Facultative anaerobic   Aerotolerant   Anaerobic

FIGURE 18.1 Procedure for determination of oxygen requirements

Name   Date

1. Observe each of the experimental cultures for the distribution of growth in each tube.
2. Record your observations and your determination of the oxygen requirements for each of the experimental species in the chart.

Species	Distribution of Growth	Classification According to Oxygen Requirement
S. aureus		
C. xerosis		
E. faecalis		
A. niger		
S. cerevisiae		
C. sporogenes		

REVIEW QUESTIONS

1. Why is it necessary to place the inoculated molten agar cultures in an ice-water bath for their rapid solidification?

2. As indicated by its oxygen requirements, which group of microorganisms has the most extensive bioenergetic enzyme system? Explain.

3. Account for the inability of aerobes to grow in the absence of O_2.

4. Account for the subsurface growth of microaerophiles in a shake-tube culture.

5. Consider the culture type in which growth was distributed throughout the entire medium and explain why the growth was more abundant toward the surface of the medium in some cultures, whereas other cultures showed an equal distribution of growth throughout the tubes.

6. Account for the fact that the *C. sporogenes* culture showed a separation within the medium or an elevation of the medium from the bottom of the test tube.

7. Your instructor asks you to explain why the *Streptococcus* species that are catalase negative are capable of growth in the presence of oxygen. How would you respond?

Techniques for the Cultivation of Anaerobic Microorganisms

LEARNING OBJECTIVES

Once you have completed this experiment, you should be familiar with

1. The methods for cultivation of anaerobic organisms.

PRINCIPLE

Microorganisms differ in their abilities to use oxygen for cellular respiration. **Respiration** involves the oxidation of substrates for energy necessary to life. A substrate is **oxidized** when it loses a hydrogen ion and its electron H^+e^-. Since the H^+e^- cannot remain free in the cell, it must immediately be picked up by an electron acceptor, which becomes reduced. Therefore reduction is the gain of the H^+e^-. These are termed **oxidation-reduction (redox)** reactions. Some microorganisms have enzyme systems in which oxygen can serve as an electron acceptor, thereby being reduced to water. These cells have high oxidation-reduction potentials; others have low potentials and must use other substances as electron acceptors.

The enzymatic differences in microorganisms are explained more fully in the section dealing with metabolism (see Part V). This discussion is limited to cultivation of the strict anaerobes, which cannot be cultivated in the presence of atmospheric oxygen (Figure 19.1). The procedure is somewhat more difficult because it involves sophisticated equipment and media enriched with substances that lower the redox potential. Figure 19.2 shows some of the methods available for anaerobic cultivation.

The following experiment uses fluid thioglycollate medium and the GasPak™ anaerobic system (Figure 19.3).

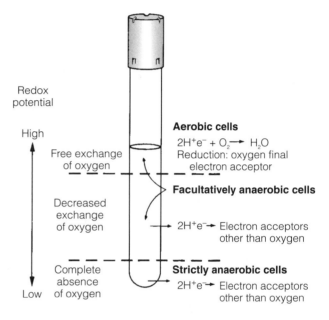

FIGURE 19.1 Illustration of redox potentials in an agar deep tube

MATERIALS

Cultures

24- to 48-hour nutrient broth cultures of *Bacillus cereus, Escherichia coli,* and *Micrococcus luteus;* and 48-hour thioglycollate broth culture of *Clostridium sporogenes.*

Media

Per designated student group: four screw-cap tubes of fluid thioglycollate medium and four nutrient agar plates.

Equipment

Bunsen burner, inoculating loop, GasPak anaerobic system, test tube rack, and glassware marking pencil.

Evacuation and replacement of oxygen atmosphere in sealed jars

Brewer jar: High-vacuum pump evacuates $O_2\uparrow$, which is replaced with a mixture of 95% $N_2\uparrow$ + 5% $CO_2\uparrow$.
Platinum catalyst in jar lid results in binding of residual $O_2\uparrow$ with $H_2\uparrow$, causing formation of H_2O.

GasPak system: Disposable $H_2\uparrow$ + $CO_2\uparrow$ envelope generator. Requires no evacuation of jar, no high-vacuum pumping equipment. Room-temperature catalyst that requires no electrical activation is used. Evolved $H_2\uparrow$ reacts with $O_2\uparrow$ to yield H_2O. (See Figure 19.3.)

Chromium–sulfuric acid method: $H_2\uparrow$ is generated in a desiccator jar following the reaction of 15% H_2SO_4 with chromium powder. $H_2SO_4 + Cr^{++} \rightarrow CrSO_4 + H_2\uparrow$. As $H_2\uparrow$ is evolved, $O_2\uparrow$ is forced out of desiccator jar and replaced with $H_2\uparrow$.

Specialized methods not requiring the use of sealed jars

Solid medium

Shake-culture technique: Molten and cooled nutrient agar is inoculated with loopful of organism. The tube is shaken, cooled rapidly, and incubated. Position of growth in tube is an index of gaseous requirement of organism. (See Figure 18.1 on page 119.)

Pyrogallic acid technique: Streak cultures on nutrient agar slants. Push a cotton plug into tube until it nearly touches slant. Fill space above cotton with pyrogallic acid crystals and add sodium hydroxide. Insert stopper tightly. Invert and incubate. Chemicals absorb $O_2\uparrow$, producing anaerobic environment.

Liquid medium

Paraffin plug technique: Any medium containing reducing substances (such as brain–heart infusion, liver–veal, cystine, or ascorbic acid) may be used. The medium is heated to drive off $O_2\uparrow$, rapidly cooled, and inoculated with a loopful of culture. This is immediately sealed with a half-inch of melted paraffin and incubated.

Fluid thioglycollate: This medium contains sodium thioglycollate, which binds to $O_2\uparrow$, thus acting as a reducing compound. Also present is a redox potential indicator, such as resazurin, that produces a pink coloration in an oxidized environment.

FIGURE 19.2 Methods for the cultivation of anaerobic microorganisms

PROCEDURE

Fluid Thioglycollate Medium

1. For the performance of this procedure, the fluid thioglycollate medium must be fresh. Freshness is indicated by the absence of a pink color in the upper one-third of the medium. If this coloration is present, loosen the screw caps and place the tubes in a boiling water bath for 10 minutes to drive off the dissolved O_2 from the medium. Cool the tubes to 45°C before inoculation.

2. Aseptically inoculate the appropriately labeled tubes of thioglycollate with their respective test organisms by means of loop inoculations *to the depths of the media.*

3. Incubate the cultures for 24 to 48 hours at 37°C.

GasPak Anaerobic Technique

The GasPak system as shown in Figure 19.3 is a contemporary method for the exclusion of oxygen from a sealed jar used for incubation of anaerobic cultures in a nonreducing medium. This system uses a GasPak generator that consists of a foil package that generates hydrogen and carbon dioxide upon the addition of water. A palladium catalyst in the lid of the jar combines the evolved hydrogen with residual oxygen to form water, thereby creating a carbon dioxide environment within the jar that is conducive for anaerobic growth. The establishment of anaerobic conditions is verified by the color change of a methylene blue indicator strip in the jar. This blue indicator becomes colorless in the absence of oxygen.

1. With a glassware marking pencil, divide the bottom of each nutrient agar plate into two sections.

2. Label each section on two plates with the name of the organism to be inoculated.

3. Repeat Step 2 to prepare a duplicate set of cultures.

4. Using sterile technique, make a single-line streak inoculation of each test organism in its respectively labeled section on both sets of plates.

5. Tear off the corner of the hydrogen and carbon dioxide gas generator and insert this inside the GasPak jar.

6. Place one set of plate cultures in an inverted position inside the GasPak chamber.

7. Expose the anaerobic indicator strip and place it inside the anaerobic jar so that the wick is visible from the outside.

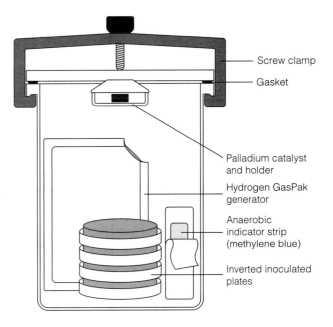

FIGURE 19.3 GasPak system (BBL Microbiology Systems, Division of Becton Dickinson and Company)

8. With a pipette, add the required 10 ml of water to the gas generator and quickly seal the chamber with its lid.

9. Place the sealed jar in an incubator at 37°C for 24 to 48 hours. After several hours of incubation, observe the indicator strip for a color change to colorless, which is indicative of anaerobic conditions.

10. Incubate the duplicate set of plates in an inverted position for 24 to 48 hours at 37°C under aerobic conditions.

Name   Date

1. Observe the fluid thioglycollate cultures, GasPak system, and aerobically incubated plate cultures for the presence of growth. Record your results in the chart.

2. Based on your observation, indicate the oxygen requirement classification of each test organism as anaerobe, facultative anaerobe, or aerobe.

Refer to photo number 26 in the color-plate insert for illustration of growth patterns.

Bacterial Species	Fluid Thioglycollate	GasPak Anaerobic Incubation	Aerobic Incubation	Oxygen Requirement Classification
M. luteus				
B. cereus				
E. coli				
C. sporogenes				

REVIEW QUESTIONS

1. Why can media such as brain–heart infusion and thioglycollate be used for the cultivation of anaerobes?

2. What are the purposes of the indicator strip and the gas generator in the GasPak system?

3. Heroin addicts have a high incidence of *Clostridium tetani* infections. Discuss the reasons for the development of this type of infection in these IV drug users.

4. ○ While you are working in your garden, a tine of the pitchfork accidentally produces a deep puncture wound in the calf of your leg. Discuss the type of infectious process you would be primarily concerned about and why.

5. ○ The physician who treats your puncture wound opts to insert a drain before applying the dressing. What is the rationale for the insertion of the drain?

Serial Dilution–Agar Plate Procedure to Quantitate Viable Cells

LEARNING OBJECTIVES

Once you have completed this experiment, you should understand

1. The diverse methods used to determine the number of cells in a bacterial culture.

2. How to determine quantitatively the number of viable cells in a bacterial culture.

PRINCIPLE

Studies involving the analysis of materials such as food, water, milk, and, in some cases, air require quantitative enumeration of microorganisms in the substances. Many methods have been devised to accomplish this, including direct microscopic counts, use of an electronic cell counter such as the Coulter counter, chemical methods for estimating cell mass or cellular constituents, turbidimetric measurements for increases in cell mass, and the serial dilution-agar plate method.

Direct Microscopic Counts

These require the use of a specialized counting chamber called the **Petroff-Hauser chamber** (Figure 20.1), in which an aliquot of a cell suspension is counted and the total cell number is determined mathematically.

1. The total number of cells counted in the five squares designated in the chamber equals the number of organisms/mm^2 of surface. If the suspension was diluted, multiply by the inverse of the dilution.

2. Cells/mm$^2 \times 10$ = number of cells/mm^3

3. Cells/mm$^3 \times 1000$ = converts cells/mm^3 to cells/ml of suspension

Although rapid, it has the disadvantages that both living and dead cells are counted and

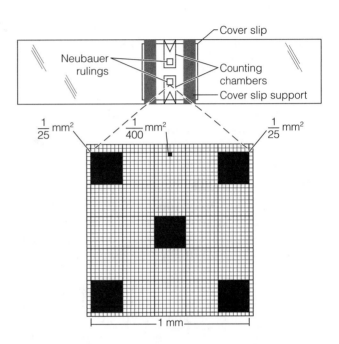

FIGURE 20.1 The Petroff-Hauser chamber

that it is not sensitive to populations of fewer than 1 million cells.

Breed smears are used mainly to quantitate bacterial cells in milk. Using stained smears confined to a 1-square-millimeter ruled area of the slide, the total population is determined mathematically. This method also fails to discriminate between viable and dead cells.

Electronic Cell Counters

The **Coulter counter**® is an example of an instrument capable of rapidly counting the number of cells suspended in a conducting fluid that passes through a minute orifice through which an electric current is flowing. Cells, which are nonconductors, increase the electrical resistance of the conducting fluid, and the resistance is electronically recorded, enumerating the number of organisms flowing

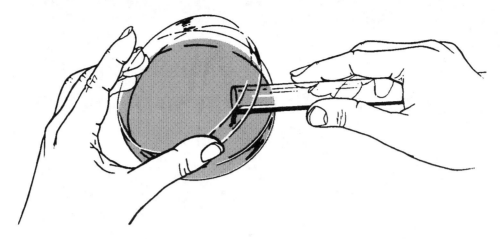

FIGURE 20.2 Pour-plate technique

through the orifice. In addition to its inability to distinguish between living and dead cells, the apparatus is also unable to differentiate inert particulate matter from cellular material.

Chemical Methods

While not considered means of direct quantitative analysis, chemical methods may be used to indirectly measure increases both in protein concentration and in DNA production. In addition, cell mass can be estimated by dry weight determination of a specific aliquot of the culture. Measurement of certain metabolic parameters may also be used to quantitate bacterial populations. The amount of oxygen consumed (oxygen uptake) is directly proportional to the increasing number of vigorously growing aerobic cells, and the rate of carbon dioxide production is related to increased growth of anaerobic organisms.

Spectrophotometric Analysis

Increased turbidity in a culture is another index of growth. With turbidimetric instruments, the amount of transmitted light decreases as the cell population increases, and the decrease in radiant energy is converted to electrical energy and indicated on a galvanometer. This method is rapid but limited because sensitivity is restricted to microbial suspensions of 10 million cells or greater.

Serial Dilution/Pour-Plate Analysis

While all these methods may be used to enumerate the number of cells in a bacterial culture, the major disadvantage common to all

is that the total count includes dead as well as living cells. Sanitary and medical microbiology at times require determination of viable cells. To accomplish this, the serial dilution–agar plate technique is used. Briefly, this method involves serial dilution of a bacterial suspension in sterile water blanks, which serve as a diluent of known volume. Once diluted, the suspensions are placed on suitable nutrient media. The **pour-plate technique,** illustrated in Figure 20.2, is the procedure usually employed. Molten agar, cooled to 45°C, is poured into a Petri dish containing a specified amount of the diluted sample. Following addition of the molten-then-cooled agar, the cover is replaced, and the plate is gently rotated in a circular motion to achieve uniform distribution of microorganisms. This procedure is repeated for all dilutions to be plated. Dilutions should be plated in duplicate for greater accuracy, incubated overnight, and counted on a **Quebec® colony counter** either by hand or by an electronically modified version of this instrument.

Plates suitable for counting must contain not fewer than 30 or more than 300 colonies. The total count of the suspension is obtained by multiplying the number of cells per plate by the dilution factor, which is the reciprocal of the dilution.

Advantages of the serial dilution–agar plate technique are as follows:

1. Only viable cells are counted.

2. It allows isolation of discrete colonies that can be subcultured into pure cultures, which may then be easily studied and identified.

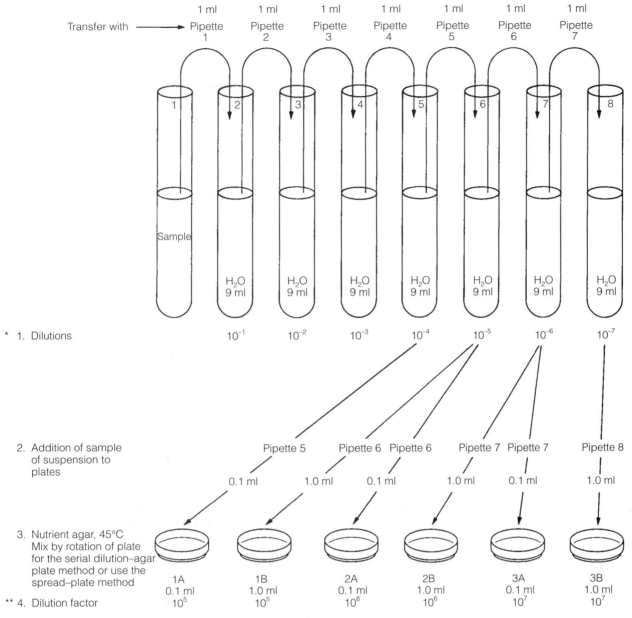

Transfer with

* 1. Dilutions $\quad 10^{-1} \quad 10^{-2} \quad 10^{-3} \quad 10^{-4} \quad 10^{-5} \quad 10^{-6} \quad 10^{-7}$

2. Addition of sample of suspension to plates

Pipette 5   Pipette 6   Pipette 6   Pipette 7   Pipette 7   Pipette 8

0.1 ml   1.0 ml   0.1 ml   1.0 ml   0.1 ml   1.0 ml

3. Nutrient agar, 45°C
Mix by rotation of plate for the serial dilution–agar plate method or use the spread–plate method

** 4. Dilution factor

	1A	1B	2A	2B	3A	3B
	0.1 ml	1.0 ml	0.1 ml	1.0 ml	0.1 ml	1.0 ml
	10^5	10^5	10^6	10^6	10^7	10^7

5. Incubate 24 hr at 37°C

6. Enumerate using Quebec colony counter

*   *Dilution* refers to varying the concentration of a substance.
**   *Dilution factor* is expressed mathematically as the reciprocal of the dilution.
For example, a dilution of 10^{-3} has a dilution factor of 10^3.

FIGURE 20.3 Serial dilution–agar plate procedure

Disadvantages of this method are as follows:

1. Overnight incubation is necessary before colonies develop on the agar surface.

2. It is necessary to use more glassware in this procedure.

3. The need for greater manipulation may result in erroneous counts due to errors in dilution or plating.

The following experiment uses the pour-plate technique for plating serially diluted culture samples. The procedure to be followed is illustrated in Figure 20.3.

MATERIALS

Culture

24- to 48-hour nutrient broth culture of *Escherichia coli*.

Media

Per designated student group: six 20-ml nutrient agar deep tubes and seven sterile 9-ml water blanks.

Equipment

Hot plate, water bath, thermometer, test tube rack, Bunsen burner, sterile 1-ml serological pipettes, mechanical pipetting device, sterile Petri dishes, Quebec colony counter, manual hand counter, disinfectant solution in a 500-ml beaker, glassware marking pencil, turntable, bent glass rod, test tube rack, and beaker with 95% alcohol.

PROCEDURE

1. Liquefy six agar deep tubes in an autoclave or by boiling. Cool the molten agar tubes and maintain in a water bath at 45°C.

2. Label the *E. coli* culture tube as Number 1 and the seven 9-ml water blanks as Numbers 2 through 8. Place the labeled tubes in a test tube rack. Label the Petri dishes 1A, 1B, 2A, 2B, 3A, and 3B.

3. Mix the *E. coli* culture (tube Number 1) by rolling the tube between the palms of your hands to ensure even dispersal of cells in the culture.

4. With a sterile pipette, aseptically transfer 1 ml from the bacterial suspension tube Number 1 to water blank tube Number 2. Discard the pipette in the beaker of disinfectant. The culture has been diluted 10 times to 10^{-1}.

5. Mix tube Number 2 and, with a fresh pipette, transfer 1 ml to tube Number 3. Discard the pipette. The culture has been diluted 100 times to 10^{-2}.

6. Mix tube Number 3 and, with a fresh pipette, transfer 1 ml to tube Number 4. Discard the pipette. The culture has been diluted 1000 times to 10^{-3}.

7. Mix tube Number 4 and, with a fresh pipette, transfer 1 ml to tube Number 5. Discard the pipette. The culture has been diluted 10,000 times to 10^{-4}.

8. Mix tube Number 5 and, with a fresh pipette, transfer 0.1 ml of this suspension to Plate 1A. Return the pipette to tube Number 5 and transfer 1 ml to tube Number 6. Discard the pipette. The culture has been diluted 100,000 times to 10^{-5}.

9. Mix tube Number 6 and, with a fresh pipette, transfer 1 ml of this suspension to Plate 1B. Return the pipette to tube Number 6 and transfer 0.1 ml to Plate 2A. Return the pipette to tube Number 6 and transfer 1 ml to tube Number 7. Discard the pipette. The culture has been diluted 1,000,000 times to 10^{-6}.

10. Mix tube Number 7 and, with a fresh pipette, transfer 1 ml of this suspension to Plate 2B. Return the pipette to tube Number 7 and transfer 0.1 ml to Plate 3A. Return the pipette to tube Number 7 and transfer 1 ml to tube Number 8. Discard the pipette. The culture has been diluted 10,000,000 times to 10^{-7}.

11. Mix tube Number 8 and, with a fresh pipette, transfer 1 ml of this suspension to Plate 3B. Discard the pipette. The dilution procedure is now complete.

12. Check the temperature of the molten agar medium to be sure the temperature is 45°C. Remove a tube from the water bath and wipe the outside surface dry with a paper towel. Using the pour-plate technique, pour the agar into Plate 1A as shown in Figure 20.2 and rotate the plate gently to ensure uniform distribution of the cells in the medium.

13. Repeat Step 12 for the addition of molten nutrient agar to Plates 1B, 2A, 2B, 3A, and 3B.

14. Once the agar has solidified, incubate the plates in an inverted position for 24 hours at 37°C.

Note: If desired, the spread-plate technique (Experiment 2) may be substituted for the agar pour-plate method described in this experiment. In this case, the dilutions may be placed on the surface of the hardened agar with a sterile pipette and distributed over the surface by means of a bent glass rod and turntable illustrated in Figure 2.2. Following incubation, cell counts may be made as described in Experiment 20. The resultant cell counts should be the same with either system. The main difference is that there will be no subsurface colonies in the spread-plate method.

Name _____   Date _____

1. Using a Quebec colony counter and a mechanical hand counter, observe all colonies on plates. Statistically valid plate counts are only obtained from bacterial cell dilutions that yield between 30 and 300 colonies. Plates with more than 300 colonies cannot be counted and are designated as **too numerous to count—TNTC;** plates with fewer than 30 colonies are designated as **too few to count—TFTC.** Count only plates containing between 30 and 300 colonies. Remember to count all subsurface as well as surface colonies.

2. The number of organisms per ml of original culture is calculated by multiplying the number of colonies counted by the dilution factor:

    Number of cells per ml = number of colonies × dilution factor

    Examples:

    a. Colonies per plate = 50
        Dilution factor = $1 : 1 \times 10^6$
                (1:1,000,000)
        Volume of dilution added to plate = 1 ml
        $50 \times 1,000,000 = 50,000,000$ or
                (5×10^7) CFUs/ml
                (colony-forming units)

    b. Colonies per plate = 50
        Dilution factor = $1 : 1 \times 10^5$ (1:100,000)
        Volume of dilution added to plate = 0.1 ml
        $50 \times 100,000 = 5,000,000$ (5×10^6)
                cells/0.1 ml
        $5,000,000 \times 10 = 50,000,000$
                (5×10^7) CFUs/ml

3. Record your observations and calculated bacterial counts per ml of sample in the chart.

4. Since the dilutions plated are replicates of each other, determine the average of the duplicate bacterial counts per ml of sample and record in the chart.

Refer to photo numbers 27–31 in the color-plate insert for illustration.

Plate	Tube Dilution	ml of Dilution Plated	Final Dilution on Plate	Number of Colonies	Bacterial Count per ml of Sample (CFU/ml)	Average Count per ml of Sample (CFU/ml)
1A						
1B						
2A						
2B						
3A						
3B						

REVIEW QUESTIONS

1. What is the major disadvantage of microbial counts performed by methods other than the serial dilution–agar plate procedure?

2. Distinguish between dilution and dilution factor.

3. What are the advantages and disadvantages of the serial dilution–agar plate procedure?

4. If 0.1 ml of a 1×10^{-6} dilution plate contains 56 colonies, calculate the number of cells per ml of the original culture.

5. How would you record your observation of a plate containing 305 colonies? A plate with 15 colonies?

6. Explain the chemical methods for measuring cell growth.

7. Your instructor asks you to determine the number of organisms in a water sample. Observation of your dilution plates reveals the presence of spreading colonial forms on some of the culture plates. What is the rationale for the elimination of these plate counts from your experimental data?

The Bacterial Growth Curve

LEARNING OBJECTIVES

Once you have completed this experiment, you should be able to

1. Understand the population growth dynamics of bacterial cultures.

2. Plot a bacterial growth curve.

3. Determine the generation time of a bacterial culture from the bacterial growth curve.

PRINCIPLE

Bacterial population growth studies require inoculation of viable cells into a sterile broth medium and incubation of the culture under optimum temperature, pH, and gaseous conditions. Under these conditions, the cells will reproduce rapidly and the dynamics of the microbial growth can be charted by means of a population growth curve, which is constructed by plotting the increase in cell numbers versus time of incubation. The curve can be used to delineate stages of the growth cycle. It also facilitates measurement of cell numbers and the rate of growth of a particular organism under standardized conditions as expressed by its **generation time,** the time required for a microbial population to double.

The stages of a typical growth curve (Figure 21.1) are:

1. **Lag phase:** During this stage the cells are adjusting to their new environment. Cellular metabolism is accelerated, resulting in rapid biosynthesis of cellular macromolecules, primarily enzymes, in preparation for the next phase of the cycle. Although the cells are increasing in size, there is no cell division and therefore no increase in numbers.

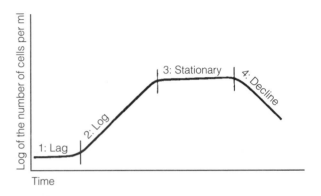

FIGURE 21.1 Population growth curve

2. **Logarithmic (log) phase:** Under optimum nutritional and physical conditions, the physiologically robust cells reproduce at a uniform and rapid rate by binary fission. Thus there is a rapid exponential increase in population, which doubles regularly until a maximum number of cells is reached. The time required for the population to double is the generation time. The length of the log phase varies, depending on the organisms and the composition of the medium. The average may be estimated to last 6 to 12 hours.

3. **Stationary phase:** During this stage, the number of cells undergoing division is equal to the number of cells that are dying. Therefore there is no further increase in cell number, and the population is maintained at its maximum level for a period of time. The primary factors responsible for this phase are the depletion of some essential metabolites and the accumulation of toxic acidic or alkaline end products in the medium.

4. **Decline, or death, phase:** Because of the continuing depletion of nutrients and buildup of metabolic wastes, the microorganisms die at a rapid and uniform rate.

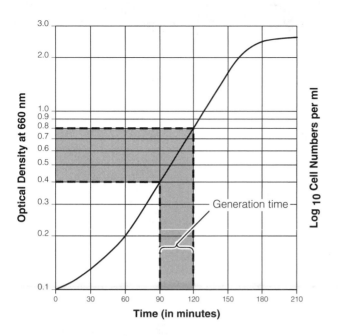

FIGURE 21.2 Indirect method of determining generation time

The decrease in population closely parallels its increase during the log phase. Theoretically, the entire population should die during a time interval equal to that of the log phase. This does not occur, however, since a small number of highly resistant organisms persist for an indeterminate length of time.

Construction of a complete bacterial growth curve requires that aliquots of a 24-hour shake-flask culture be measured for population size at intervals during the incubation period. Such a procedure does not lend itself to a regular laboratory session. Therefore this experiment follows a modified procedure designed to demonstrate only the lag and log phases. The curve will be plotted on semilog paper by using two values for the measurement of growth. The direct method requires enumeration of viable cells in serially diluted samples of the test culture taken at 30-minute intervals as described in Experiment 20. The indirect method uses spectrophotometric measurement of the developing turbidity at the same 30-minute intervals, as an index of increasing cellular mass.

You will determine generation time with indirect and direct methods by using data on the growth curve. Indirect determination is made by simple extrapolation from the log phase as illustrated in Figure 21.2. Select two points on the optical density scale, such as 0.2 and 0.4, that represent a doubling of turbidity.

Using a ruler, extrapolate by drawing a line between each of the selected optical densities on the ordinate (x-axis) and the plotted line of the growth curve. Then draw perpendicular lines from these end points on the plotted line of the growth curve to their respective time intervals on the abscissa (y-axis). With this information, determine the generation time (GT) as follows:

$$GT = t_{(O.D.\ 0.4)} - t_{(O.D.\ 0.2)}$$

$$GT = 90\ \text{minutes} - 60\ \text{minutes} = 30\ \text{minutes}$$

The generation time may be calculated directly using the log of cell numbers scale on a growth curve. The following example uses information from a hypothetical growth curve to calculate the generation time directly.

C_O = number of cells at time zero

C_E = number of cells at the end of a specified time (t)

N = number of generations (doublings)

To describe logarithmic growth, the following equation is used:

$$N = (\log C_E - \log C_O)/ \log 2$$

Using this formula, the logarithmic tables to the base 10, and the following supplied information, we may now solve for the generation time:

C_E = 52,000,000 cells	$\log C_E$ = 7.7218
C_O = 25,000 cells	$\log C_O$ = 4.4048
	$\log 2$ = 0.301

$$N = (7.7218 - 4.4048)/0.301 = 11\ \text{generations}$$

$$\text{Generation time (GT)} = \frac{\text{the specified time (t)}}{\text{number of generations (N)}}$$

$$t = 180\ \text{minutes}$$

$$GT = 180/11 = 16\ \text{minutes}$$

MATERIALS

Cultures

5- to 10-hour (log phase) brain–heart infusion broth culture of *Escherichia coli* with O.D. of 0.08–0.10 at 600 nm.

Media

Per designated student group: 100 ml of brain–heart infusion in a 250-ml Erlenmeyer flask; 18 99-ml sterile water blanks; and four 100-ml bottles of nutrient agar.

Equipment

37°C waterbath shaker incubator, Bausch & Lomb Spectronic 20 spectrophotometer, 13×100 mm cuvettes, Quebec colony counter, 24 sterile Petri dishes, 1-ml and 10-ml sterile pipettes, mechanical pipetting device, glassware marking pencil, 1000-ml beaker, and Bunsen burner.

PROCEDURE

1. Separate the 18 99-ml sterile water blanks into six sets of three water blanks each. Label each set as to time of inoculation (t_0, t_{30}, t_{60}, t_{90}, t_{120}, t_{150}) and the dilution to be effected in each water blank (10^{-2}, 10^{-4}, 10^{-6}).

2. Label six sets of four Petri dishes as to time of inoculation and dilution to be plated (10^{-4}, 10^{-5}, 10^{-6}, 10^{-7}).

3. Liquefy the four bottles of nutrient agar in an autoclave. Cool and maintain at 45°C.

4. With a sterile pipette, add approximately 5 ml of the log phase *E. coli* culture to the flask containing 100 ml of brain-heart infusion broth. The approximate initial O.D. (t_0) should be 0.08 to 0.1 at 600 nm. Refer to Experiment 14 for proper use of the spectrophotometer.

5. After the t_0 O.D. has been determined, shake the culture flask and aseptically transfer 1 ml to the 99-ml water blank labeled t_0 10^{-2} and continue to dilute serially to 10^{-4} and 10^{-6}. *Note: A new pipette must be used for each subsequent dilution.*

6. Place the culture flask in a waterbath shaker set at 120 rpm at 37°C, and time for the required 30-minute intervals.

7. Shake the t_0 dilution bottle as illustrated in Figure 21.3. Plate the t_0 dilutions on the

FIGURE 21.3 Method for mixing sample in a dilution bottle

appropriately labeled t_0 plates as shown in Figure 21.4. Aseptically pour 15 ml of the molten agar into each plate and mix by gentle rotation.

8. Thereafter, at each 30-minute interval, shake and aseptically transfer a 5-ml aliquot of the culture to a cuvette and determine its optical density. Also, aseptically transfer a 1-ml aliquot of the culture into the 10^{-2} water blank of the set labeled with the appropriate time, complete the serial dilution, and plate in the respectively labeled Petri dishes. *Note: A new pipette must be used for each subsequent dilution.*

9. When the pour-plate cultures harden, incubate them in an inverted position for 24 hours at 37°C.

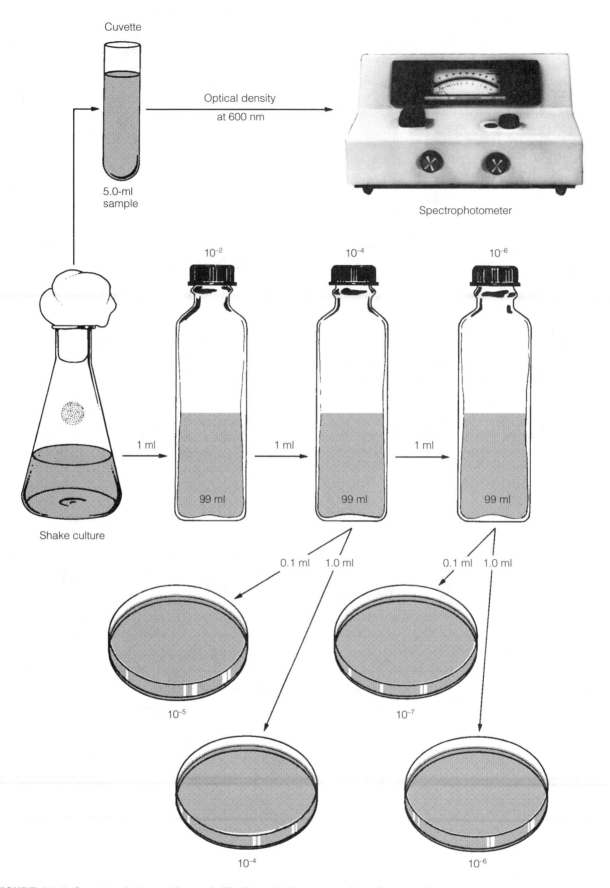

FIGURE 21.4 Spectrophotometric and dilution-plating procedure for use in bacterial growth curves

Name   Date

1. Perform cell counts on all plates as described in Experiment 20. Cell counts are often referred to as colony-forming units (CFUs) because each single cell in the plate becomes visible as a colony, which can then be counted.
2. Record the optical densities and corresponding cell counts in the chart.

Incubation Time (minutes)	Optical Density @ 600 nm	Plate Counts (CFU/ml)	Log of CFU/ml
0			
30			
60			
90			
120			
150			

3. On the semilog paper provided on pages 141 and 142:
   a. Plot a curve relating the optical densities on the ordinate versus incubation time on the abscissa as shown in Figure 21.2.
   b. Plot a population curve with the log of the viable cells/ml on the ordinate and the incubation time on the abscissa. On both graphs, use a ruler to draw the best line connecting the plotted points. The straight-line portion of the curve represents the log phase.
4. Calculate the generation time for this culture by the direct method (using the mathematical formula) and by the indirect method (extrapolating from the O.D. scale on the plotted curve). Show calculations, and record the generation time.

   a. Direct method:

   b. Indirect method:

REVIEW QUESTIONS

1. Does the term *growth* convey the same meaning when applied to bacteria and to multicellular organisms? Explain.

2. Why do variations in generation time exist:
   a. Among different species of microorganisms?

   b. Within a single microbial species?

3. The generation time and growth rate of an organism grown in the laboratory can be easily determined by constructing a typical growth curve.
   a. Would you expect the growth rate of the infectious organisms found in an abscess that developed from a wound to mimic the growth curve obtained in the laboratory? Explain.

   b. Would you expect antibiotic therapy to be effective without any other concurrent treatment of the abscess?

4. Is generation time a useful parameter to indicate the types of media best suited to support the growth of a specific organism? Explain.

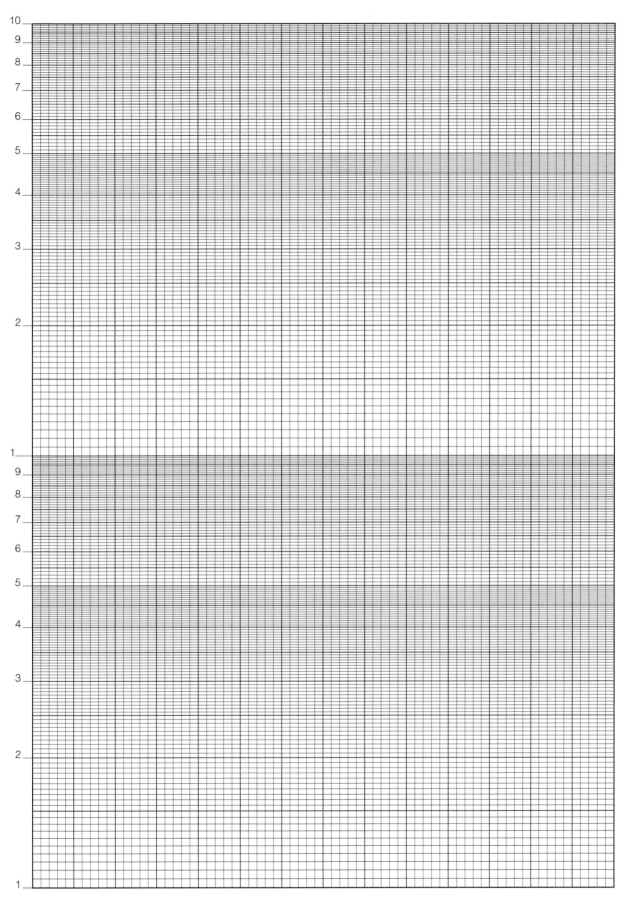

Optical Density versus Incubation Time

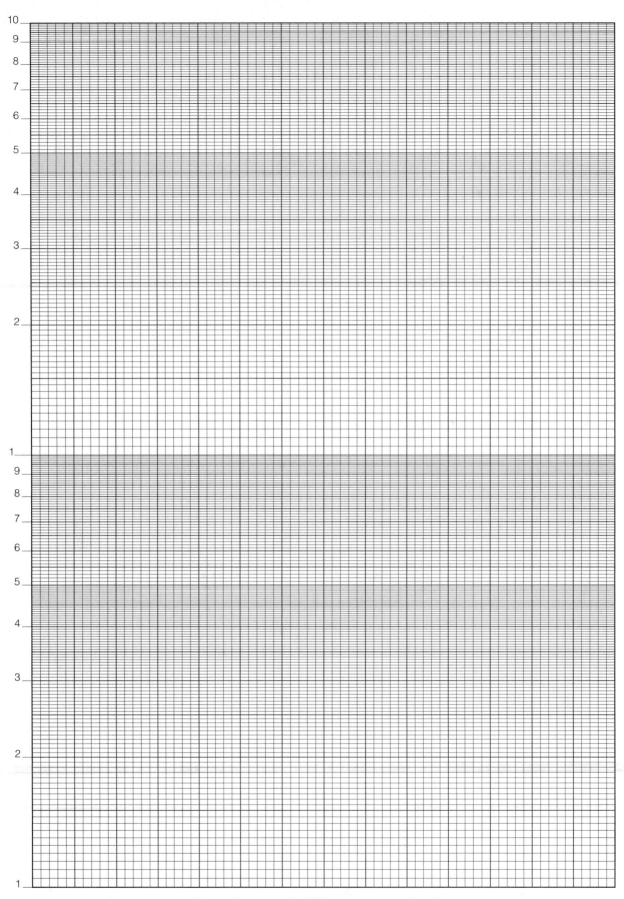

Colony Forming Units (CFUs) versus Incubation Time

Biochemical Activities
of Microorganisms

LEARNING OBJECTIVES

When you have completed the experiments in this section, you should understand

1. The nature and activities of exoenzymes and endoenzymes.

2. Experimental procedures for differentiation of enteric microorganisms.

3. Biochemical test procedures for identification of microorganisms.

INTRODUCTION

Microorganisms must be separated and identified for a wide variety of reasons, such as:

1. Determination of pathogens responsible for infectious diseases.
2. Selection and isolation of strains of fermentative microorganisms necessary for the industrial production of alcohols, solvents, vitamins, organic acids, antibiotics, and industrial enzymes.
3. Isolation and development of suitable microbial strains necessary for the manufacture and the enhancement of quality and flavor in certain food materials such as yogurt, cheeses, and other milk products.
4. Comparison of biochemical activities for taxonomic purposes.

To accomplish these tasks, the microbiologist is assisted by the fact that, just as human beings possess a characteristic and specific set of fingerprints, microorganisms all have their own identifying biochemical characteristics. These so-called biochemical fingerprints are the properties controlled by the cells' enzymatic activity, and they are responsible for bioenergetics, biosynthesis, and biodegradation.

The sum of all these chemical reactions is defined as **cellular metabolism,** and the biochemical transformations that occur both outside and inside the cell are governed by biological catalysts called **enzymes.**

Extracellular Enzymes (Exoenzymes)

Exoenzymes act on substances outside of the cell. Most high-molecular-weight substances are not able to pass through cell membranes, and therefore these raw materials—foodstuffs such as polysaccharides, lipids, and proteins—must be degraded to low-molecular-weight materials—nutrients—before they can be transported into the cell. Because of the reactions involved, exoenzymes are mainly **hydrolytic enzymes** that reduce high-molecular-weight materials into their building blocks by introducing water into the molecule. This liberates smaller molecules, which may then be transported into the cell and assimilated.

Intracellular Enzymes (Endoenzymes)

Endoenzymes function inside the cell and are mainly responsible for synthesis of new protoplasmic requirements and production of cellular energy from assimilated materials. The ability of cells to act on nutritional substrates permeating cell membranes indicates the presence of many endoenzymes capable of transforming the chemically specific substrates into essential materials.

This transformation is necessary for cellular survival and function, and it is the basis of cellular metabolism. As a result of these metabolic processes, metabolic products are formed and excreted by the cell into the environment.

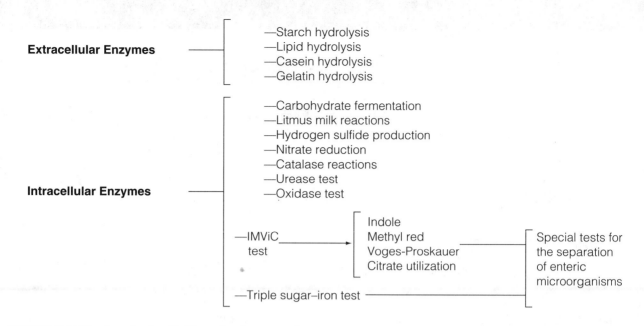

Extracellular Enzymes
- Starch hydrolysis
- Lipid hydrolysis
- Casein hydrolysis
- Gelatin hydrolysis

Intracellular Enzymes
- Carbohydrate fermentation
- Litmus milk reactions
- Hydrogen sulfide production
- Nitrate reduction
- Catalase reactions
- Urease test
- Oxidase test
- IMViC test → Indole / Methyl red / Voges-Proskauer / Citrate utilization → Special tests for the separation of enteric microorganisms
- Triple sugar–iron test

FIGURE V.1 Biochemical activities of microorganisms

Assay of these end products not only aids in identification of specific enzyme systems but also serves to identify, separate, and classify microorganisms. Figure V.1 represents a simplified schema of experimental procedures used to acquaint students with the intracellular and extracellular enzymatic activities of microorganisms.

The experiments you will carry out in this section can be performed in either of two ways. A short version uses a limited number of organisms to illustrate the possible end product(s) that may result from enzyme action on a substrate. The organisms for this version are designated in the individual exercises.

The alternative, or long, version involves the use of 13 microorganisms. This version provides a complete overview of the biochemical fingerprints of the organisms and supplies the format for their separation and identification.

These organisms were chosen to serve as a basis for identification of an unknown microorganism in Experiment 33. If this alternative version is selected, the following organisms are recommended for use:

Escherichia coli
Enterobacter aerogenes
Klebsiella pneumoniae
Shigella dysenteriae
Salmonella typhimurium
Proteus vulgaris
Pseudomonas aeruginosa
Alcaligenes faecalis
Micrococcus luteus
Lactococcus lactis
Staphylococcus aureus
Bacillus cereus
Corynebacterium xerosis

Extracellular Enzymatic Activities of Microorganisms

LEARNING OBJECTIVES

Once you have completed this experiment, you should be able to

1. Understand the function of microbial extracellular enzymes.
2. Determine the ability of microorganisms to excrete hydrolytic extracellular enzymes capable of degrading the polysaccharide starch, the lipid tributyrin, and the proteins casein and gelatin.

PRINCIPLE

Because of their large sizes, high-molecular-weight nutrients such as polysaccharides, lipids, and proteins are not capable of permeating the cell membrane. These macromolecules must first be hydrolyzed by specific extracellular enzymes into their respective basic building blocks. These low-molecular-weight substances can then be transported into the cells and used for the synthesis of protoplasmic requirements and energy production. The following procedures are designed to investigate the exoenzymatic activities of different microorganisms.

Starch Hydrolysis

Starch is a high-molecular-weight, branching polymer composed of **glucose** molecules linked together by **glycosidic bonds.** The degradation of this macromolecule first requires the presence of the extracellular enzyme **amylase** for its hydrolysis into shorter polysaccharides, namely **dextrins,** and ultimately into **maltose** molecules. The final hydrolysis of this disaccharide, which is catalyzed by **maltase,** yields low-molecular-weight, soluble **glucose** molecules that can be transported into the cell and used for energy production through the process of glycolysis.

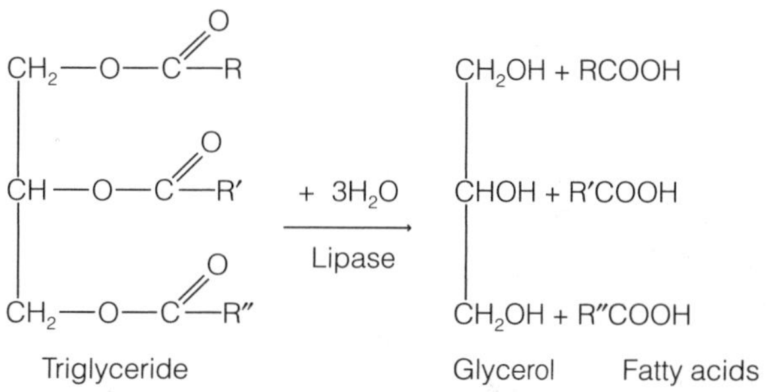

FIGURE 22.1 Lipid hydrolysis

In this experimental procedure, starch agar is used to demonstrate the hydrolytic activities of these exoenzymes. The medium is composed of nutrient agar supplemented with starch, which serves as the polysaccharide substrate. The detection of the hydrolytic activity following the growth period is made by performing the starch test to determine the presence or absence of starch in the medium. Starch in the presence of iodine will impart a blue-black color to the medium, indicating the absence of starch-splitting enzymes and representing a negative result. If the starch has been hydrolyzed, a clear zone of hydrolysis will surround the growth of the organism. This is a positive result.

Lipid Hydrolysis

Lipids are high-molecular-weight compounds possessing large amounts of energy. The degradation of lipids such as **triglycerides** is accomplished by extracellular hydrolyzing enzymes, called **lipases** (esterases), that cleave the **ester bonds** in this molecule by the addition of water to form the building blocks **glycerol** (an alcohol) and **fatty acids.** Figure 22.1 shows this reaction. Once assimilated into the cell, these basic components can be further metabolized through aerobic respiration to

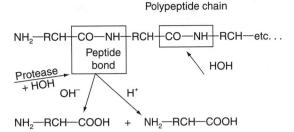

Polypeptide chain

$$NH_2—RCH—CO—NH—RCH—CO—NH—RCH—etc. . .$$

Peptide bond

Protease + HOH

HOH

OH^- H^+

$$NH_2—RCH—COOH + NH_2—RCH—COOH$$

FIGURE 22.2 Protein hydrolysis

produce cellular energy, adenosine triphosphate (ATP). The components may also enter other metabolic pathways for the synthesis of other cellular protoplasmic requirements.

In this experimental procedure, tributyrin agar is used to demonstrate the hydrolytic activities of the exoenzyme lipase. The medium is composed of nutrient agar supplemented with the triglyceride tributyrin as the lipid substrate. Tributyrin forms an emulsion when dispersed in the agar, producing an opaque medium that is necessary for observing exoenzymatic activity.

Following inoculation and incubation of the agar plate cultures, organisms excreting lipase will show a zone of **lipolysis,** which is demonstrated by a clear area surrounding the bacterial growth. This loss of opacity is the result of the hydrolytic reaction yielding soluble glycerol and fatty acids and represents a positive reaction for lipid hydrolysis. In the absence of lipolytic enzymes, the medium retains its opacity. This is a negative reaction.

Casein Hydrolysis

Casein, the major milk protein, is a macromolecule composed of **amino acid** subunits linked together by **peptide bonds** (CO—NH). Before their assimilation into the cell, proteins must undergo step-by-step degradation into **peptones, polypeptides, dipeptides,** and ultimately into their building blocks, **amino acids.** This process is called peptonization, or **proteolysis,** and it is mediated by extracellular enzymes called **proteases.** The function of these proteases is to cleave the peptide bond CO—NH by introducing water into the molecule. The reaction then liberates the amino acids, as illustrated in Figure 22.2.

The low-molecular-weight soluble amino acids can now be transported through the cell membrane into the intracellular amino acid pool for use in the synthesis of structural and functional cellular proteins.

In this experimental procedure, milk agar is used to demonstrate the hydrolytic activity of these exoenzymes. The medium is composed of nutrient agar supplemented with milk that contains the protein substrate casein. Similar to other proteins, milk protein is a colloidal suspension that gives the medium its color and opacity, because it deflects light rays rather than transmitting them.

Following inoculation and incubation of the agar plate cultures, organisms secreting proteases will exhibit a zone of proteolysis, which is demonstrated by a clear area surrounding the bacterial growth. This loss of opacity is the result of a hydrolytic reaction yielding soluble, noncolloidal amino acids, and it represents a positive reaction. In the absence of protease activity, the medium surrounding the growth of the organism remains opaque, which is a negative reaction.

Gelatin Hydrolysis

Although the value of gelatin as a nutritional source is questionable (it is an incomplete protein, lacking the essential amino acid tryptophan), its value in identifying bacterial species is well-established. Gelatin is a protein produced by hydrolysis of collagen, a major component of connective tissue and tendons in humans and other animals. Below temperatures of 25°C, gelatin will maintain its gel properties and exist as a solid; at temperatures above 25°C, gelatin is liquid.

Liquefaction is accomplished by some microorganisms capable of producing a proteolytic extracellular enzyme called **gelatinase,** which acts to hydrolyze this protein to **amino acids.** Once this degradation occurs, even very low temperatures of 4°C will not restore the gel characteristic.

In this experimental procedure, you will use nutrient gelatin deep tubes to demonstrate the hydrolytic activity of gelatinase. The medium consists of nutrient broth supplemented with 12% gelatin. This high gelatin concentration results in a stiff medium and also serves as the substrate for the activity of gelatinase.

Following inoculation and incubation for 48 hours, the cultures are placed in a refrigerator at 4°C for 30 minutes. Cultures that remain liquefied produce gelatinase and demonstrate *rapid* gelatin hydrolysis. Re-incubate all solidified cultures for an additional 5 days.

Refrigerate for 30 minutes and observe for liquefaction. Cultures that remain liquefied are indicative of *slow* gelatin hydrolysis.

MATERIALS

Cultures

24- to 48-hour trypticase soy broth cultures of *Escherichia coli, Bacillus cereus, Pseudomonas aeruginosa,* and *Staphylococcus aureus* for the short version. 24- to 48-hour trypticase soy broth cultures of the 13 organisms listed on page 144 for the long version.

Media

Short version: Two plates each of starch agar, tributyrin agar, and milk agar, and three nutrient gelatin deep tubes per designated student group. Long version: Four plates each of starch agar, tributyrin agar, and milk agar, and 14 nutrient gelatin deep tubes per designated student group.

Reagent

Gram's iodine solution.

Equipment

Bunsen burner, inoculating loop and needle, glassware marking pencil, test tube rack, and refrigerator.

PROCEDURE

1. Prepare the starch agar, tributyrin agar, and milk agar plates for inoculation as follows:

   a. Short procedure: Using two plates per medium, divide the bottom of each Petri dish into two sections. Label the sections as *E. coli, B. cereus, P. aeruginosa,* and *S. aureus,* respectively.

   b. Long procedure: Repeat Step 1a, dividing three plate bottoms into three sections and one plate bottom into four sections for each of the required media, to accommodate the 13 test organisms.

2. Using sterile technique, make a single-line streak inoculation of each test organism on the agar surface of its appropriately labeled section on the agar plates.

3. Using sterile technique, inoculate each experimental organism in its appropriately labeled gelatin deep tube by means of a stab inoculation.

4. Incubate all plates in an inverted position for 24 to 48 hours at 37°C. Incubate the gelatin deep tube cultures for 48 hours. Re-incubate all negative cultures for an additional 5 days.

Name _____   Date _____

Starch Hydrolysis

1. Flood the starch agar plate cultures with Gram's iodine solution, allow the iodine to remain in contact with the medium for 30 seconds, and pour off the excess.
2. Examine the cultures for the presence or absence of a blue-black color surrounding the growth of each test organism. Record your results in the chart.
3. Based on your observations, determine and record which organisms were capable of hydrolyzing the starch.

📷 *Refer to photo numbers 32 and 33 in the color-plate insert for illustration of these reactions.*

Lipid Hydrolysis

1. Examine the tributyrin agar plate cultures for the presence or absence of a clear area, or zone of lipolysis, surrounding the growth of each of the organisms. Record your results in the chart.
2. Based on your observations, determine and record which organisms were capable of hydrolyzing the lipid.

📷 *Refer to photo number 34 in the color-plate insert for illustration of these reactions.*

Bacterial Species	Starch Hydrolysis		Tributyrin Hydrolysis	
	Appearance of Medium	Result (+) or (−)	Appearance of Medium	Result (+) or (−)
E. coli				
E. aerogenes				
K. pneumoniae				
S. dysenteriae				
S. typhimurium				
P. vulgaris				
P. aeruginosa				
A. faecalis				
M. luteus				
L. lactis				
S. aureus				
B. cereus				
C. xerosis				

Casein Hydrolysis

1. Examine the milk agar plate cultures for the presence or absence of a clear area, or zone of proteolysis, surrounding the growth of each of the bacterial test organisms. Record your results in the chart.
2. Based on your observations, determine and record which of the organisms were capable of hydrolyzing the milk protein casein.

Gelatin Hydrolysis

1. Place all gelatin deep tube cultures into a refrigerator at 4°C for 30 minutes.
2. Examine all the cultures to determine whether the medium is solid or liquid. Record your results in the chart.
3. Based on your observations following the 2-day and 7-day incubation periods, determine and record (a) which organisms were capable of hydrolyzing gelatin and (b) the rate of hydrolysis.

Bacterial Species	Casein Hydrolysis		Gelatin Hydrolysis		
	Appearance of Medium	Result (+) or (−)	Liquefaction (+) or (−)		Rate of Hydrolysis (Slow or Rapid)
			2 days	7 days	
E. coli					
E. aerogenes					
K. pneumoniae					
S. dysenteriae					
S. typhimurium					
P. vulgaris					
P. aeruginosa					
A. faecalis					
M. luteus					
L. lactis					
S. aureus					
B. cereus					
C. xerosis					

Carbohydrate Fermentation

LEARNING OBJECTIVES

Once you have completed this experiment, you should

1. Understand the difference between cellular respiration and fermentation.
2. Be able to determine the ability of microorganisms to degrade and ferment carbohydrates with the production of acid and gas.

PRINCIPLE

Most microorganisms obtain their energy through a series of orderly and integrated enzymatic reactions leading to the biooxidation of a substrate, frequently a carbohydrate. The major pathways by which this is accomplished are shown in Figure 23.1.

Organisms use carbohydrates differently depending on their enzyme complement. Some organisms are capable of fermenting sugars such as glucose anaerobically, while others use the aerobic pathway. Still others, facultative anaerobes, are enzymatically competent to use both aerobic and anaerobic pathways, and some organisms lack the ability to oxidize glucose by either. In this exercise the fermentative pathways are of prime concern.

In fermentation, substrates such as carbohydrates and alcohols undergo anaerobic dissimilation and produce an organic acid (for example, lactic, formic, or acetic acid) that may be accompanied by gases such as hydrogen or carbon dioxide. Facultative anaerobes are usually the so-called fermenters of carbohydrates. Fermentation is best described by considering the degradation of glucose by way of the **Embden-Meyerhof pathway,** also known as the **glycolytic pathway,** illustrated in Figure 23.2.

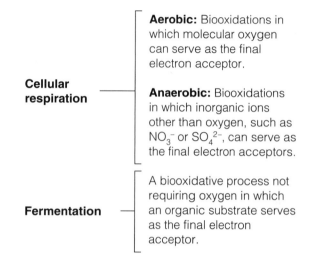

Aerobic: Biooxidations in which molecular oxygen can serve as the final electron acceptor.

Anaerobic: Biooxidations in which inorganic ions other than oxygen, such as NO_3^- or SO_4^{2-}, can serve as the final electron acceptors.

A biooxidative process not requiring oxygen in which an organic substrate serves as the final electron acceptor.

FIGURE 23.1 Biooxidative pathways

As the diagram shows, one mole of glucose is converted into two moles of pyruvic acid, which is the major intermediate compound produced by glucose degradation. Subsequent metabolism of pyruvate is not the same for all organisms, and a variety of end products result that define their different fermentative capabilities. This can be seen in Figure 23.3.

Fermentative degradation under anaerobic conditions is carried out in a fermentation broth tube containing a Durham tube, an inverted inner vial for the detection of gas production as illustrated in Figure 23.4. A typical carbohydrate fermentation medium contains:

1. Nutrient broth ingredients for the support of the growth of all organisms.
2. A specific carbohydrate that serves as the substrate for determining the organism's fermentative capabilities.

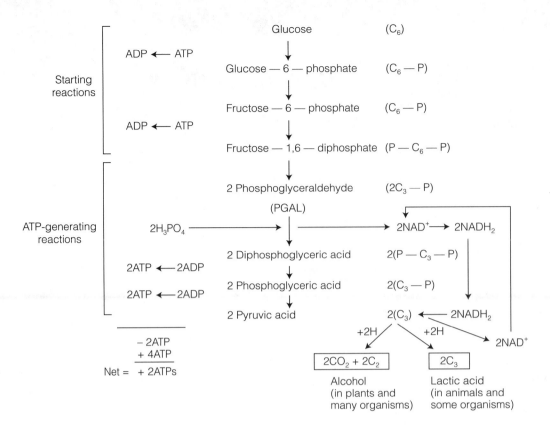

FIGURE 23.2 The Embden-Meyerhof pathway

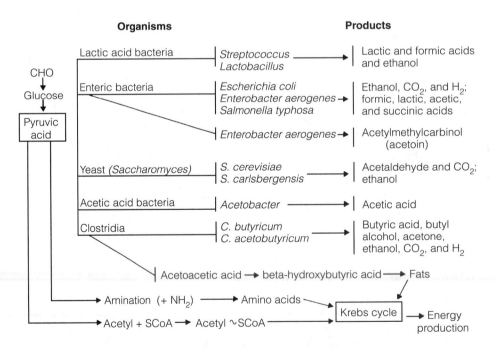

FIGURE 23.3 Variations in the use of pyruvic acid

3. The pH indicator phenol red, which is red at a neutral pH (7) and changes to yellow at a slightly acidic pH of 6.8, indicating that slight amounts of acid will cause a color change.

The critical nature of the fermentation reaction and the activity of the indicator make it imperative that all cultures should be observed within 48 hours. Extended incubation may mask acid-producing reactions by production of alkali because of enzymatic action on substrates other than the carbohydrate.

Following incubation, carbohydrates that have been fermented with the production of acidic wastes will cause the phenol red to turn yellow, thereby indicating a positive reaction. In some cases, acid production is accompanied by the evolution of a gas (CO_2) that will be visible as a bubble in the inverted tube. Cultures that are not capable of fermenting a carbohydrate substrate will not change the indicator, and the tubes will appear red; there will not be a concomitant evolution of gas. This is a negative reaction.

The lack of carbohydrate fermentation by some organisms should not be construed as absence of growth. The organisms use other nutrients in the medium as energy sources. Among these nutrients are peptones present in nutrient broth. Peptones can be degraded by microbial enzymes to amino acids that are in turn enzymatically converted by oxidative deamination to ketoamino acids. These are then metabolized through the Krebs cycle for energy production. These reactions liberate ammonia, which accumulates in the medium, forming ammonium hydroxide (NH_4OH) and producing an alkaline environment. When this occurs, the phenol red turns to a deep red in the now basic medium. This alternative pathway of aerobic respiration is illustrated in Figure 23.5.

MATERIALS
Cultures

24- to 48-hour trypticase soy broth cultures of *Escherichia coli, Alcaligenes faecalis, Salmonella typhimurium,* and *Staphylococcus aureus*

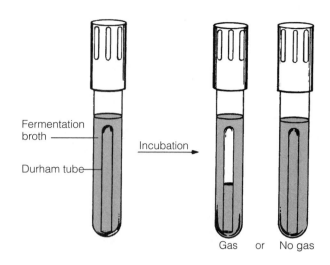

FIGURE 23.4 Detection of gas production

for the short version. 24- to 48-hour trypticase soy broth cultures of the 13 organisms listed on page 144 for the long version.

Media

Per designated student group: phenol red lactose, dextrose (glucose), and sucrose broths: five of each for the short version, 14 of each for the long version.

Equipment

Bunsen burner, inoculating loop, and glassware marking pencil.

PROCEDURE

1. Using sterile technique, inoculate each experimental organism into its appropriately labeled medium by means of loop inoculation. *Note: Take care during this step not to shake the fermentation tube;* shaking the tube may accidentally force a bubble of air into the inverted gas vial, displacing the medium and possibly rendering a false-positive result. The last tube will serve as a control.

2. Incubate all tubes for 24 hours at 37°C.

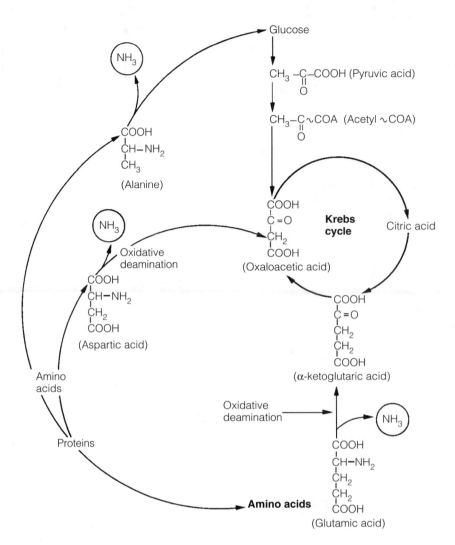

FIGURE 23.5 Proteins as energy sources for microbes

Name _____   Date _____

1. Examine all carbohydrate broth cultures for color and the presence or absence of a gas bubble. Record your results in the chart.

2. Based on your observations, determine and record whether or not each organism was capable of fermenting the carbohydrate substrate with the production of acid or acid and gas.

📷 *Refer to photo number 35 in the color-plate insert for illustration of these reactions.*

Bacterial Species	Lactose Observation (color of medium, bubble in fermentation tube)	Result (A), (A/G), or (−)	Dextrose Observation (color of medium, bubble in fermentation tube)	Result (A), (A/G), or (−)	Sucrose Observation (color of medium, bubble in fermentation tube)	Result (A), (A/G), or (−)
E. coli						
E. aerogenes						
K. pneumoniae						
S. dysenteriae						
S. typhimurium						
P. vulgaris						
P. aeruginosa						
A. faecalis						
M. luteus						
L. lactis						
S. aureus						
B. cereus						
C. xerosis						
Control						

REVIEW QUESTIONS

1. Distinguish between respiration and fermentation.

2. Do all microorganisms use pyruvic acid in the same way? Explain.

3. Describe a pathway used for the degradation of carbohydrates by strict anaerobes.

4. From your experimental data, you know that *P. aeruginosa* did not utilize any of the carbohydrates in the test media. In view of this, how do these organisms generate energy to sustain their viability?

5. *Clostridium perfringens,* an obligate anaerobe, is capable of utilizing the carbohydrates released from injured tissues as an energy source. During the infectious process, large amounts of gas accumulate in the infected tissues. Would you expect this gas to be CO_2? Explain.

Triple Sugar—Iron Agar Test

LEARNING OBJECTIVES

Once you have completed this experiment, you should understand a rapid screening procedure that will

1. Differentiate among members of the Enterobacteriaceae.

2. Distinguish between the Enterobacteriaceae and other groups of intestinal bacilli.

PRINCIPLE

The **triple sugar–iron (TSI) agar test** is designed to differentiate among the different groups or genera of the Enterobacteriaceae, which are all gram-negative bacilli capable of fermenting glucose with the production of acid, and to distinguish the Enterobacteriaceae from other gram-negative intestinal bacilli. This differentiation is made on the basis of differences in carbohydrate fermentation patterns and hydrogen sulfide production by the various groups of intestinal organisms.

To facilitate observation of carbohydrate utilization patterns, the TSI agar slants contain lactose and sucrose in 1% concentrations and glucose in a concentration of 0.1%, which permits detection of the utilization of this substrate only. The acid–base indicator phenol red is also incorporated to detect carbohydrate fermentation that is indicated by a change in color of the medium from orange-red to yellow in the presence of acids. The slant is inoculated by means of a stab-and-streak procedure. This requires the insertion of a sterile, straight needle from the base of the slant into the butt. Upon withdrawal of the needle, the slanted surface of the medium is streaked. Following incubation, you will determine the fer-

mentative activities of the organisms as described below.

1. **Alkaline slant (red) and acid butt (yellow) with or without gas production (breaks in the agar butt):** Only glucose fermentation has occurred. The organisms preferentially degrade glucose first. Since this substrate is present in minimal concentration, the small amount of acid produced on the slant surface is oxidized rapidly. The peptones in the medium are also used in the production of alkali. In the butt the acid reaction is maintained because of reduced oxygen tension and slower growth of the organisms.

2. **Acid slant (yellow) and acid butt (yellow) with or without gas production.** Lactose and/or sucrose fermentation has occurred. Since these substances are present in higher concentrations, they serve as substrates for continued fermentative activities with maintenance of an acid reaction in both slant and butt.

3. **Alkaline slant (red) and alkaline butt (red) or no change (orange-red) butt.** No carbohydrate fermentation has occurred. Instead, peptones are catabolized under anaerobic and/or aerobic conditions, resulting in an alkaline pH due to production of ammonia. If only aerobic degradation of peptones occurs, the alkaline reaction is evidenced only on the slant surface. If there is aerobic and anaerobic utilization of peptone, the alkaline reaction is present on the slant and the butt.

For you to obtain accurate results, it is absolutely essential to observe the cultures within 18 to 24 hours following incubation. Doing so will ensure that the carbohydrate

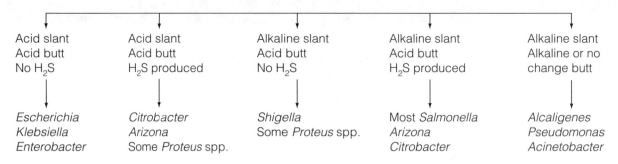

Acid slant	Acid slant	Alkaline slant	Alkaline slant	Alkaline slant
Acid butt	Acid butt	Acid butt	Acid butt	Alkaline or no
No H$_2$S	H$_2$S produced	No H$_2$S	H$_2$S produced	change butt
Escherichia	*Citrobacter*	*Shigella*	Most *Salmonella*	*Alcaligenes*
Klebsiella	*Arizona*	Some *Proteus* spp.	*Arizona*	*Pseudomonas*
Enterobacter	Some *Proteus* spp.		*Citrobacter*	*Acinetobacter*

FIGURE 24.1 TSI reactions for differentiation of enteric microorganisms

substrates have not been depleted and that degradation of peptones yielding alkaline end products has not taken place.

The TSI agar medium also contains sodium thiosulfate, a substrate for hydrogen sulfide (H$_2$S) production, and ferrous sulfate for detection of this colorless end product. Following incubation, only cultures of organisms capable of producing H$_2$S will show an extensive blackening in the butt because of the precipitation of the insoluble ferrous sulfide. (Refer to Experiment 26 for a more detailed biochemical explanation of H$_2$S production.)

Figure 24.1 is a schema for the differentiation of intestinal bacilli on the basis of the TSI agar reactions.

MATERIALS

Cultures

24-hour trypticase soy broth cultures of *Pseudomonas aeruginosa, Escherichia coli, Salmonella typhimurium, Shigella dysenteriae, Proteus vulgaris,* and *Alcaligenes faecalis* for the

short version. 24-hour trypticase soy broth cultures of the 13 organisms listed on page 144 for the long version.

Media

Per designated student group: triple sugar–iron agar slants: seven for the short version, 14 for the long version.

Equipment

Bunsen burner, inoculating needle, test tube rack, and glassware marking pencil.

PROCEDURE

1. Using sterile technique, inoculate each experimental organism into its appropriately labeled tube by means of a stab-and-streak inoculation. *Note: Do not fully tighten screw cap.* The last tube will serve as a control.

2. Incubate for 18 to 24 hours at 37°C.

Name _____   Date _____

1. Examine the color of both the butt and slant of all agar slant cultures. Based on your observations, determine the type of reaction that has taken place (acid, alkaline, or none) and the carbohydrate that has been fermented (dextrose, lactose, and/or sucrose, all, or none) in each culture. Record your observations and results in the chart.

2. Examine all cultures for the presence or absence of blackening within the medium. Based on your observations, determine whether or not each organism was capable of H_2S production. Record your observations and results in the chart.

📷 *Refer to photo number 36 in the color-plate insert for illustration of these reactions.*

Bacterial Species	Carbohydrate Fermentation			H_2S Production	
	Butt Color and Reaction	Slant Color and Reaction	Carbohydrate Fermented	Blackening	H_2S (+) or (−)
E. coli					
E. aerogenes					
K. pneumoniae					
S. dysenteriae					
S. typhimurium					
P. vulgaris					
P. aeruginosa					
A. faecalis					
M. luteus					
L. lactis					
S. aureus					
B. cereus					
C. xerosis					
Control					

REVIEW QUESTIONS

1. What is the purpose of the TSI test?

2. Explain why the TSI medium contains a lower concentration of glucose than of lactose and sucrose.

3. Explain the purpose of the phenol red in the medium.

4. Explain the purpose of thiosulfate in the medium.

5. Explain why the test observations must be made between 18 and 24 hours after inoculation.

IMViC Test

Identification of enteric bacilli is of prime importance in controlling intestinal infections by preventing contamination of food and water supplies. The groups of bacteria that can be found in the intestinal tract of humans and lower mammals are classified as members of the family **Enterobacteriaceae.** They are short, gram-negative, nonspore-forming bacilli. Included in this family are:

1. **Pathogens** such as members of the genera *Salmonella* and *Shigella*.

2. **Occasional pathogens** such as members of the genera *Proteus* and *Klebsiella*.

3. **Normal intestinal flora** such as members of the genera *Escherichia* and *Enterobacter*, which are saprophytic inhabitants of the intestinal tract.

Differentiation of the principal groups of Enterobacteriaceae can be accomplished on the basis of their biochemical properties and enzymatic reactions in the presence of specific substrates. The **IMViC** series of tests (**indole, methyl-red, Voges-Proskauer,** and **citrate utilization**) can be used.

Figure 25.7 on page 166 shows the biochemical reactions that occur during the IMViC tests. It is designed to assist you in the execution and interpretation of each test.

The following experiments are designed for either a short or long version. The short version uses selected members of the enteric family. The long procedure makes use of bacterial species that do not belong solely to the Enterobacteriaceae. Nonenteric forms are included to acquaint you with the biochemical activities of other organisms grown in these media and to enable you to use these data for further comparisons of both types of bacteria.

Selected organisms to be used in the long-version procedures are listed below. The enteric organisms are subdivided as lactose fermenters and nonfermenters.

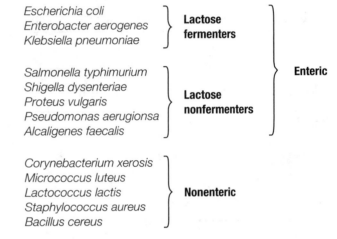

PART A: Indole Production Test

LEARNING OBJECTIVE

Once you have completed this test, you should be able to

1. Determine the ability of microorganisms to degrade the amino acid tryptophan.

PRINCIPLE

Tryptophan is an essential amino acid that can undergo oxidation by way of the enzymatic activities of some bacteria. Conversion of tryptophan into metabolic products is mediated by the enzyme **tryptophanase**. This reaction is illustrated in Figure 25.1. This ability to hydrolyze tryptophan with the production

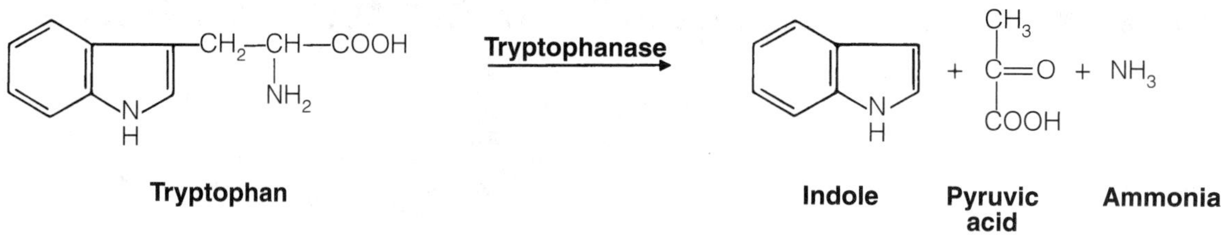

FIGURE 25.1 Enzymatic degradation of tryptophan

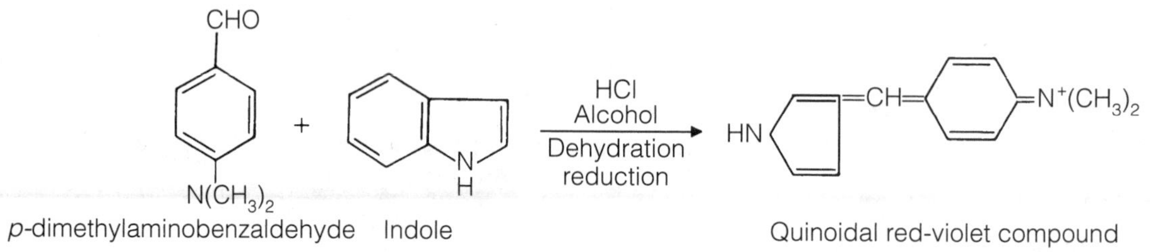

FIGURE 25.2 Indole reaction with Kovac's reagent

of indole is not a characteristic of all microorganisms and therefore serves as a biochemical marker.

In this experiment, SIM agar, which contains the substrate tryptophan, is used. The presence of indole is detectable by adding Kovac's reagent, which produces a cherry red reagent layer. This color is produced by the reagent, which is composed of p-dimethylaminobenzaldehyde, butanol, and hydrochloric acid. Indole is extracted from the medium into the reagent layer by the acidified butyl alcohol component and forms a complex with the p-dimethylaminobenzaldehyde, yielding the cherry red color. This reaction is illustrated in Figure 25.2.

Cultures producing a red reagent layer following addition of Kovac's reagent are indole-positive. The absence of red coloration demonstrates that the substrate tryptophan was not hydrolyzed and indicates an indole-negative reaction.

MATERIALS

Cultures

24- to 48-hour trypticase soy broth cultures of *Escherichia coli*, *Proteus vulgaris*, and *Enterobacter aerogenes* for the short version. 24- to 48-hour trypticase soy broth cultures of the 13 organisms listed on page 161 for the long version.

Media

SIM agar deep tubes per designated student group: four for the short version, 14 for the long version.

Reagent

Kovac's reagent.

Equipment

Bunsen burner, inoculating needle, test tube rack, and glassware marking pencil.

PROCEDURE

1. Using sterile technique, inoculate each experimental organism into its appropriately labeled deep tube by means of a stab inoculation. The last tube will serve as a control.

2. Incubate tubes for 24 to 48 hours at 37°C.

$$\text{Glucose} + H_2O \longrightarrow \begin{bmatrix} \text{Lactic acid} \\ \text{Acetic acid} \\ \text{Formic acid} \end{bmatrix} + CO_2 + H_2 \text{ (pH 4.0)}$$

Methyl red indicator turns red color

FIGURE 25.3 Glucose fermentation reaction with methyl red pH reagent

PART B: Methyl Red Test

LEARNING OBJECTIVES

Once you have completed this experiment, you should be able to

1. Determine the ability of microorganisms to oxidize glucose with the production and stabilization of high concentrations of acid end products.

2. Differentiate between all glucose-oxidizing enteric organisms, particularly *E. coli* and *E. aerogenes*.

PRINCIPLE

The hexose monosaccharide **glucose** is the major substrate oxidized by all enteric organisms for energy production. The end products of this process will vary depending on the specific enzymatic pathways present in the bacteria. In this test the pH indicator methyl red detects the presence of large concentrations of acid end products. Although all enteric microorganisms ferment glucose with the production of organic acids, this test is of value in the separation of *E. coli* and *E. aerogenes*.

Both of these organisms initially produce organic acid end products during the early incubation period. The low acidic pH (4) is stabilized and maintained by *E. coli* at the end of incubation. During the later incubation period, *E. aerogenes* enzymatically converts these acids to nonacidic end products such as 2,3-butanediol and acetoin (acetylmethylcarbinol), resulting in an elevated pH of approximately 6. The glucose fermentation reaction generated by *E. coli* is illustrated in Figure 25.3.

As shown, the methyl red indicator in the pH range of 4 will turn red, which is indicative of a positive test. At a pH of 6, still indicating

the presence of acid but with a lower hydrogen ion concentration, the indicator turns yellow and is a negative test. Production and detection of the nonacidic end products from glucose fermentation by *E. aerogenes* is amplified in Part C of this exercise, the Voges-Proskauer test, which is performed simultaneously with the methyl red test.

MATERIALS

Cultures

24- to 48-hour trypticase soy broth cultures of *E. coli*, *E. aerogenes*, and *Klebsiella pneumoniae* for the short version. 24- to 48-hour trypticase soy broth cultures of the 13 organisms listed on page 161 for the long version. Aliquots of these experimental cultures must be set aside from the Voges-Proskauer test.

Media

MR-VP broth per designated student group: four for the short version, 14 for the long version.

Reagent

Methyl red indicator.

Equipment

Bunsen burner, inoculating loop, test tubes, and glassware marking pencil.

PROCEDURE

1. Using sterile technique, inoculate each experimental organism into its appropriately labeled tube of medium by means of a loop inoculation. The last tube will serve as a control.

2. Incubate all cultures for 24 to 48 hours at 37°C.

$$\text{Glucose} + O_2 \longrightarrow \underset{\text{acid}}{\text{Acetic}} \longrightarrow \left[\begin{array}{c}\text{2,3-butanediol}\\\text{acetylmethylcarbinol}\end{array}\right] + CO_2 + H_2 \text{ (pH 6.0)}$$

FIGURE 25.4 Glucose fermentation by *E. aerogenes*

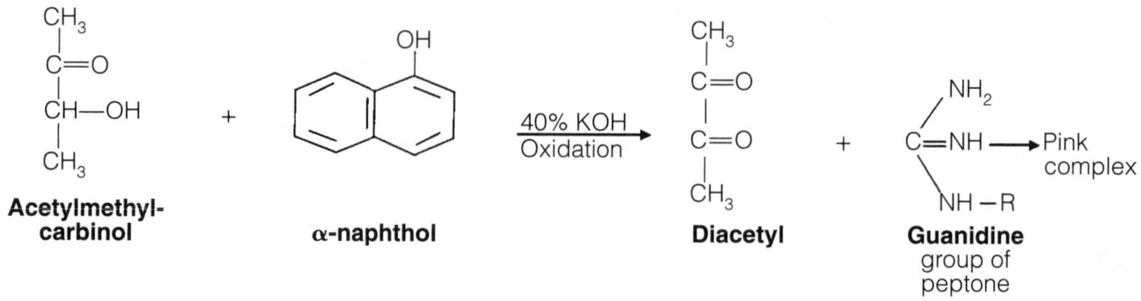

FIGURE 25.5 Acetylmethylcarbinol reaction with Barritt's reagent

PART C: Voges-Proskauer Test

LEARNING OBJECTIVE

Once you have completed this experiment, you should be able to

1. Differentiate further among enteric organisms such as *E. coli, E. aerogenes,* and *K. pneumoniae.*

PRINCIPLE

The Voges-Proskauer test determines the capability of some organisms to produce nonacidic or neutral end products, such as acetylmethylcarbinol, from the organic acids that result from glucose metabolism. This glucose fermentation, which is characteristic of *E. aerogenes,* is illustrated in Figure 25.4.

The reagent used in this test, Barritt's reagent, consists of a mixture of alcoholic α-naphthol and 40% potassium hydroxide solution. Detection of acetylmethylcarbinol requires this end product to be oxidized to a diacetyl compound. This reaction will occur in the presence of the α-naphthol catalyst and a guanidine group that is present in the peptone of the MR-VP medium. As a result, a pink complex is formed, imparting a rose color to the medium. The chemistry of this reaction is illustrated in Figure 25.5.

Development of a deep rose color in the culture 15 minutes following the addition of Barritt's reagent is indicative of the presence of acetylmethylcarbinol and represents a positive result. The absence of rose coloration is a negative result.

MATERIALS
Cultures

24- to 48-hour trypticase soy broth cultures of *E. coli, E. aerogenes,* and *K. pneumoniae* for the short version. 24- to 48-hour trypticase soy broth cultures of the 13 organisms listed on page 161 for the long version. *Note: Aliquots of these experimental cultures must be set aside from the methyl red test.*

Reagent
Barritt's reagents A and B.

Equipment
Bunsen burner, inoculating loop, and glassware marking pencil.

PROCEDURE
Refer to the methyl red test in Part B of this exercise.

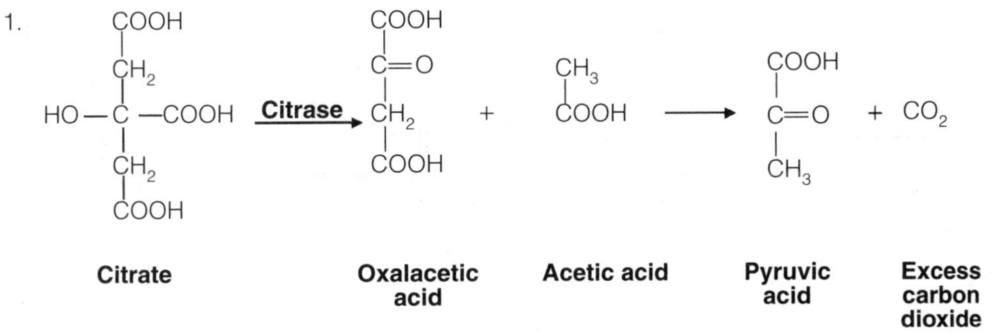

1.

$$\text{Citrate} \xrightarrow{\text{Citrase}} \text{Oxalacetic acid} + \text{Acetic acid} \longrightarrow \text{Pyruvic acid} + CO_2 \text{ (Excess carbon dioxide)}$$

Citrate:
```
      COOH
       |
      CH₂
       |
 HO—C —COOH
       |
      CH₂
       |
      COOH
```

Oxalacetic acid:
```
      COOH
       |
      C=O
       |
      CH₂
       |
      COOH
```

Acetic acid:
```
      CH₃
       |
      COOH
```

Pyruvic acid:
```
      COOH
       |
      C=O
       |
      CH₃
```

| Citrate | Oxalacetic acid | Acetic acid | Pyruvic acid | Excess carbon dioxide |

2. $CO_2 + 2Na^+ + H_2O \longrightarrow Na_2CO_3 \longrightarrow$ Alkaline pH $\longrightarrow$ Color change from green to blue

FIGURE 25.6 Enzymatic degradation of citrate

PART D: Citrate Utilization Test

LEARNING OBJECTIVE

Once you have completed this experiment, you should be able to

1. Differentiate among enteric organisms on the basis of their ability to ferment citrate as a sole source of carbon.

PRINCIPLE

In the absence of fermentable glucose or lactose, some microorganisms are capable of using **citrate** as a carbon source for their energy. This ability depends on the presence of a **citrate permease** that facilitates the transport of citrate in the cell. Citrate is the first major intermediate in the Krebs cycle and is produced by the condensation of active acetyl with oxalacetic acid. Citrate is acted on by the enzyme **citrase,** which produces oxalacetic acid and acetate. These products are then enzymatically converted to pyruvic acid and carbon dioxide. During this reaction the medium becomes alkaline—the carbon dioxide that is generated combines with sodium and water to form sodium carbonate, an alkaline product. The presence of sodium carbonate changes the bromthymol blue indicator incorporated into the medium from green to deep Prussian blue. This reaction is illustrated in Figure 25.6.

Following incubation, citrate-positive cultures are identified by the presence of growth on the surface of the slant, which is accompanied by blue coloration. Citrate-negative cultures will show no growth, and the medium will remain green.

MATERIALS

Cultures

24- to 48-hour trypticase soy broth cultures of *E. coli, E. aerogenes,* and *K. pneumoniae* for the short version. 24- to 48-hour trypticase soy broth cultures of the 13 organisms listed on page 161 for the long version.

Media

Simmons citrate agar slants per designated student group: four for the short version, 14 for the long version.

Equipment

Bunsen burner, inoculating needle, test tube rack, and glassware marking pencil.

PROCEDURE

1. Using sterile technique, inoculate each organism into its appropriately labeled tube by means of streak inoculation. The last tube will serve as a control.

2. Incubate all cultures for 24 to 48 hours at 37°C.

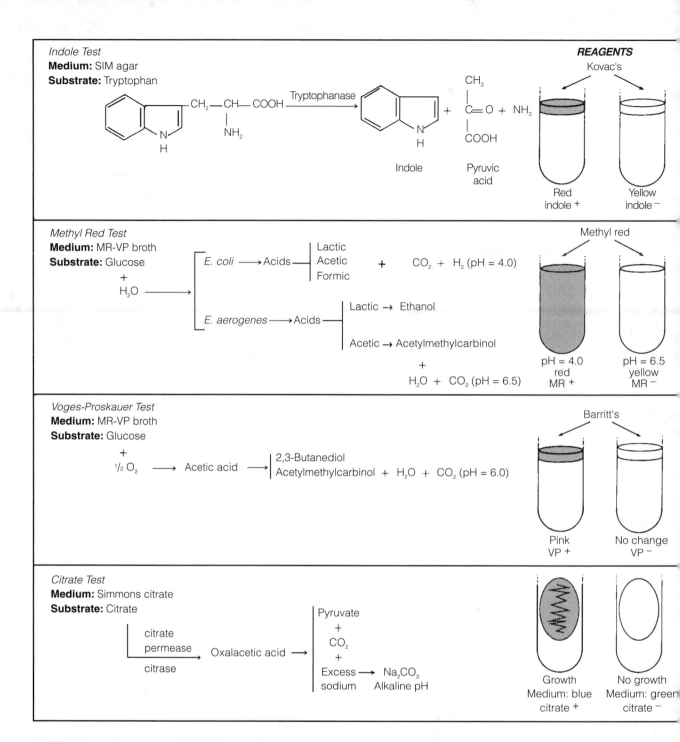

FIGURE 25.7 Summary of IMViC reactions

Name   Date

PART A: Indole Production Test

1. Add 10 drops of Kovac's reagent to all deep tube cultures and agitate the cultures gently.
2. Examine the color of the reagent layer in each culture. Record your results in the chart.
3. Based on your observations, determine and record whether or not each organism was capable of hydrolyzing the tryptophan.

Refer to photo number 37 in the color-plate insert for illustration of these reactions.

Bacterial Species	Color of Reagent Layer	Tryptophan Hydrolysis (+) or (−)
E. coli		
E. aerogenes		
K. pneumoniae		
S. dysenteriae		
S. typhimurium		
P. vulgaris		
P. aeruginosa		
A. faecalis		
M. luteus		
L. lactis		
S. aureus		
B. cereus		
C. xerosis		
Control		

PART B: Methyl Red Test

1. Transfer approximately one-third of each culture into an empty test tube and set these tubes aside for the Voges-Proskauer test.
2. Add five drops of the methyl red indicator to the remaining aliquot of each culture.
3. Examine the color of all cultures. Record the results in the chart on the following page.
4. Based on your observations, determine and record whether or not each organism was capable of fermenting glucose with the production and maintenance of a high concentration of acid.

Refer to photo number 38 in the color-plate insert for illustration of these reactions.

PART C: Voges-Proskauer Test

1. To the aliquots of each broth culture separated during the methyl red test, add 10 drops of Barritt's reagent A and shake the cultures. Immediately add 10 drops of Barritt's reagent B and shake. Reshake the cultures every 3 to 4 minutes.
2. Examine and record the color of the cultures 15 minutes after the addition of Barritt's reagent.
3. Based on your observations, determine and record whether or not each organism was capable of fermenting glucose with ultimate production of acetylmethylcarbinol.

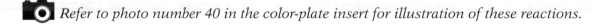 *Refer to photo number 39 in the color-plate insert for illustration of these reactions.*

Bacterial Species	Methyl Red Test		Voges-Proshauer Test	
	Color of Medium	(+) or (−)	Color of Medium	(+) or (−)
E. coli				
E. aerogenes				
K. pneumoniae				
S. dysenteriae				
S. typhimurium				
P. vulgaris				
P. aeruginosa				
A. faecalis				
M. luteus				
L. lactis				
S. aureus				
B. cereus				
C. xerosis				
Control				

PART D: Citrate Utilization Test

1. Examine all agar slant cultures for the presence or absence of growth and coloration of the medium. Record your results in the chart on the following page.
2. Based on your observations, determine and record whether or not each organism was capable of using citrate as its sole source of carbon.

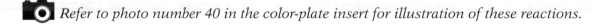 *Refer to photo number 40 in the color-plate insert for illustration of these reactions.*

Bacterial Species	Presence or Absence of Growth (+) or (−)	Color of Medium	Citrate Utilization (+) or (−)
E. coli			
E. aerogenes			
K. pneumoniae			
S. dysenteriae			
S. typhimurium			
P. vulgaris			
P. aeruginosa			
A. faecalis			
M. luteus			
L. lactis			
S. aureus			
B. cereus			
C. xerosis			
Control			

REVIEW QUESTIONS

1. Discuss the medical significance of the IMViC series of tests.

2. Explain the chemical mechanism for detecting indole in a bacterial culture.

3. Account for the development of alkalinity in cultures capable of using citrate as their sole carbon source.

4. In the carbohydrate fermentation test, we found that both *E. coli* and *E. aerogenes* produced the end products acid and gas. Account for the fact that *E. coli* is methyl red-positive and *E. aerogenes* is methyl red-negative.

5. The end products of tryptophan degradation are indole and pyruvic acid. Why do we test for the presence of indole rather than pyruvic acid as the indicator of tryptophanase activity?

6. Simmons citrate medium contains primarily inorganic ammonium, potassium, and sodium salts, plus organic citrate. What is the rationale for using a medium with this type of composition for the performance of the citrate utilization test?

Hydrogen Sulfide Test

LEARNING OBJECTIVES

Once you have completed this experiment, you will be able to determine

1. The ability of microorganisms to produce hydrogen sulfide from substances such as the sulfur-containing amino acids or inorganic sulfur compounds.

2. Mobility of microorganisms in SIM agar.

PRINCIPLE

There are two major fermentative pathways by which some microorganisms are able to produce hydrogen sulfide (H_2S).

Pathway 1: Gaseous H_2S may be produced by the reduction (hydrogenation) of organic sulfur present in the amino acid cysteine, which is a component of peptones contained in the medium. These peptones are degraded by microbial enzymes to amino acids, including the sulfur-containing amino acid cysteine. This amino acid in the presence of a **cysteine desulfurase** loses the sulfur atom and is then reduced by the addition of hydrogen from water to form hydrogen sulfide gas as illustrated:

$$CH_2 - SH \quad \xrightarrow[\text{desulfurase}]{\text{Cysteine}} \quad CH_3$$

$$CH - NH_2 \qquad\qquad C - O + H_2S \uparrow + NH_3$$

$$COOH \qquad\qquad\quad COOH$$

Cysteine | **Pyruvic acid** **Hydrogen sulfide gas** **Ammonia**

Pathway 2: Gaseous H_2S may also be produced by the reduction of inorganic sulfur compounds such as the thiosulfates ($S_2O_3^{2-}$),

sulfates (SO_4^{2-}), or sulfites (SO_3^{2-}). The medium contains sodium thiosulfate, which certain microorganisms are capable of reducing to sulfite with the liberation of hydrogen sulfide. The sulfur atoms act as hydrogen acceptors during oxidation of the inorganic compound as illustrated in the following:

$$3S_2O_3^{2-} + 4H^+ + 4e^- \xrightarrow[\text{reductase}]{\text{Thiosulfate}} 2SO_3^{2-} + 2H_2S \uparrow$$

Thiosulfate | **Sulfite** **Hydrogen sulfide gas**

In this experiment the SIM medium contains peptone and sodium thiosulfate as the sulfur substrates; ferrous sulfate ($FeSO_4$), which behaves as the H_2S indicator; and sufficient agar to make the medium semisolid and thus enhance anaerobic respiration. Regardless of which pathway is used, the hydrogen sulfide gas is colorless and therefore not visible. Ferrous ammonium sulfate in the medium serves as an indicator by combining with the gas, forming an insoluble black ferrous sulfide precipitate that is seen along the line of the stab inoculation and is indicative of H_2S production. Absence of the precipitate is evidence of a negative reaction. The overall reactions for both pathways and their interpretation are illustrated in Figure 26.1.

Motility

SIM agar may also be used to detect motile organisms. Motility is recognized when culture growth (turbidity) of flagellated organisms is not restricted to the line of inoculation. Growth of nonmotile organisms is confined to the line of inoculation.

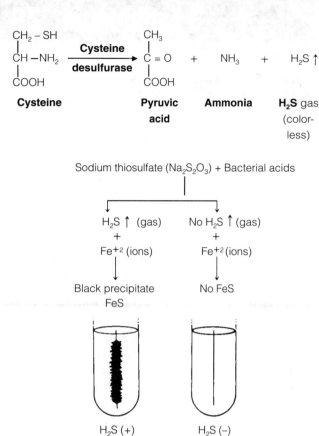

FIGURE 26.1 Detection of hydrogen sulfide

Cultures

24- to 48-hour trypticase soy broth cultures of *Enterobacter aerogenes, Shigella dysenteriae, Proteus vulgaris,* and *Salmonella typhimurium* for the short version. 24- to 48-hour trypticase soy broth cultures of the 13 organisms listed on page 144 for the long version.

Media

SIM agar deep tubes per designated student group: five for the short version, 14 for the long version.

Equipment

Bunsen burner, inoculating needle, test tube rack, and glassware marking pencil.

PROCEDURE

1. Aseptically inoculate each experimental organism into its appropriately labeled tube by means of stab inoculation. The last tube will serve as a control.
2. Incubate all cultures for 24 to 48 hours at 37°C.

Name_____ Date_____

1. Examine all SIM cultures for the presence or absence of black coloration along the line of the stab inoculation. Record your results in the chart.
2. Based on your observations, determine and record whether or not each organism was capable of producing hydrogen sulfide.
3. Observe all cultures for the presence (+) or absence (−) of motility. Record your results in the chart.

Refer to photo number 41 in the color-plate insert for illustration of these reactions.

Bacterial Species	Color of Medium	H_2S Production (+) or (−)	Motility (+) or (−)
E. coli			
E. aerogenes			
K. pneumoniae			
S. dysenteriae			
S. typhimurium			
P. vulgaris			
P. aeruginosa			
A. faecalis			
M. luteus			
L. lactis			
S. aureus			
B. cereus			
C. xerosis			
Control			

REVIEW QUESTIONS

1. Distinguish between the types of substrates available to cells for H_2S production.

2. Explain how SIM medium is used to detect motility.

3. Explain the function of the ferrous ammonium sulfate in SIM agar.

4. Why is *P. vulgaris* H_2S-positive and *E. aerogenes* H_2S-negative?

5. A stool specimen of a patient with severe diarrhea was cultured in a series of specialized media for isolation of enteric organisms. The three isolates included species of *Salmonella, Shigella,* and *Escherichia.* Explain why the H_2S production test would be diagnostically significant.

Urease Test

LEARNING OBJECTIVE

Once you have completed this experiment, you should be able to

1. Determine the ability of microorganisms to degrade urea by means of the enzyme urease.

PRINCIPLE

Urease, which is produced by some microorganisms, is an enzyme that is especially helpful in the identification of *Proteus vulgaris*. Although other organisms may produce urease, their action on the substrate urea tends to be slower than that seen with *Proteus* species. Therefore this test serves to rapidly distinguish members of this genus from other lactose-nonfermenting enteric microorganisms.

Urease is a hydrolytic enzyme that attacks the nitrogen and carbon bond in amide compounds such as urea and forms the alkaline end product ammonia. This chemical reaction is illustrated in Figure 27.1.

The presence of urease is detectable when the organisms are grown in a urea broth medium containing the pH indicator phenol red. As the substrate urea is split into its products, the presence of ammonia creates an alkaline environment that causes the phenol red to turn to a deep pink. This is a positive reaction for the presence of urease. Failure of a deep pink color to develop is evidence of a negative reaction.

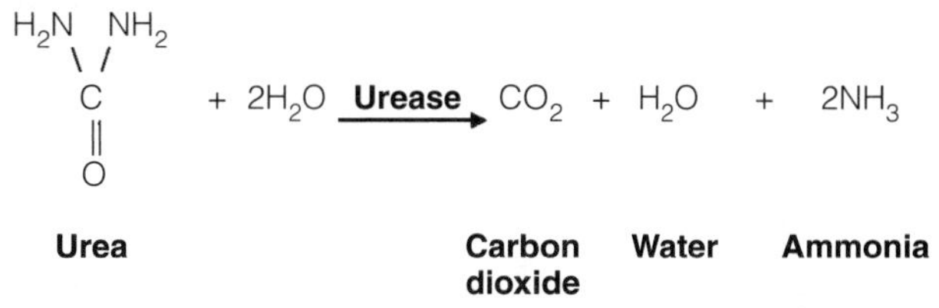

MATERIALS
Cultures

24- to 48-hour trypticase soy broth cultures of *Escherichia coli, Proteus vulgaris, Klebsiella*

FIGURE 27.1 Enzymatic degradation of urea

pneumoniae, and *Salmonella typhimurium* for the short version. 24- to 48-hour trypticase soy broth cultures of the 13 organisms listed on page 144 for the long version.

Media

Urea broth per designated student group: five for the short version, 14 for the long version.

Equipment

Bunsen burner, inoculating loop, test tube rack, and glassware marking pencil.

PROCEDURE

1. Using sterile technique, inoculate each experimental organism into its appropriately labeled tube by means of loop inoculation. The last tube will serve as a control.
2. Incubate cultures 24 to 48 hours at 37°C.

Name _____ Date _____

1. Examine all urea broth cultures for color. Record your results in the chart.
2. Based on your observations, determine and record whether or not each organism was capable of hydrolyzing the substrate urea.

📷 *Refer to photo number 42 in the color-plate insert for illustration of these reactions.*

Bacterial Species	Color of Medium	Urea Hydrolysis (+) or (−)
E. coli		
E. aerogenes		
K. pneumoniae		
S. dysenteriae		
S. typhimurium		
P. vulgaris		
P. aeruginosa		
A. faecalis		
M. luteus		
L. lactis		
S. aureus		
B. cereus		
C. xerosis		
Control		

REVIEW QUESTIONS

1. Explain the mechanism of urease activity.

2. Explain the function of phenol red in the urea broth medium.

3. Explain how the urease test is useful for identifying members of the genus *Proteus*.

4. A swollen can of chicken soup is examined by the public health laboratory and found to contain large numbers of gram-negative H_2S-positive bacilli. Which biochemical tests would you perform to identify the genus of the contaminant? Justify your test choices.

Litmus Milk Reactions

LEARNING OBJECTIVE

Once you have completed this experiment, you should be able to

1. Differentiate among microorganisms that enzymatically transform different milk substrates into varied metabolic end products.

PRINCIPLE

The major milk substrates capable of transformation are the milk sugar lactose and the milk proteins casein, lacto-albumin, and lactoglobulin. To distinguish among the metabolic changes produced in milk, a pH indicator, the oxidation–reduction indicator litmus, is incorporated into the medium. Litmus milk now forms an excellent differential medium in which microorganisms can metabolize milk substrates depending on their enzymatic complement. A variety of different biochemical changes result, as follows:

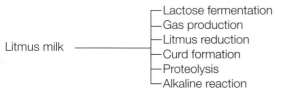

Litmus milk
— Lactose fermentation
— Gas production
— Litmus reduction
— Curd formation
— Proteolysis
— Alkaline reaction

Lactose Fermentation

Organisms capable of using **lactose** as a carbon source for energy production utilize the inducible enzyme **β-galactosidase** and degrade lactose as follows:

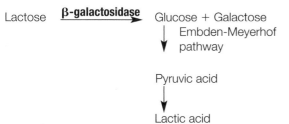

Lactose $\xrightarrow{\text{β-galactosidase}}$ Glucose + Galactose
                        Embden-Meyerhof
                        ↓ pathway

Pyruvic acid
↓
Lactic acid

The presence of **lactic acid** is easily detected because litmus is purple at a neutral pH and turns pink when the medium is acidified to an approximate pH of 4.

Gas Formation

The end products of the microbial fermentation of lactose is likely to include the **gases $CO_2\uparrow + H_2\uparrow$.** The presence of gas may be seen as separations of the curd or by the development of tracks or fissures within the curd as gas rises to the surface.

Litmus Reduction

Fermentation is an anaerobic process involving biooxidations that occur in the absence of molecular oxygen. These oxidations may be visualized as the removal of hydrogen (dehydrogenation) from a substrate. Since hydrogen ions cannot exist in the free state, there must be an immediate and concomitant electron acceptor available to bind these hydrogen ions, or else oxidation–reduction reactions are not possible and cells cannot manufacture energy. In the litmus milk test, **litmus** acts as such an acceptor. While in the oxidized state, the litmus is purple; when it accepts hydrogen from a substrate, it will become reduced and turn white or milk-colored. This oxidation of lactose, which produces lactic acid, butyric acid, $CO_2\uparrow$, and $H_2\uparrow$, is as follows:

Lactose → Glucose → Pyruvic acid →
— Lactic acid
— Butyric acid
— $CO_2 + H_2$

The excess hydrogen is now accepted by the hydrogen acceptor litmus, which turns white and is said to be reduced.

Curd Formation

The biochemical activities of different microorganisms grown in litmus milk may result in the production of two distinct types of curds (clots). Curds are designated as either acid or rennet, depending on the biochemical mechanism responsible for their formation.

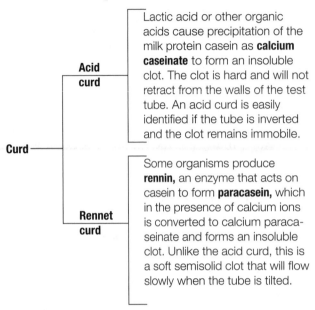

Curd

Acid curd: Lactic acid or other organic acids cause precipitation of the milk protein casein as **calcium caseinate** to form an insoluble clot. The clot is hard and will not retract from the walls of the test tube. An acid curd is easily identified if the tube is inverted and the clot remains immobile.

Rennet curd: Some organisms produce **rennin,** an enzyme that acts on casein to form **paracasein,** which in the presence of calcium ions is converted to calcium paracaseinate and forms an insoluble clot. Unlike the acid curd, this is a soft semisolid clot that will flow slowly when the tube is tilted.

Proteolysis (Peptonization)

The inability of some microorganisms to obtain their energy by way of lactose fermentation means they must use other nutritional sources such as proteins for this purpose (see Figure 23.5 on page 154). By means of **proteolytic enzymes,** these organisms hydrolyze the milk proteins, primarily **casein,** into their basic building blocks, namely **amino acids.** This digestion of proteins is accompanied by the evolution of large quantities of ammonia, resulting in an alkaline pH in the medium. The litmus turns deep purple in the upper portion of the tube, while the medium begins to lose body and produces a translucent, brown, wheylike appearance as the protein is hydrolyzed to amino acids.

Alkaline Reaction

An alkaline reaction is evident when the color of the medium remains unchanged or changes to a deeper blue. This reaction is indicative of the partial degradation of **casein** into **shorter polypeptide chains,** with the simultaneous release of alkaline end products that are responsible for the observable color change.

Figure 28.1 is a summary of the possible litmus milk reactions and their appearance following the appropriate incubation of the cultures.

MATERIALS

Cultures

24- to 48-hour trypticase soy broth cultures of *Escherichia coli, Alcaligenes faecalis, Lactococcus lactis,* and *Pseudomonas aeruginosa* for the short version. 24- to 48-hour trypticase soy broth cultures of the 13 organisms listed on page 144 for the long version.

Media

Litmus milk broth per designated student group: five for the short version, 14 for the long version.

Equipment

Bunsen burner, inoculating loop, test tube rack, and glassware marking pencil.

PROCEDURE

1. Using sterile technique, inoculate each experimental organism into its appropriately labeled tube by means of a loop inoculation. The last tube will serve as a control.

2. Incubate all cultures for 24 to 48 hours at 37°C.

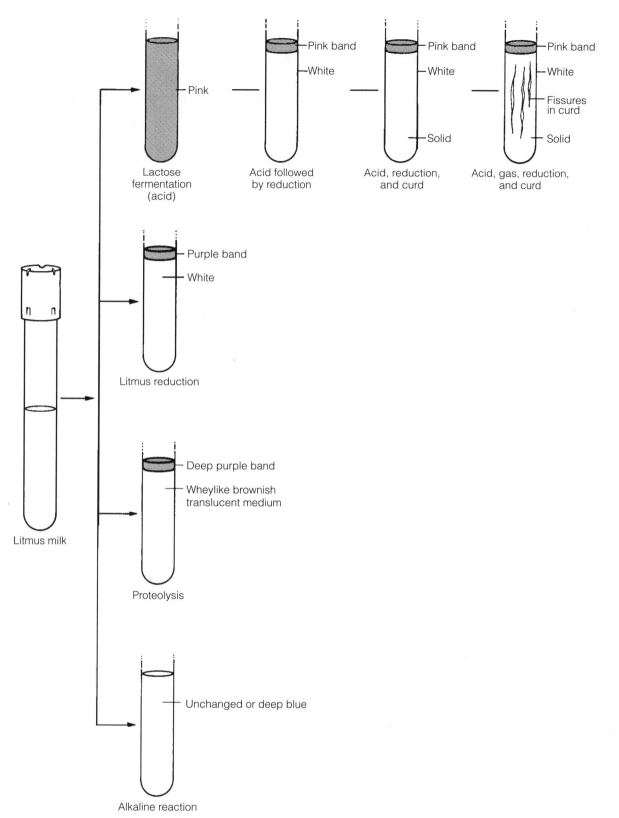

FIGURE 28.1 Litmus milk reactions (📷 **Refer to photo number 43 in the color-plate insert.)**

Name   Date

1. Examine all the litmus milk cultures for color and consistency of the medium. Record your results in the chart.
2. Based on your observations, determine and record the type(s) of reaction(s) that have taken place in each culture.

Refer to photo number 43 in the color-plate insert for illustration of these reactions.

Bacterial Species	Appearance of Medium	Litmus Milk Reactions
E. coli		
E. aerogenes		
K. pneumoniae		
S. dysenteriae		
S. typhimurium		
P. vulgaris		
P. aeruginosa		
A. faecalis		
M. luteus		
L. lactis		
S. aureus		
B. cereus		
C. xerosis		
Control		

REVIEW QUESTIONS

1. Distinguish between acid and rennet curds.

2. Describe the litmus milk reactions that may occur when proteins are metabolized as an energy source.

3. Explain how the litmus in the litmus milk acts as a redox indicator.

4. Can a litmus milk culture show a pink band at the top and a brownish translucent layer at the bottom? Explain.

5. Explain why litmus milk is considered a good differential medium.

Nitrate Reduction Test

LEARNING OBJECTIVE

Once you have completed this experiment, you should be able to

1. Determine the ability of some microorganisms to reduce nitrates (NO_3^-) to nitrites (NO_2^-) or beyond the nitrite stage.

PRINCIPLE

The reduction of nitrates by some aerobic and facultative anaerobic microorganisms occurs in the absence of molecular oxygen, an anaerobic process. In these organisms anaerobic respiration is an oxidative process whereby the cell uses inorganic substances such as nitrates (NO_3^-) or sulfates (SO_4^{2-}) to supply oxygen that is subsequently utilized as a final hydrogen acceptor during energy formation. The biochemical transformation may be visualized as follows:

$$NO_3^- + 2H^+ + 2e^- \xrightarrow[\text{Reductase}]{\text{Nitrate}} NO_2^- + H_2O$$

| Nitrate | Hydrogen electrons | | Nitrite | Water |

Some organisms possess the enzymatic capacity to act further on nitrites to reduce them to ammonia (NH_3^+) or molecular nitrogen (N_2). These reactions may be described as follows:

$$NO_2^- \longrightarrow NH_3^+$$

| Nitrite | Ammonia |

or

$$2NO_3^- + 12H^+ + 10e^- \longrightarrow N_2 + 6H_2O$$

| Nitrate | | Molecular nitrogen |

Nitrate reduction can be determined by cultivating organisms in a nitrate broth medium. The medium is basically a nutrient broth supplemented with 0.1% potassium nitrate (KNO_3) as the nitrate substrate. In addition, the medium is made into a semisolid by the addition of 0.1% agar. The semisolidity impedes the diffusion of oxygen into the medium, thereby favoring the anaerobic requirement necessary for nitrate reduction.

Following incubation of the cultures, an organism's ability to reduce nitrates to nitrites is determined by the addition of two reagents: Solution A, which is sulfanilic acid, followed by Solution B, which is α-naphthylamine. *Note: This should not be confused with Barritt's reagent.* Following reduction, the addition of Solutions A and B will produce an immediate cherry red color.

$$NO_3^- \xrightarrow[\text{Reductase}]{\text{Nitrate}} NO_2^- \text{ (Red color on addition of Solutions A and B)}$$

Cultures not producing a color change suggest one of two possibilities: (1) nitrates were not reduced by the organism, or (2) the organism possessed such potent **nitrate reductase** enzymes that nitrates were rapidly reduced beyond nitrites to ammonia or even molecular nitrogen. To determine whether or not nitrates were reduced past the nitrite stage, a small amount of zinc powder is added to the basically colorless cultures already containing Solutions A and B. Zinc reduces nitrates to nitrites. The development of red color therefore verifies that nitrates were not reduced to nitrites by the organism. If nitrates were not reduced, a negative nitrate reduction reaction has occurred. If the addition of zinc does not produce a color change, the nitrates in the medium were reduced beyond nitrites to ammonia or nitrogen gas. This is a positive reaction, as shown in Figure 29.1.

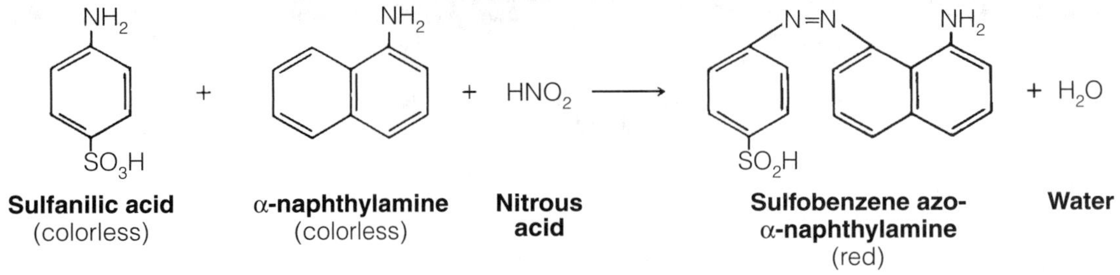

FIGURE 29.1 Formation of colored complex indicative of NO_3^- reduction

MATERIALS

Cultures

24- to 48-hour trypticase soy broth cultures of *Escherichia coli, Alcaligenes faecalis,* and *Pseudomonas aeruginosa* for the short version. 24- to 48-hour trypticase soy broth cultures of the 13 organisms listed on page 144 for the long version.

Media

Trypticase nitrate broth per designated student group: four for the short version, 14 for the long version.

Reagents

Solution A (sulfanilic acid), Solution B (α-naphthylamine), and zinc powder.

Equipment

Bunsen burner, inoculating loop, test tube rack, and glassware marking pencil.

PROCEDURE

1. Using sterile technique, inoculate each experimental organism into its appropriately labeled tube by means of a loop inoculation. The last tube will serve as a control.

2. Incubate all cultures for 24 to 48 hours at 37°C.

Name _____   Date _____

1. Add five drops of Solution A and then five drops of Solution B to all nitrate broth cultures. Observe and record in the chart whether or not a red coloration develops in each of the cultures.
2. Add a minute quantity of zinc to the cultures in which no red color developed. Observe and record whether or not red coloration develops in each of the cultures.
3. On the basis of your observations, determine and record in the chart whether or not each organism was capable of nitrate reduction. Identify the end product (NO_2^- or NH_3^+/N_2), if any, that is present.

Refer to photo number 44 in the color-plate insert for illustration of these reactions.

Bacterial Species	Red Coloration with Solutions A and B (+) or (−)	Red Coloration with Zinc (+) or (−)	Nitrate Reductions (+) or (−)	End Products
E. coli				
E. aerogenes				
K. pneumoniae				
S. dysenteriae				
S. typhimurium				
P. vulgaris				
P. aeruginosa				
A. faecalis				
M. luteus				
L. lactis				
S. aureus				
B. cereus				
C. xerosis				
Control				

REVIEW QUESTIONS

1. Explain the function of the 0.1% agar in the nitrate medium.

2. Explain the functions of Solutions A and B.

3. If a culture does not undergo a color change on the addition of Solutions A and B, explain how you would interpret this result.

4. Explain why the development of a red color on the addition of zinc is a negative test.

5. Discuss the relationship between an organism's ability to reduce nitrate past the nitrite stage and that organism's proteolytic activity.

Catalase Test

LEARNING OBJECTIVE

Once you have completed this experiment, you should be able to

1. Determine the ability of some microorganisms to degrade hydrogen peroxide by producing the enzyme catalase.

PRINCIPLE

During aerobic respiration, microorganisms produce hydrogen peroxide and, in some cases, an extremely toxic superoxide. Accumulation of these substances will result in death of the organism unless they can be enzymatically degraded. These substances are produced when aerobes, facultative anaerobes, and microaerophiles use the aerobic respiratory pathway, in which oxygen is the final electron acceptor, during degradation of carbohydrates for energy production. Organisms capable of producing **catalase** rapidly degrade hydrogen peroxide as illustrated:

$$2H_2O_2 \xrightarrow{\text{Catalase}} 2H_2O + O_2 \uparrow$$

Hydrogen peroxide   **Water**   **Free oxygen**

Aerobic organisms that lack catalase can degrade especially toxic superoxides using the enzyme **superoxide dismutase;** the end product of a superoxide dismutase is H_2O_2, but this is less toxic to the bacterial cells than are the superoxides.

The inability of strict anaerobes to synthesize catalase, peroxidase, or superoxide dismutase may explain why oxygen is poisonous to these microorganisms. In the absence of these enzymes, the toxic concentration of H_2O_2 cannot be degraded when these organisms are cultivated in the presence of oxygen.

Catalase production can be determined by adding the substrate H_2O_2 to an appropriately incubated trypticase soy agar slant culture. If catalase is present, the chemical reaction mentioned is indicated by bubbles of free oxygen gas ($O_2\uparrow$). This is a positive catalase test; the absence of bubble formation is a negative catalase test.

MATERIALS

Cultures

24- to 48-hour trypticase soy broth cultures of *Staphylococcus aureus, Micrococcus luteus,* and *Lactococcus lactis* for the short version. 24- to 48-hour trypticase soy broth cultures of the 13 organisms listed on page 144 for the long version.

Media

Trypticase soy agar slants per designated student group: four for the short version, 14 for the long version.

Reagent

3% hydrogen peroxide.

Equipment

Bunsen burner, inoculating loop, test tube rack, and glassware marking pencil.

PROCEDURE

1. Using sterile technique, inoculate each experimental organism into its appropriately labeled tube by means of a streak inoculation. The last tube will serve as a control.

2. Incubate all cultures for 24 to 48 hours at 37°C.

Experiment 30 / Observations and Results

Name _____   Date _____

1. Allow three or four drops of the 3% hydrogen peroxide to flow over the entire surface of each slant culture.
2. Examine each culture for the presence or absence of bubbling or foaming. Record your results in the chart.
3. Based on your observations, determine and record whether or not each organism was capable of catalase activity.

📷 *Refer to photo numbers 45 and 46 in the color-plate insert for illustration of these reactions.*

Bacterial Species	Presence or Absence of Bubbling	Catalase Production (+) or (−)
E. coli		
E. aerogenes		
K. pneumoniae		
S. dysenteriae		
S. typhimurium		
P. vulgaris		
P. aeruginosa		
A. faecalis		
M. luteus		
L. lactis		
S. aureus		
B. cereus		
C. xerosis		
Control		

REVIEW QUESTIONS

1. Explain the toxic effect of O_2 on strict anaerobes.

2. Illustrate the chemical reaction involved in the degradation of hydrogen peroxide in the presence of catalase.

3. Would catalase be classified as an endoenzyme or an exoenzyme? Explain.

4. Account for the ability of streptococci to tolerate O_2 in the absence of catalase activity.

Oxidase Test

LEARNING OBJECTIVE

Once you have completed this experiment, you should be able to

1. Perform an experimental procedure that is designed to distinguish among groups of bacteria on the basis of cytochrome oxidase activity.

PRINCIPLE

Oxidase enzymes play a vital role in the operation of the electron transport system during aerobic respiration. **Cytochrome oxidase** catalyzes the oxidation of a reduced cytochrome by molecular oxygen (O_2), resulting in the formation of H_2O or H_2O_2. Aerobic bacteria, as well as some facultative anaerobes and microaerophiles, exhibit oxidase activity. The oxidase test aids in differentiation among members of the genera *Neisseria* and *Pseudomonas*, which are oxidase-positive, and Enterobacteriaceae, which are oxidase-negative.

The ability of bacteria to produce cytochrome oxidase can be determined by the addition of the test reagent, *p*-aminodimethylaniline oxalate, to colonies grown on a plate medium. This light pink reagent serves as an artificial substrate, donating electrons and thereby becoming oxidized to a blackish compound in the presence of the oxidase and free oxygen. Following the addition of the test reagent, the development of pink, then maroon, and finally black coloration on the surface of the colonies is indicative of cytochrome oxidase production and represents a positive test. No color change, or a light pink coloration on the colonies, is indicative of the absence of oxidase activity and is a negative test.

MATERIALS

Cultures

24- to 48-hour trypticase soy broth cultures of *Escherichia coli, Pseudomonas aeruginosa,* and *Alcaligenes faecalis* for the short version. 24- to 48-hour trypticase soy broth cultures of the 13 organisms listed on page 144 for the long version.

Media

Trypticase soy agar plates per designated student group: one for the short version, four for the long version.

Reagent

p-aminodimethylaniline oxalate (Difco 0329-13-9).

Equipment

Bunsen burner, inoculating loop, and glassware marking pencil.

PROCEDURE

1. Prepare the trypticase soy agar plate(s) for inoculation as follows:
   a. Short procedure: With a glassware marking pencil, divide the bottom of a Petri dish into three sections and label each section with the name of the test organism to be inoculated.
   b. Long procedure: Repeat Step 1a, dividing three plates into three sections and one plate into four sections to accommodate the 13 test organisms.
2. Using sterile technique, make a single-line streak inoculation of each test organism on the agar surface of its appropriate section of the plate(s).
3. Incubate the plate(s) in an inverted position for 24 to 48 hours at 37°C.

Experiment 31 / Observations and Results

Name ____ Date _______________

1. Add two or three drops of the *p*-aminodimethylaniline oxalate to the surface of the growth of each test organism.
2. Observe the growth for the presence or absence of a color change from pink, to maroon, and finally to purple. Positive test (+), color change in 10–30 seconds; negative test (−), no color change, or light pink color. Record the results on the chart below.
3. Based on your observations, determine and record whether or not each organism was capable of producing cytochrome oxidase.

Refer to photo number 47 in the color-plate insert for illustration of these reactions.

Bacterial Species	Color of Colonies	Oxidase Production (+) or (−)
E. coli		
E. aerogenes		
K. pneumoniae		
S. dysenteriae		
S. typhimurium		
P. vulgaris		
P. aeruginosa		
A. faecalis		
M. luteus		
L. lactis		
S. aureus		
B. cereus		
C. xerosis		
Control		

REVIEW QUESTIONS

1. What is the function of cytochrome oxidase?

2. Why are strict aerobes oxidase-positive?

3. The oxidase test is used to differentiate among which groups of bacteria?

4. What is the function of the test reagent in this procedure?

5. Your instructor asks you to isolate and identify the organisms in an unknown culture. You find that the culture contains two gram-negative bacilli that produce swarming colonies. What biochemical test would you use to identify the bacilli? Justify your answer.

Utilization of Amino Acids

PART A: Decarboxylase Test

LEARNING OBJECTIVE

Once you have completed this experiment, you should be able to

1. Identify and differentiate organisms based on their ability to enzymatically degrade amino acid substrates.

PRINCIPLE

Every biologically active protein is composed of the 20 essential amino acids. Structurally amino acids are composed of an alpha carbon (—C—), an amino group (—NH$_2$), a carboxyl group (—COOH), and a hydrogen atom (—H). Also attached to the alpha carbon is a side group or an atom designated by an (—R), which differs in each of the amino acids.

$$NH_2 - \underset{\underset{H}{|}}{\overset{\overset{R}{|}}{C}} - COOH$$

Decarboxylation is a process whereby some microorganisms that possess decarboxylase enzymes are capable of removing the carboxyl group to yield end products consisting of an **amine** or **diamine** plus **carbon dioxide.** Decarboxylated amino acids play an essential role in cellular metabolism since the amines produced may serve as end products for the synthesis of other molecules required by the cell. Decarboxylase enzymes are designated as adaptive (or induced) enzymes and are produced in the presence of specific amino acid substrates upon which they act. These amino acid substrates must possess at least one chemical group other than an amine (—NH$_2$)

or a carboxyl group (—COOH). In the process of decarboxylation, organisms are cultivated in an acid environment and in the presence of a specific substrate. The produced end product (amines) results in a shift to a more alkaline pH.

In the clinical or diagnostic microbiology laboratory, three decarboxylase enzymes are used to differentiate members of the Enterobacteriaceae: lysine, ornithine, and arginine. Decarboxylase activity is determined by cultivating the organism in a nutrient medium containing glucose, the specific amino acid substrate, and bromthymol blue (the pH indicator). If decarboxylation occurs, the pH of the medium becomes alkaline despite the fermentation of glucose since the end products (amines or diamines) are alkaline. The function of the glucose in the medium is to insure good microbial growth and thus more reliable results in the presence of the pH indicator. The presence of each decarboxylase enzyme can be tested for by supplementing decarboxylase broth with the specific amino acid substrate, namely, lysine, arginine, and ornithine. For example, **lysine decarboxylase** degrades L-lysine, forming the diamine end product **cadaverine** plus **carbon dioxide** as illustrated in Figure 32.1.

In the experiment that follows, the decarboxylation of L-lysine will be studied. It should be noted that decarboxylation reactions occur under anaerobic conditions that are satisfied by sealing the culture tubes with sterile mineral oil. In the sealed tubes, all of the unbound oxygen is utilized during the organisms' initial growth phase, and the pH of the medium becomes alkaline as carbon dioxide (CO_2) is produced in the culture tube. A pH indicator such as bromcresol purple is usually incorporated into the medium for the easy detection of pH changes. The production of acid end products

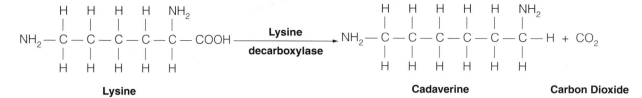

FIGURE 32.1 Degradation of lysine

will cause the bromcresol purple to change color from purple to yellow, indicating that acid has formed, the medium has been acidified, and the decarboxylase enzymes have been activated. The activated enzyme responds with the production of the alkalinizing diamine (cadaverine) and carbon dioxide, which will produce a final color change from yellow back to purple, thereby indicating that L-lysine has been decarboxylated. The development of a turbid purple color verifies a positive test for amino acid decarboxylation. The absence of a purple color indicates a negative result.

MATERIALS

Cultures

24-hour nutrient broth cultures of *Proteus vulgaris, Escherichia coli,* and *Citrobacter freundii* for the short version. 24- to 48-hour nutrient broth cultures of the 13 organisms listed on page 144 for the long version.

Media

Per designated student group: three tubes of Moeller's decarboxylase broth supplemented with L-lysine (10 gm/l) (labeled LD+), three tubes of Moeller's decarboxylase broth without lysine (labeled LD−).

Equipment

Bunsen burner, glassware marking pencil, inoculating loop and needle, sterile Pasteur pipettes, rubber bulbs, test tube rack, and sterile mineral oil.

PROCEDURE

1. With a glassware marking pencil, label three tubes of the LD+ medium with the name of the organism to be inoculated. Similarly label three tubes of LD− medium. The use of (LD−) control tubes is essential since some bacterial strains are capable of

turning substrate-free media positive. *Note: Control tubes should remain yellow after incubation, denoting that only glucose was fermented.* The presence of a positive control tube invalidates the test, and no interpretation is possible.

2. Using sterile technique, inoculate each experimental organism into its appropriately labeled tube using a loop inoculation.

3. Place a rubber bulb onto a sterile Pasteur pipette and overlay the surface of the inoculated culture tubes with 1 ml of sterile mineral oil. Hold the tubes in a slanted position while adding the mineral oil. *Note: Do not let the tip of the pipette touch the inoculated medium or the sides of the test tube walls.*

4. Repeat the above procedure for the remaining test cultures.

5. Incubate all tubes at 37°C for 24 to 48 hours.

PART B: Phenylalanine Deaminase Test

LEARNING OBJECTIVE

Once you have completed this experiment, you should be able to

1. Demonstrate the ability of some organisms to remove the amino group ($-NH_2$) from amino acids.

PRINCIPLE

Microorganisms that contain deaminase enzymes are capable of removing the amino group ($-NH_2$) from amino acids and other NH_2-containing chemical compounds. During this process the amino acid, under the auspices of its specific deaminase, will produce keto acids and ammonia as end products. In the experiment to follow, the amino acid phenylalanine will be deaminated by **phenylalanine deaminase** and converted to the **keto**

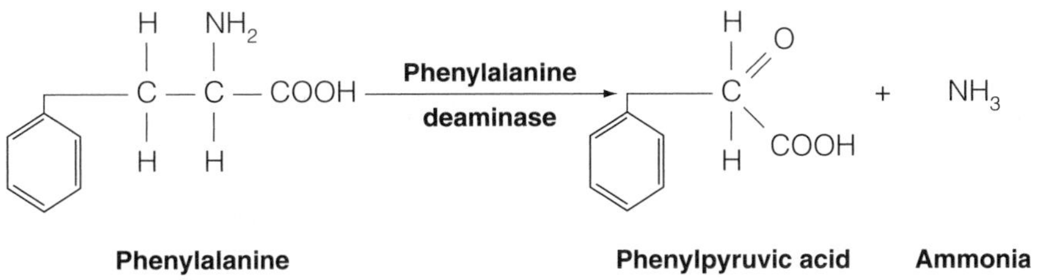

FIGURE 32.2 Deamination of phenylalanine

acid phenylpyruvic acid and **ammonia.** The organisms are cultured on a medium incorporating phenylalanine as the substrate. This reaction is illustrated in Figure 32.2.

If the organism possesses phenylalanine deaminase, phenylpyruvic acid will be released into the medium and can be detected by the addition of a 10 to 12% ferric chloride solution to the surface of the medium. If a green color develops, the enzymatic deamination of the substrate has occurred and is indicative of a positive result. No color change indicates a negative result. The resultant green color produced upon the addition of ferric chloride ($FeCl_3$) is due to the formation of a keto acid (phenylpyruvic acid). It has been shown that α- and β-keto acids give a positive color reaction with either alcoholic or aqueous solutions of $FeCl_3$. Phenylpyruvic acid is a α-keto acid. The results should be read immediately following the addition of the reagent since the color produced fades quickly. When not in use, the ferric chloride reagent should be refrigerated and kept in a dark bottle to avoid exposure to light. The stability of this reagent varies and should be checked weekly with known positive cultures.

MATERIALS

Cultures

24-hour nutrient broth cultures of *Escherichia coli* and *Proteus vulgaris* for the short version. 24-hour nutrient broth cultures of the 13 organisms listed on page 144 in long version.

Media

Two phenylalanine agar slants.

Reagents

10 to 12% ferric chloride solution.

Equipment

Bunsen burner, glassware marking pencil, Pasteur pipettes, rubber bulbs, test tube racks, and inoculating loop.

PROCEDURE

1. Using sterile technique, inoculate each experimental organism into its appropriately labeled tube using a streak inoculation.
2. Incubate cultures at 37°C for 24 to 48 hours.

Name   Date

PART A: Decarboxylase Test

1. Examine each culture tube for the presence of a color change.
2. Based on your observations, determine whether or not each organism was capable of performing decarboxylation of lysine.
3. Record your results in the chart below.

📷 *Refer to photo number 49 in the color-plate insert for illustration of this reaction.*

Bacterial Species	Color of Medium		Lysine decarboxylase (+) or (−)	
	LD+	LD−	LD+	LD−
E. coli				
P. vulgaris				
C. freundii				

PART B: Phenylalanine Deaminase Test

1. Add 5 to 10 drops of the ferric chloride solution to each agar slant and mix gently. Ferric chloride is a chelating agent and binds to the phenylpyruvic acid to produce a green color on the slant.
2. Based on your observations, determine and record whether or not each organism was capable of amino acid deamination.
3. *Note: Results should be read immediately following the addition of ferric chloride because the green color fades rapidly.*

📷 *Refer to photo number 48 in the color-plate insert for illustration of this reaction.*

Bacterial Species	Color after FeCl$_3$	Deamination (+) or (−)
E. coli		
P. vulgaris		

REVIEW QUESTIONS

1. A negative decarboxylase test is indicated by the production of a yellow color in the medium. Explain the reason for the development of this color.

2. Explain why deaminase activity must be determined immediately following the addition of ferric chloride.

3. What is the function of ferric chloride in the detection of deaminase activity?

4. Explain why the anaerobic environment is essential for decarboxylation of the substrate to occur.

5. Following a normal delivery, a nurse observes that the urine of the infant has a peculiar odor resembling that of burnt sugar or maple syrup. Subsequent examination by the pediatrician reveals that this child has maple syrup urine disease.

   a. What is this disease?

   b. How is it treated?

Genus Identification of Unknown Bacterial Cultures

LEARNING OBJECTIVE

Once you have completed this experiment, you should be able to

1. Use previously studied staining, cultural characteristics, and biochemical procedures for independent genus identification of an unknown bacterial culture.

PRINCIPLE

Identification of unknown bacterial cultures is one of the major responsibilities of the microbiologist. Samples of blood, tissue, food, water, and cosmetics are examined daily in laboratories throughout the world for the presence of contaminants. In addition, industrial organizations are constantly screening materials to isolate new antibiotic-producing organisms or organisms that will increase the yield of marketable products such as vitamins, solvents, and enzymes. Once isolated, these unknown organisms must be identified and classified.

The science of classification is called **taxonomy** and deals with the separation of living organisms into interrelated groups. *Bergey's Manual* has been the official, internationally accepted reference for bacterial classification since 1923. The current edition, *Bergey's Manual of Systematic Bacteriology,* arranges related bacteria into 33 groups called sections rather than into the classical taxonomic groupings of phylum, class, order, and family. The interrelationship of the organisms in each section is based on characteristics such as morphology, staining reactions, nutrition, cultural characteristics, physiology, cellular chemistry, and biochemical test results for specific metabolic end products.

At this point you have developed sufficient knowledge of staining methods, isolation techniques, microbial nutrition, biochemical activities, and characteristics of microorganisms to be able to work independently in attempting to identify the genus of an unknown culture. Characteristics of the major organisms that have been used in experiments thus far are given in Table 33.1. You are to use this table for the identification of the unknown cultures. The observations and results obtained following the experimental procedures are the basis of this identification. However, you should note that your biochemical results may not be identical to those shown in Table 33.1; they may vary because of variations in bacterial strains (subgroups of a species). Therefore, it becomes imperative to recall the specific biochemical tests that differentiate among the different genera of the test organisms.

Identification of an unknown culture using a more extensive procedure to differentiate bacterial species is presented in Experiment 70. The rationale for the performance of this exercise later in the semester is twofold. First, you will have acquired expanded knowledge of microbial activities and will be more proficient in laboratory skills. Second, and more important, you will be more cognizant of and more critical in your approach to species identification using dichotomous keys supplemented with *Bergey's Manual.*

MATERIALS

Cultures

Number-coded 24- to 48-hour trypticase soy agar slant cultures of the 13 bacterial species used in the long experimental procedure and listed on page 144. You will be provided with one unknown pure culture.

TABLE 33.1 Cultural and Biochemical Characteristics of Unknown Organisms

Organism	Gram Stain	Agar Slant Cultural Characteristics	Litmus Milk Reaction	Fermentation Lactose	Dextrose	Sucrose	H_2S Production	NO_3 Reduction	Indole Production	MR Reaction	VP Reaction	Citrate Use	Urease Activity	Catalase Activity	Oxidase Activity	Gelatin Liquefaction	Starch Hydrolysis	Lipid Hydrolysis
Escherichia coli	Rod −	White, moist, glistening growth	Acid, curd ±, gas ±, reduction ±	AG	AG	A±	−	+	+	+	−	−	−	+	−	−	−	−
Enterobacter aerogenes	Rod −	Abundant, thick, white, glistening growth	Acid	AG	AG	AG±	−	+	−	−	+	+	−	+	−	−	−	−
Klebsiella pneumoniae	Rod −	Slimy, white, somewhat translucent, raised growth	Acid, gas, curd ±	AG	AG	AG	−	+	−	±	±	+	+	+	−	−	−	−
Shigella dysenteriae	Rod −	Thin, even, grayish growth	Alkaline	−	A	A±	−	+	±	+	−	−	−	+	−	−	−	−
Salmonella typhimurium	Rod −	Thin, even, grayish growth	Alkaline	−	AG±	A±	+	+	−	+	−	+	−	+	−	−	−	−
Proteus vulgaris	Rod −	Thin, blue-gray, spreading growth	Alkaline	−	AG	AG±	+	+	+	+	−	±	+	+	−	+	−	−
Pseudomonas aeruginosa	Rod −	Abundant, thin, white growth, with medium turning green	Rapid peptonization	−	−	−	−	+	−	−	−	+	−	+	+	+ Rapid	−	+
Alcaligenes faecalis	Rod* −	Thin, white, spreading, viscous growth	Alkaline	−	−	−	−	−	−	−	−	±	−	+	+	−	−	−
Staphylococcus aureus	Cocci +	Abundant, opaque, golden growth	Acid, reduction ±	A	A	A	−	+	−	+	±	−	−	+	−	+	−	+
Lactococcus lactis	Cocci +	Thin, even growth	Acid, rapid reduction with curd	A	A	A	−	−	−	+	−	−	−	−	−	−	−	−
Micrococcus luteus	Cocci +	Soft, smooth, yellow growth	Alkaline	−	−	−	−	±	−	−	−	−	+	+	−	+ Slow	−	−
Corynebacterium xerosis	Rod +	Grayish, granular, limited growth	Alkaline	−	A±	A±	−	+	−	−	−	−	−	+	−	−	−	−
Bacillus cereus	Rod +	Abundant, opaque, white waxy growth	Peptonization	−	A	A	−	+	−	−	±	−	−	+	−	+ Rapid	+	±

Note: AG = Acid and gas; ± = Variable reaction; Rod* = Coccobacillus

Media

Two trypticase soy agar slants, and one each of the following per student: phenol red sucrose broth, phenol red lactose broth, phenol red dextrose broth, SIM agar deep tube, MR-VP broth, tryptic nitrate broth, Simmons citrate agar slant, urea broth, litmus milk, trypticase soy agar plate, nutrient gelatin deep tube, starch agar plate, and tributyrin agar plate.

Reagents

Crystal violet, Gram's iodine, 95% ethyl alcohol, safranin, methyl red, 3% hydrogen peroxide, Barritt's reagent, Solutions A and B, Kovac's reagent, zinc powder, *p*-aminodimethylaniline oxalate.

Equipment

Bunsen burner, inoculating loop and needle, staining tray, immersion oil, lens paper, bibulous paper, microscope, and glassware marking pencil.

PROCEDURE

1. Perform a Gram stain of the unknown organism. Observe and record the reaction and the morphology and arrangement of the cells.

2. Using sterile inoculating technique, inoculate two trypticase soy agar slants by means of a streak inoculation. Following incubation, you will use one slant culture to determine the cultural characteristics of the unknown microorganism. You will use the second as a stock subculture should it be necessary to repeat any of the tests.

3. Exercising care in sterile technique so as not to contaminate cultures and thereby obtain spurious results, inoculate the media for the following biochemical tests:

Medium	Test
a. Phenol red lactose broth	
b. Phenol red dextrose broth	Carbohydrate fermentation
c. Phenol red sucrose broth	
d. Litmus milk	Litmus milk reactions
e. SIM medium	Indole production H$_2$S production
f. Tryptic nitrate broth	Nitrate reduction
g. MR-VP broth	Methyl red test Voges-Proskauer test
h. Simmons citrate agar slant	Citrate utilization
i. Urea broth	Urease activity
j. Trypticase soy agar slant	Catalase activity
k. Starch agar plate	Starch hydrolysis
l. Tributyrin agar plate	Lipid hydrolysis
m. Nutrient gelatin deep tube	Gelatin liquefaction
n. Trypticase soy agar plate	Oxidase test

4. Incubate all cultures for 24 to 72 hours at 37°C.

Name _____ Date _____

1. Examine a trypticase soy agar slant culture and determine the cultural characteristics of your unknown organism. Record your results in the chart.
2. Perform biochemical tests on the remaining cultures, making reference to the specific laboratory exercise for each test. Record your observations and results.
3. Based on your results, identify the genus and species of the unknown organism. *Note: Results may vary depending on the strains of each species used and the length of time the organism has been maintained in stock culture. The observed results may not be identical to the expected results. Therefore choose the organism that best fits the results summarized in Table 33.1.*

Description of Unknown's Characteristics	Student	_____
	Culture no.	_____
	Organism	_____

Experimental Procedure	Observations	Results
Grain stain		
Acid-fast stain		
Shape and arrangement		
Cultural characteristics		
Litmus milk reactions		
Carbohydrate fermentations: Lactose		
Dextrose		
Sucrose		
H_2S production		
Nitrate reduction		
Indole production		
Methyl red test		
Voges-Proskauer test		
Citrate utilization		
Urease activity		
Catalase activity		
Starch hydrolysis		
Gelatin liquefaction		
Oxidase test		

The Protozoa

LEARNING OBJECTIVES

Once you have completed the experiments in this section, you should be

1. Familiar with the distinguishing characteristics of the members of the phylum protozoa.
2. Able to perform experimental procedures to identify free-living and parasitic protozoans.

INTRODUCTION

The protozoa are a large and diverse group of unicellular, eukaryotic organisms. Most are free-living, but some are parasites. Their major distinguishing characteristics are:

1. The absence of a cell wall; some, however, possess a flexible layer, a pellicle, or a rigid shell of inorganic materials outside of the cell membrane.
2. The ability during their entire life cycle or part of it to move by locomotor organelles or by a gliding mechanism.
3. Heterotrophic nutrition whereby the free-living forms ingest particulates such as bacteria, yeast, and algae, while the parasitic forms derive nutrients from the body fluids of their hosts.
4. Primarily asexual means of reproduction, although sexual modes occur in some groups.

Taxonomically, the classification of protozoa depends on their means of locomotion in the mature stage. The phylum is subdivided into the following four classes (or subphyla, according to some taxonomists):

1. **Sarcodina:** Motility results from the streaming of ectoplasm, producing proto-plasmic projections called pseudopods (false feet). Prototypic amoebas include the free-living *Amoeba proteus* and its parasitic congener *Entamoeba histolytica*.
2. **Mastigophora:** Locomotion is effected by one or more whiplike, thin structures called flagella. Free-living members include the genera *Cercomonas*, *Heteronema*, and *Euglena*, which are photosynthetic protists that may be classified as flagellated algae. The parasitic forms include *Trichomonas vaginalis*, *Giardia lamblia*, and the *Trypanosoma* species.
3. **Ciliophora:** Locomotion is carried out by means of short hairlike projections called cilia, whose synchronous beating propels the organisms. The characteristic example of free-living members of this group is *Paramecium caudatum*, and the parasitic example is *Balantidium coli*.
4. **Sporozoa:** Unlike other members of this phylum, sporozoa do not have locomotor organelles in their mature stage; however, immature forms exhibit some type of movement. All the members of this group are parasites. The most significant members belong to the genus *Plasmodium*, the malarial parasites of animals and humans.

Free-Living Protozoa

LEARNING OBJECTIVE

Once you have completed this experiment, you should be familiar with

1. The protozoan flora of pond water.

PRINCIPLE

There are more than 20,000 known species of free-living protozoa. It is not within the scope of this manual to present an in-depth study of this large and diverse population. Therefore in this procedure you will use Table 34.1 to become familiar with the general structural characteristics of representative protozoa, and you will identify these in a sample of pond water.

MATERIALS

Cultures

Stagnant pond water and prepared slides of amoebas, paramecia, euglena, and stentor.

Reagent

Methyl cellulose.

Equipment

Microscope, glass slides, coverslips, and Pasteur pipettes.

PROCEDURE

1. Obtain a drop of pond water from the bottom of the culture and place it in the center of a clean slide.

2. Add a drop of methyl cellulose to the culture to slow down the movement of the protozoa.

3. Apply a coverslip in the following manner to prevent formation of air bubbles:

   a. Place one edge of the coverslip against the outer edge of the drop of culture.

   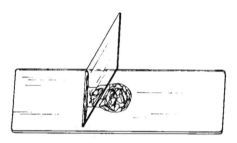

   b. After the drop of culture spreads along the inner aspect of the edge of the coverslip, gently lower the coverslip onto the slide.

   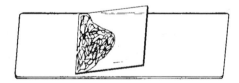

4. Examine your slide preparation under scanning, low-power, and high-power objectives with diminished light.

TABLE 34.1 Structural Characteristics of Free-Living Protozoa

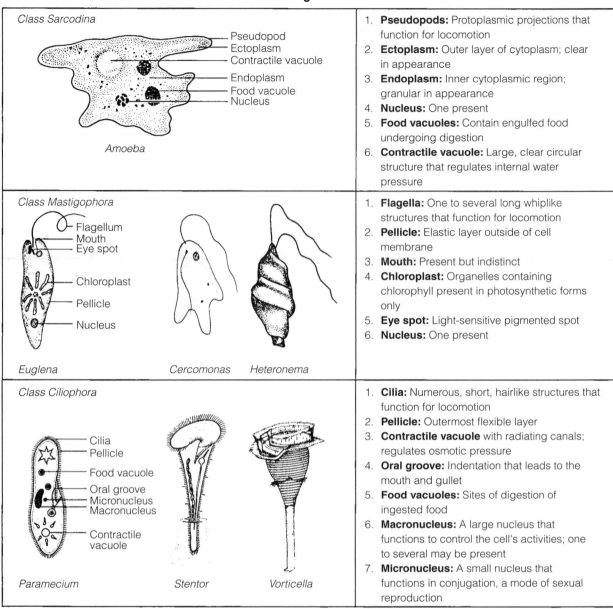

Class Sarcodina Pseudopod Ectoplasm Contractile vacuole Endoplasm Food vacuole Nucleus *Amoeba*	1. **Pseudopods:** Protoplasmic projections that function for locomotion 2. **Ectoplasm:** Outer layer of cytoplasm; clear in appearance 3. **Endoplasm:** Inner cytoplasmic region; granular in appearance 4. **Nucleus:** One present 5. **Food vacuoles:** Contain engulfed food undergoing digestion 6. **Contractile vacuole:** Large, clear circular structure that regulates internal water pressure
Class Mastigophora Flagellum Mouth Eye spot Chloroplast Pellicle Nucleus *Euglena* *Cercomonas* *Heteronema*	1. **Flagella:** One to several long whiplike structures that function for locomotion 2. **Pellicle:** Elastic layer outside of cell membrane 3. **Mouth:** Present but indistinct 4. **Chloroplast:** Organelles containing chlorophyll present in photosynthetic forms only 5. **Eye spot:** Light-sensitive pigmented spot 6. **Nucleus:** One present
Class Ciliophora Cilia Pellicle Food vacuole Oral groove Micronucleus Macronucleus Contractile vacuole *Paramecium* *Stentor* *Vorticella*	1. **Cilia:** Numerous, short, hairlike structures that function for locomotion 2. **Pellicle:** Outermost flexible layer 3. **Contractile vacuole** with radiating canals; regulates osmotic pressure 4. **Oral groove:** Indentation that leads to the mouth and gullet 5. **Food vacuoles:** Sites of digestion of ingested food 6. **Macronucleus:** A large nucleus that functions to control the cell's activities; one to several may be present 7. **Micronucleus:** A small nucleus that functions in conjugation, a mode of sexual reproduction

Experiment 34 / Observations and Results

Name ______________________________ Date ______________

1. Under the low- and high-power objectives, observe your slide preparation for the different protozoa present.
2. In the space provided, draw a representative sketch of several of the observed protozoa in stagnant pond water, indicate the magnifications used, and label their structural components. Identify each organism according to its class based on its mode of locomotion and its genus.

Magnification:		
Organelles of locomotion:		
Class:		
Genus:		

Magnification:		
Organelles of locomotion:		
Class:		
Genus:		

3. Using the low- and high-power objectives, observe the prepared slides of different protozoa. Draw representative sketches, indicate magnification, and label the structural components. Identify each organism according to its class based on locomotion and genus.

Amoeba

Paramecium

Magnification: _____   _____

Organelles of locomotion: _____   _____

Class: _____   _____

Genus: _____   _____

Euglena

Stentor

Magnification: _____   _____

Organelles of locomotion: _____   _____

Class: _____   _____

Genus: _____   _____

REVIEW QUESTIONS

1. What are the distinguishing characteristics of the free-living members of Sarcodina, Mastigophora, and Ciliophora?

2. Identify and give the function of the following:
   a. Pseudopods:

   b. Contractile vacuole:

   c. Eye spot:

   d. Micronucleus:

   e. Pellicle:

   f. Oral groove:

3. 🔍 People with AIDS are vulnerable to pneumonia caused by the protozoan *Pneumocystis carinii*. This organism is known to be a member of the normal flora in most humans. Why then are these individuals so susceptible to this generally rare form of pneumonia?

Parasitic Protozoa

LEARNING OBJECTIVE

Once you have completed this experiment, you should be familiar with

1. Parasitic protozoan forms.

PRINCIPLE

Unlike the life cycles of the free-living forms, the life cycles of parasitic protozoa vary greatly in complexity. Knowledge of the various developmental stages in these life cycles is essential in the diagnosis, clinical management, and chemotherapy of parasitic infections.

Parasites with the simplest or most direct life cycles not requiring an intermediate host are:

1. *Entamoeba histolytica:* A pseudopodian parasite of the class Sarcodina that causes amebic dysentery. Infective, resistant cysts are released from the lumen of the intestine through the feces and are deposited in water, in soil, or on vegetation. On ingestion, the mature quadrinucleated cyst wall disintegrates and the nuclei divide, producing eight active trophozoites (metabolically active cells) that move to the colon, where they establish infection.

2. *Balantidium coli:* The ciliated parasitic protozoan exhibits a life cycle similar to that of *E. histolytica* except that no multiplication occurs within the cyst. This organism resides primarily in the lumen and submucosa of the large intestine. It causes intestinal ulceration and alternating constipation and diarrhea.

3. *Giardia lamblia:* The intestinal mastigophoric flagellate exhibits a life cycle comparable to those of the above parasites. This organism is responsible for the induction of abdominal discomfort and severe diarrhea. Diagnosis is made by finding

cysts in the formed stool and both cysts and trophozoites in the diarrhetic stool.

The mastigophoric hemoflagellate responsible for various forms of African sleeping sickness has a more complex life cycle. The *Trypanosoma* must have two hosts to complete its cyclic development: a vertebrate, and an invertebrate, blood-sucking insect host. Humans are the definitive hosts harboring the sexually mature forms; the tsetse fly (*Glossina*) and the reduviid bug are the invertebrate hosts in which the developmental forms occur.

Table 35.1 illustrates the morphological characteristics of prototypic members of the parasitic protozoa except the Sporozoa.

Protozoa demonstrating the greatest degree of cyclic complexity are found in the class Sporozoa. They are composed of exclusively obligate parasitic forms, such as members of the genus *Plasmodium,* and are responsible for malaria in both humans and animals. The life cycle requires two hosts, a human being and the female *Anopheles* mosquito. It is significant to note that in this life cycle, the mosquito, and not the human, is the definitive host harboring the sexually mature parasite.

Malaria is initiated when a person is bitten by an infected mosquito, during which time infective sexually mature sporozoites are injected with the insect's saliva. These parasites pass rapidly from the blood into the liver, where they infect the parenchymal cells. This is the **pre-erythrocytic stage.** The parasites develop asexually within the liver cells by a process called **schizogony,** producing **merozoites.** This cycle may be repeated or the merozoites that are released from the ruptured liver cells may now infect red blood cells and initiate the **erythrocytic stage.** During this asexual development, the parasite undergoes a series of morphological changes that are of diagnostic

TABLE 35.1 Characteristics of Representative Parasitic Protozoa

Class, Organism, and Infection	Structural Characteristics	Locomotor Organelles	Site of Infection	Isolation of Parasitic Form
Sarcodina *ENTAMOEBA HISTOLYTICA* INFECTION: AMEBIC DYSENTERY Peripheral chromatin Central karyosome Uniform cytoplasm Red blood cell Nuclei Chromatoid body Early Late	**Trophozoite:** Shape: Variable Nucleus: Discrete nuclear membrane with central karyosome and peripheral chromatin granules Cytoplasm: Clear, red blood cells may be present **Cyst:** Shape: Round to oval with thick wall Nuclei: 1–4 present; mature cyst is quadrinucleated Chromatoid bodies: Sausage-shaped with rounded ends, present in young cysts only	Pseudopods None	Large intestine by ingestion of mature cysts	Diarrhetic stool Formed stool
Mastigophora *TRYPANOSOMA GAMBIENSE* INFECTION: AFRICAN SLEEPING SICKNESS Flagellum Undulating membrane Nucleus Volutin granules Kinetoplast	**Trophozoite:** Shape: Crescent Nucleus: Large, central, and polymorphic Cytoplasm: Granular **Cyst:** None	Single flagellum along undulating membrane	Peripheral blood stream by means of tsetse fly vector	Peripheral blood

value. These forms are designated as **signet rings, trophozoites, schizonts, segmenters, merozoites,** and **gametocytes.** The merozoites are capable of reinfecting other blood cells or liver cells. Ingestion of the **microgametocytes** (♂) and **macrogametocytes** (♀) by another mosquito during a blood meal initiates the sexual cycle called **sporogamy.** Male and female gametes give rise to a zygote in the insect's gut. The zygote is then transformed into an **ookinete** that burrows through the gut wall to form an **oocyst** in which the sexually mature **sporozoites** develop, thereby completing the life cycle.

In this experiment, you will study the parasitic protozoa using prepared slides and the diagnostic characteristics shown in Figure 35.1 on page 220 and Table 35.1. The purpose of the experiment is to help you understand life cycles of parasitic protozoa.

MATERIALS
Prepared Slides

Entamoeba histolytica trophozoite and cyst, *Giardia lamblia* trophozoite and cyst, *Balantidium coli* trophozoite and cyst, *Trypanosoma gambiense,* and *Plasmodium vivax* in human blood smears.

Equipment

Microscope, immersion oil, and lens paper.

TABLE 35.1 (continued)

Class, Organism, and Infection	Structural Characteristics	Locomotor Organelles	Site of Infection	Isolation of Parasitic Form
Mastigophora *GIARDIA LAMBLIA* INFECTION: DYSENTERY Nucleus Karyosome Median bodies Axonemes Nuclei Median bodies Retracted protoplasm Axonemes	**Trophozoite:** Shape: Pear-shaped with concave sucking disc Nuclei: 2 bilaterally located with central karyosome and no peripheral chromatin Cytoplasm: Uniform and clear **Cyst:** Shape: Oval to ellipsoidal Nuclei: 2–4 present and protoplasm retracted from cyst wall Axostyle Parabasal body	4 pairs of flagella 4 pairs of flagella within cyst	Small intestine through ingestion of cysts	Diarrhetic stool Formed stool
Ciliophora *BALANTIDIUM COLI* INFECTION: DYSENTERY Cytostome Cilia Micronucleus Macronucleus Cyst wall Macronucleus	**Trophozoite:** Shape: Oval Nuclei: Kidney-shaped macronucleus and a micronucleus Cytoplasm: Vacuolated **Cyst:** Shape: Round and thick-walled Nuclei: 1 macronucleus and a micronucleus that is not visible	Cilia Cilia within cyst	Large intestine by the ingestion of cysts	Diarrhetic stool Formed stool

PROCEDURE

Examine all available slides under the oil-immersion objective. Use Table 35.1 and Figure 35.1 to identify the distinguishing microscopic characteristics of each parasite studied.

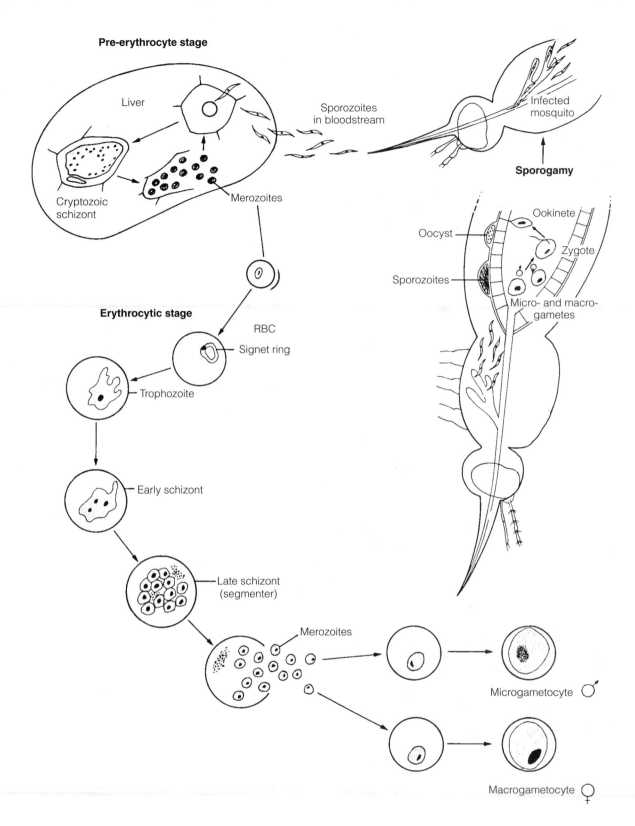

FIGURE 35.1 Life cycle of *Plasmodium vivax* (From *Basic Clinical Parasitology*, 4th ed., by Harold S. Brown. New York: Appleton-Century-Crofts, 1975. By permission.)

Name   Date

Draw representative sketches of the parasitic organisms that you studied, and label the distinguishing structural characteristics you were able to observe.

Refer to photo numbers 50–56 in the color-plate insert for illustration of these organisms.

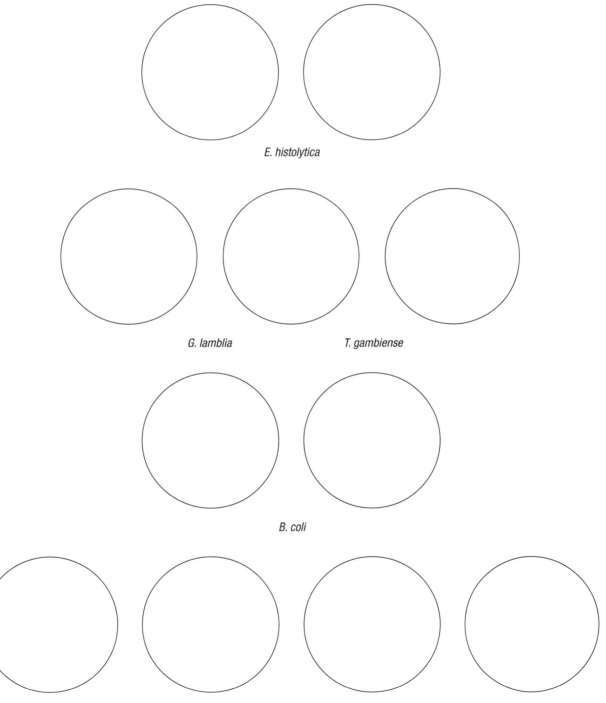

E. histolytica

G. lamblia   T. gambiense

B. coli

P. vivax: Erythrocytic stages

REVIEW QUESTIONS

1. Describe the developmental stages of the malarial parasite during sporogamy and schizogony.

2. What role does the invertebrate host play in the life cycle of the trypanosomes? Explain.

3. Distinguish between the pre-erythrocytic and erythrocytic stages in the life cycle of the malarial parasite.

4. In malarial infections, the sexually mature parasite is found in which host? Is this true for all other protozoan parasitic infections? Explain.

5. On returning from a trip overseas, an individual with persistent diarrhea is diagnosed as having an *E. histolytica* infection. Fecal examination reveals the presence of blood in the stool, suggesting damage to the intestinal mucosa. Explain why and how the mucosa was compromised by this parasite.

The Fungi

LEARNING OBJECTIVES

When you have completed the experiments in this section, you should be familiar with

1. The macroscopic and microscopic structures of yeast and molds.

2. The basic mycological culturing and staining procedures.

3. The ability to identify common fungal organisms.

INTRODUCTION

The branch of microbiology that deals with the study of fungi (yeasts and molds) is called **mycology.** True fungi are separated into the following four classes on the basis of their sexual modes of reproduction:

1. **Phycomycetes:** Water, bread, and terrestrial molds. Reproductive spores are external and uncovered.

2. **Ascomycetes:** Yeasts and molds. Sexual spores, called ascospores, are produced in a saclike structure called an ascus.

3. **Basidiomycetes:** Fleshy fungi, toadstools, mushrooms, puffballs, and bracket fungi. Reproductive spores, basidiospores, are separate from specialized stalks called basidia.

4. **Deuteromycetes:** Also called **Fungi Imperfecti** because no sexual reproductive phase has been observed.

The major characteristics of the four classes of these fungi are shown in Table VII.1.

Nutritionally, the fungi are heterotrophic, eukaryotic microorganisms that are enzymatically capable of metabolizing a wide variety of organic substrates. Fungi can have beneficial or detrimental effects on humans. Those that inhabit the soil play a vital role in decomposing dead plant and animal tissues, thereby maintaining a fertile soil environment. The fermentative fungi are of industrial importance in producing beer and wine, bakery products, cheeses, industrial enzymes, and antibiotics. The detrimental activities of some fungi include spoilage of foods by rots, mildews, and rusts found on fruit, vegetables, and grains. Some species are capable of producing toxins (aflatoxin) and hallucinogens. A few fungal species are of medical significance because of their capacities to produce diseases in humans. Many of the pathogenic fungi are Deuteromycetes and can be divided into two groups based on site of infection. The **superficial mycoses** cause infections of the skin, hair, and nails (for example, ringworm infections). The **systemic mycoses** cause infections of the subcutaneous and deeper tissues such as those of the lungs, genital areas, and nervous system.

TABLE VII.1 Major Characteristics of the Four Classes of Fungi

Characteristics	Class			
	Phycomycetes	**Ascomycetes**	**Basidiomycetes**	**Deuteromycetes**
Mycelium	Nonseptate	Septate	Septate	Septate
Asexual spores	Found in sporangium; zoospores (motile), sporangiospores (nonmotile)	Formed on tip of conidiophore; conidiospores (nonmotile)	Same as the Ascomycetes	Same as the Ascomycetes
Sexual spores	Zygospores, found in terrestrial forms; oospores, found in aquatic forms	Ascospores, contained in a saclike structure called the ascus	Basidiospores, carried on the outer surface of a club-shaped cell called the basidium	Fungi Imperfecti— no sexual reproductive phase observed; some members of the Ascomycetes and Basidiomycetes are Fungi Imperfecti
Common species	Bread molds, mildews, potato blight, *Rhizopus* species	Cup fungi, ergot, Dutch elm, yeast species	Smuts, rusts, puffballs, toadstools, mushrooms	Aspegillus, Candida, Trichphyton, Cryptococcus, Blastomyces, Histoplasma, Micosporum, and Sporotrichium

Cultivation and Morphology of Molds

Molds are the major fungal organisms that can be seen by the naked eye. We have all seen them growing on foods such as bread or citrus fruit as a cottony, fuzzy, black, green, or orange growth, depending on the mold. Examination with a simple hand lens shows that these organisms are composed of an intertwining branching mat called a **mycelium.** The filaments that make up this mycelial mat are called **hyphae.** Most of the mat grows on or in the surface of the nutrient medium so that it can extract nutrients; the mat is therefore called **vegetative mycelium.** Some of the mycelium mat rises upward from the mat and is referred to as **aerial mycelium.** Specialized hyphae are produced from the aerial mycelium and give rise to spores that are the reproductive elements of the mold.

Figure 36.1 and color plate 58 shows the reproductive structures of some fungi.

The cultivation, growth, and observation of molds require techniques that differ from those used for bacteria. Mold cultivation requires the use of a selective media such as Sabouraud agar or potato dextrose agar. These media favor mold growth because their low acidity (pH 4.5 to 5.6) discourages the growth of bacteria, which favor a neutral (pH 7.0) environment. The temperature requirements of molds are also different from those of bacteria, in that molds grow best at room temperature (25° C). In addition, molds grow at a much slower rate than bacteria do, requiring several days to weeks before visible colonies appear on a solid agar surface.

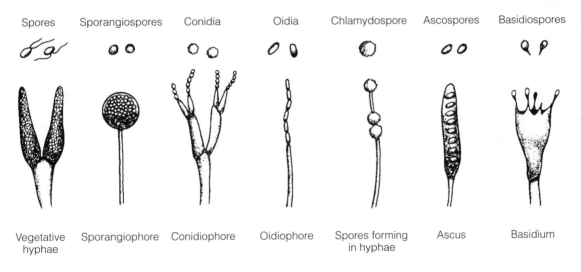

| Spores | Sporangiospores | Conidia | Oidia | Chlamydospore | Ascospores | Basidiospores |

| Vegetative hyphae | Sporangiophore | Conidiophore | Oidiophore | Spores forming in hyphae | Ascus | Basidium |

FIGURE 36.1 Spore and sporangia types

PART A: Slide Culture Technique

LEARNING OBJECTIVES

When you have completed this experiment you should be

1. Acquainted with mold cultivation on glass slides.
2. Able to visualize and identify the structural components of molds.

PRINCIPLE

Because the structural components of molds are very delicate even simple handling with an inoculating loop may result in mechanical disruption of their components. The following slide culture technique is used to avoid such disruption. A deep concave slide containing a suitable nutrient medium with an acidic pH, such as Sabouraud agar, is covered by a removable coverslip. Mold spores are deposited in the surface of the agar and incubated in a moist chamber at room temperature. Direct microscopic observation is then possible without fear of disruption or damage to anatomical components. Molds can be identified as to spore type and shape, type of sporangia, and type of mycelium, as shown in Figure 36.1 and Table 38.1 on pages 239–241.

MATERIALS

Cultures

7- to 10-day Sabouraud agar cultures of *Penicillium notatum* and *Aspergillus niger, Mucor mucedo,* and *Rhizopus stolonifer.*

Media

Per designated student group: one Sabouraud agar deep tube.

Equipment

Bunsen burner, water bath, four concave glass slides, four coverslips, petroleum jelly, sterile Pasteur pipettes, toothpicks, four sterile Petri dishes, filter paper, forceps, inoculating loop and needle, four sterile U-shaped bent glass rods, thermometer, dissecting microscope, and beaker with 95% ethyl alcohol.

PROCEDURE

1. Melt the deep tube of Sabouraud agar in a boiling water bath and cool to 45°C.
2. Place a piece of filter paper in the bottom of each Petri dish, insert a sterile bent glass rod into each dish, and replace the covers as illustrated below:

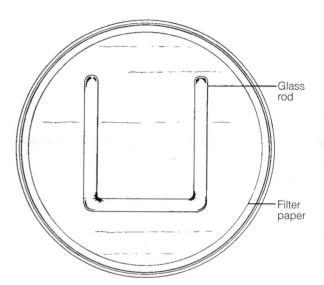

3. Using forceps, dip the concave slides and coverslips in a beaker of 95% ethanol, pass through Bunsen burner flame, remove from flame, and hold until all the alcohol has burned off the slides and coverslips.
4. Cool slides and coverslips. Place a slide, concave side up, with a coverslip to one side of the concavity, on the glass rod inside each Petri dish.
5. With a toothpick, add petroleum jelly to three sides surrounding the concave section of each slide. The fourth side will serve as a vent for air passage.

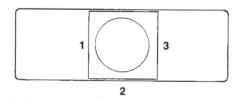

6. With a sterile Pasteur pipette, add one or two drops of cooled Sabouraud agar to the concavity of each slide.
7. Place a coverslip over the concave portion of each slide so that it is completely sealed.

8. With forceps, stand each slide upright inside its respective Petri dish until the agar solidifies. Following solidification, the slide will resemble the following illustration:

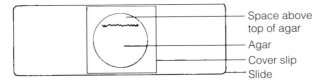

9. When agar is fully hardened, slide coverslips downward with forceps, and with a sterile needle inoculate each prepared slide with the spores from the test cultures.

10. Push the coverslips to their original positions, thereby sealing off the slide.

11. With a Pasteur pipette, moisten the filter paper with sterile water to provide a moist atmosphere. Remoisten filter paper when necessary during the incubation period.

12. Place slide on U-shaped bent rod, replace Petri dish cover, and label the covers with the names of the organism inoculated and identify each with your initials.

13. Incubate the preparations for 7 days at 25°C.

PART B: Mold Cultivation on Solid Surfaces

LEARNING OBJECTIVES

When you have completed this experiment, you should be

1. Acquainted with the technique of mold cultivation on agar plates.

2. Able to observe and identify colonial characteristics such as growth rate, texture, pigmentation on the surface and reverse side, and folds or ridges on the surface.

PRINCIPLE

Cultivating molds on solid surfaces allows you to observe the variations in gross colonial morphology among different genera of molds. These variations in colonial appearance play a major role in the identification of the filamentous fungi. Most microbiologists are familiar with the gross appearance of multicellular fungi, but even to the untrained, the macroscopic differences in colonial growths are obvious and recognizable. For example, most people have seen rotting citrus fruits (lemons and oranges) produce a blue-green velvety growth characteristic of *Penicillium* species. It is also common for stale cheese to show a grayish-white furry growth of *Mucor* species, and likewise, the black stalklike appearance of *Rhizopus* molds growing on bread is familiar to many.

In this part of the experiment, you will be able to visualize the gross appearance of the colonial growth of four different molds.

MATERIALS

Cultures

7- to 10-day Sabouraud agar cultures of *Aspergillus niger, Penicillium notatum, Mucor mucedo,* and *Rhizopus stolonifer.*

Media

Per designated student group: three Sabouraud agar plates and one potato dextrose agar plate.

Equipment

Bunsen burner, four test tubes containing 2 ml of sterile saline, dissecting microscope, and an inoculating loop.

PROCEDURE

1. Label the three Sabouraud agar plates as *Aspergillus niger, Penicillium notatum,* and *Mucor mucedo,* and label the fourth plate containing potato dextrose agar as *Rhizopus stolonifer.*

2. Prepare a saline suspension of each mold culture. Label each of the four tubes of saline with the name of the organism. Using a sterile inoculating loop, scrape two loopfuls of mold culture into the corresponding tube of 2 ml of sterile saline and mix well by tapping the tube with your finger.

3. Using sterile technique, inoculate each of the plates by placing a single loopful of mold suspension in the center of its respective agar plate. *Note: Do not spread the inoculum and do not shake or jostle the plates.*

4. Incubate all plates at room temperature, 25° C, for 7 to 10 days. *Note: Do not invert the plates.*

Name   Date

PART A: Slide Culture Technique

1. Examine each mycological slide preparation under the low and high power of a dissecting microscope. Identify the mycelial mat, vegetative and reproductive hyphae, and spores. Use Table 38.1 on pages 239–241 to aid with your identification of mold structures.

2. Draw a representative microscopic field under low-power and high-power magnification and label the structural components of each test organism in the chart.

Refer to photo numbers 57 and 58 in the color-plate insert for illustration of fungal colonial growth.

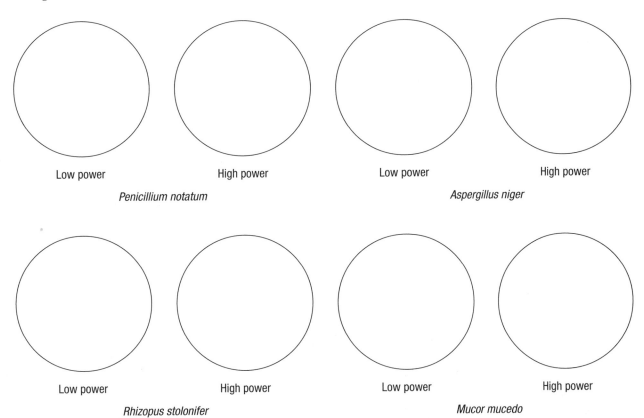

| Low power | High power | Low power | High power |

Penicillium notatum   *Aspergillus niger*

| Low power | High power | Low power | High power |

Rhizopus stolonifer   *Mucor mucedo*

PART B: Mold Cultivation on Solid Surfaces

1. Examine each mold plate under the low and high power of a dissecting microscope. Draw sketches of the mold colonies under low power, indicating the extent of growth (diameter in mm), pigmentation, and the presence or absence of aerial hyphae. Refer to Table 38.1 to aid with your identification of mold structures. *Note: Do not remove Petri dish covers.*

Refer to photo numbers 57 and 58 in the color-plate insert for illustration of fungal colony growth.

Penicillium notatum

Aspergillus niger

Colony diameter (mm): _____   _____

Pigmentation: _____   _____

Aerial hyphae (+ or −): _____   _____

Rhizopus stolonifer

Mucor mucedo

Colony diameter (mm): _____   _____

Pigmentation: _____   _____

Aerial hyphae (+ or −): _____   _____

REVIEW QUESTIONS

1. Cite some beneficial and harmful aspects of molds.

2. What is the advantage of using Sabouraud agar?

3. In the slide culture technique, what is the purpose of
   a. Moistened filter paper in the Petri dish

   b. A U-shaped glass rod in the Petri dish

4. What is the advantage of the slide culture technique over that of a simple loop inoculation onto an agar plate (as in Part B)?

5. Why would it be advantageous to observe mold colonies on an agar plate?

6. ⚲ Since dimorphism is a property of fungi, how do you account for the fact that molds grow preferentially in vitro rather than in vivo?

Yeast Morphology, Cultural Characteristics, and Reproduction

LEARNING OBJECTIVES

Once you have completed this experiment, you should

1. Know the morphology of different genera of yeast.

2. Understand the growth and fermentative properties of yeast cells.

3. Be familiar with the sexual and asexual modes of reproduction in yeast cells.

PRINCIPLE

Yeasts are nonfilamentous unicellular fungi. Yeast cultures resemble bacteria when grown on the surface of artificial laboratory media; however, they are 5 to 10 times larger than bacteria. Microscopically, yeast cells may be ellipsoidal, spherical, or in some cases, cylindrical. Unlike molds, yeast do not have aerial hyphae and supporting sporangia.

Yeast reproduce asexually by **budding** or by **fission.** In budding, an outgrowth from the parent cell (a **bud**) pinches off, producing a daughter cell (Figure 37.1a). Fission occurs in certain species of yeast, such as those in the genus *Schizosaccharomyces.* During fission, the parent cell elongates, its nucleus divides, and it splits evenly into two daughter cells.

Some yeast may also undergo sexual reproduction when two sexual spores conjugate, giving rise to a zygote, or diploid cell. The nucleus of this cell divides by meiosis, producing four new haploid nuclei (sexual spores), called **ascospores,** contained within a structure, called the **ascus** (Figure 37.1b). When the ascus ruptures, the ascospores are released and conjugate, starting the cycle again.

Yeasts are important for many reasons. *Saccharomyces cerevisiae* is referred to as baker's yeast and is used as the leavening

(a)

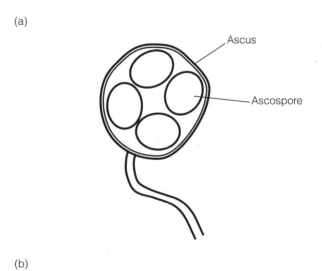

(b)

FIGURE 37.1 Asexual (a) and sexual (b) reproductive structures of yeast

agent in dough. Two major strains of yeast, *Saccharomyces carlsbergensis* and *Saccharomyces cerevisiae*, are used for brewing. The wine industry relies on wild yeast (present of the grape) for the fermentation of grape juice, which is supplemented with *Saccharomyces ellipsoideus* to begin the fermentation. Also, the high vitamin content of yeasts makes them particularly valuable as food supplements. As useful as some yeasts are, there are a few species that can create problems in the food industry or are harmful to humans. Undesired yeast must be excluded from the manufacture of fruit juices, such as grape juice or apple cider, to prevent the fermentation of fruit sugars to alcohol. The contamination of soft cheese by some forms of yeast will destroy the product. Finally, some yeast such as *Candida albicans* are pathogenic and responsible for urinary tract and vaginal infections known as **moniliasis** and infections of the mouth called **thrush.**

The cultural characteristics, the type of reproduction, and the fermentative activities used to identify the different genera of yeast will all be studied in this experiment.

MATERIALS
Cultures

7-day Sabouraud agar cultures of *Saccharomyces cerevisiae*, *Candida albicans*, *Rhodotorula rubra*, *Selenotilia intestinalis*, and *Schizosaccharomyces octosporus*.

Media

Per designated student group: five tubes each of bromcresol purple glucose broth, bromcresol purple maltose broth, bromcresol purple lactose broth, and bromcresol purple sucrose broth, containing a Durham tube; two glucose-acetate agar plates; and five test tubes (13 × 100 mm) containing 2 ml of sterile saline.

Reagents

Water-iodine solution, lactophenol–cotton-blue solution.

Equipment

Bunsen burner, inoculating loop and needle, 10 glass slides, 10 coverslips, five sterile Pasteur pipettes, glassware marking pencil, and microscope.

PROCEDURE
Part 1: Morphological Characteristics

Prepare a wet mount of each yeast culture in the following manner:

1. Suspend a loopful of yeast culture in a few drops of lactophenol–cotton-blue solution on a microscope slide and cover with a coverslip.

Part 2: Fermentation Studies

1. With a sterile loop, inoculate each experimental organism into appropriately labeled tubes of bromcresol purple glucose, maltose, lactose, and sucrose fermentation broths.

2. Incubate all cultures at 25°C for 4 to 5 days.

Part 3: Sexual Reproduction

1. With a glassware marking pencil, divide the bottom of a glucose-acetate agar plate into three sections, and divide another glucose-acetate agar plate in half.

2. Label each section with the name of a test organism.

3. Label each tube of sterile saline with the name a test organism.

4. With a sterile inoculating loop, suspend a heavy loopful of each test organism into its appropriately labeled tube of saline. Tap the tube with your finger to obtain a uniform cell suspension.

5. With a sterile Pasteur pipette, inoculate one drop of each test organism onto the surface of the appropriately labeled section on an agar plate. *Note: Allow the inoculum to diffuse into the agar for a few minutes. Do not swirl or rotate the plates.*

6. Incubate all plates at 25° for 7 days. *Note: Visit the laboratory, if possible, during the incubation period and note when sporulation begins.*

Name _____   Date _____

PART 1: Morphological Characteristics

Examine all yeast wet-mount slide preparations under low and high power and draw a representative field for each organism in the chart below. Note the shape and presence or absence of budding (+ or −).

Refer to photo numbers 59 and 60 in the color-plate insert for illustration of yeast colony growth and microscopic cell structure.

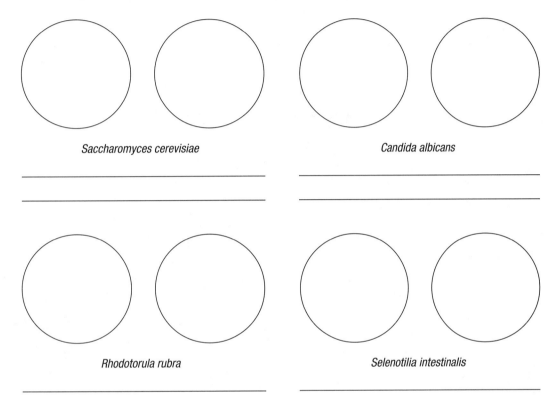

Saccharomyces cerevisiae   Candida albicans

Shape: _____   _____

Budding (+ or −): _____   _____

Rhodotorula rubra   Selenotilia intestinalis

Shape: _____   _____

Budding (+ or −): _____   _____

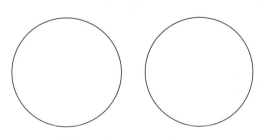

Schizosaccharomyces octosporus

Shape: _____

Budding (+ or −): _____

PART 2: Fermentation Studies

1. Examine all fermentation tubes for the presence of growth (turbidity), the presence or absence of acid (change in color of medium), and the presence or absence of gas (bubble in Durham tube). Use a plus (+) or minus (−) in the chart below to record your results.

Organism	Glucose			Maltose			Lactose			Sucrose		
	T	A	G	T	A	G	T	A	G	T	A	G
Saccharomyces cerevisiae												
Candida albicans												
Rhodotorula rubra												
Selenotilia intestinalis												
Schizosaccharomyces octosporus												

Note: T = turbidity, A = acid, and G = gas

PART 3: Sexual Reproduction

1. Examine the glucose-acetate agar plates for the presence or absence of sporulation.
2. Prepare a water-iodine wet mount using a loopful of culture from each respective section on the glucose-acetate agar plate.
3. Observe the cells using the high-dry objective.
4. In the circles below, draw representative reproductive structures and label the parts.

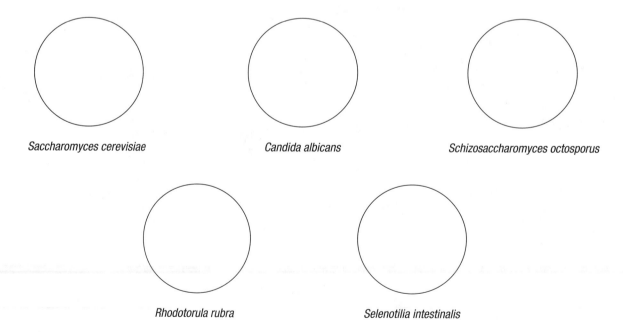

Saccharomyces cerevisiae

Candida albicans

Schizosaccharomyces octosporus

Rhodotorula rubra

Selenotilia intestinalis

REVIEW QUESTIONS

1. Indicate the significance of the following structures in the reproductive activities of yeast cells.

   a. Buds

   b. Ascus

   c. Ascospores

2. Why are yeast cells classified as fungi, and how do they differ from other fungi?

3. Why is yeast of industrial importance?

4. Why are yeasts significant from a medical persective?

5. Why is it necessary to pasteurize fruit juices?

6. A female patient develops candidiasis (moniliasis) following prolonged antibiotic therapy for a bladder infection caused by *Pseudomonas aeruginosa.* How can you account for the development of this concurrent vaginal infection?

7. With regard to the fermentation of wine, what kind of wine would be produced if you washed the grapes prior to crushing them?

Identification
of Unknown Fungi

LEARNING OBJECTIVE

When you have completed this experiment, you should be able to

1. Identify a fungal unknown based on colonial morphology and microscopic appearance.

PRINCIPLE

In this experiment, you will be provided with a number-coded pure culture of a representative fungal organism for cultivation and subsequent identification. Use Table 38.1 to aid in identification of the unknown culture.

MATERIALS

Cultures

Number-coded, 7-day Sabouraud broth spore suspensions of *Aspergillus, Mucor, Penicillium, Alternaria, Rhizopus, Cladosporium, Cephalosporium, Fusarium, Torula,* and *Candida.*

Media

One Sabouraud agar plate per student.

TABLE 38.1 Identification of Fungi

Diagram	Colonial Morphology	Microscopic Appearance
Molds Sporangium Columella Sporangiophore Mycelium Rhizoid *RHIZOPUS:* BLACK BREAD MOLD; COMMON LABORATORY CONTAMINANT	Rapidly growing white-colored fungus swarms over entire plate; aerial mycelium cottony and fuzzy	Spores are oval, colorless, or brown; nonseptate mycelium gives rise to straight sporangiophores that terminate with black sporangium containing a columella; rootlike hyphae (rhizoids) penetrate the medium
Sporangium Columella Sporangiophore Mycelium *MUCOR:* FOOD CONTAMINANT	Resembles the colonies of *Rhizopus*	Spores are oval; nonseptate mycelium gives rise to single sporangiophores with globular sporangium containing a columella; there are no rhizoids

➡

TABLE 38.1 Identification of Fungi (continued)

Diagram	Colonial Morphology	Microscopic Appearance
Molds (continued) — Conidia — Conidiophore — Mycelium *ALTERNARIA:* NORMALLY FOUND ON PLANT MATERIAL	Grayish-green or black colonies with gray edges rapidly swarming over entire plate; aerial mycelium not very dense, appears grayish to white	Multicelled spores (conidia) are pear-shaped and attached to single conidiophores arising from a septate mycelium
— Conidia — Conidiophore *FUSARIUM:* FOUND IN SOIL	Woolly, white, fuzzy colonies changing color to pink, purple, or yellow	Multicelled spores (conidia) are oval or crescent-shaped and attached to conidiophores arising from a septate mycelium
— Conidia — Sterigma — Vesicle — Conidiophore — Mycelium *ASPERGILLUS:* PLANT AND ANIMAL PATHOGENS; SOME SPECIES USED INDUSTRIALLY	White colonies become greenish-blue, black, or brown as culture matures	Single-celled spores (conidia) in chains developing at the end of the sterigma arising from the terminal bulb of the conidiophore, the vesicle; long conidiophores arise from a septate mycelium
— Conidia — Sterigma — Metula — Conidiophore — Mycelium *PENICILLIUM:* ANTIBIOTIC-PRODUCING CITRUS FRUIT CONTAMINANT; SOIL INHABITANT	Mature cultures usually greenish or blue-green	Single-celled spores (conidia) in chains develop at the end of the sterigma arising from the metula of the conidiophore; branching conidiophores arise from a septate mycelium

TABLE 38.1 (continued)

Diagram	Colonial Morphology	Microscopic Appearance
Molds (continued) — Conidia — Conidiophore — Septate mycelium *CLADOSPORIUM:* DEAD AND DECAYING PLANTS	Small, heaped colonies are greenish-black and powdery	Spores (conidia) develop at the end of complex conidiophores arising from a septate mycelium that is usually brownish
— Conidia — Conidiophore — Mycelium *CEPHALOSPORIUM:* ANTIBIOTIC PRODUCTION	Rapidly growing compact and moist colonies becoming cottony with aerial hyphae that are gray or rose colored	Single-celled conical or elliptical spores (conidia) held together in clusters at the tips of the conidiophores by a mucoid substance; erect, unbranched conidiophores arise from a septate mycelium
Yeast — Bud *TORULA:* CHEESE AND FOOD CONTAMINANT	Colonies are pink, moist, with unbroken, even edges	Cells are oval, colorless, and reproduce by budding
— Yeast cell — Pseudomycelium *CANDIDA:* HUMAN PATHOGEN	Colonies are small, round, moist, and colorless, with unbroken, even edges	Yeastlike fungus produces pseudomycelium

Reagent

Lactophenol–cotton-blue solution.

Equipment

Bunsen burner, dissecting microscope, hand lens, sterile cotton swabs, glass slides, cover-slips, inoculating loop, and glassware marking pencil.

PROCEDURE

1. With a sterile inoculating loop, inoculate an appropriately labeled Sabouraud agar plate with one of the provided unknown cultures by placing one loopful in the center of the plate. *Note: Do not spread culture.*

2. Incubate the plates in a noninverted position for 1 week at 25°C in a moist incubator.

Name _____   Date _____

1. Observe mold cultures with a hand lens, noting and recording their colonial morphologies.
2. Prepare a wet mount by suspending some of the culture in a few drops of lactophenol–cotton-blue solution. Be gentle to avoid damaging the fungal structures.
3. Examine the preparation under high-power and low-power magnifications with the aid of a dissecting microscope and record your observations.
4. Draw a representative microscopic field of your culture in the chart.
5. Using Table 38.1 and Figure 3.1 on page 22, identify your unknown fungal organism.
   a. Color pigmentation

   b. Diameter (mm)

   c. Texture (cottony, smooth, etc.)

   d. Margin (entire, undulating, lobular, etc.)

   e. Aerial hyphae (septate, nonseptate)

Diagram of microscopic appearance:

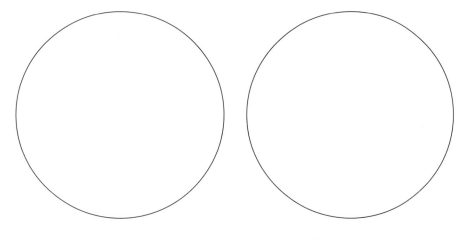

Low-power magnification   High-power magnification

Number assigned to
unknown culture: _____

Genus of fungal unknown: _____

The Viruses

LEARNING OBJECTIVES

Once you have completed the experiments in this section, you should know

1. The chemical structures, morphologies, and replicative activities of bacterial viruses (bacteriophages).

2. How to perform a phage dilution procedure for the cultivation and enumeration of bacterial viruses.

3. How to isolate bacteriophages from sewage.

INTRODUCTION

Viruses are noncellular biological entities composed solely of a single type of nucleic acid surrounded by a protein coat called the capsid. Because of their limited and simplistic structures, viruses can be chemically defined as **nucleoproteins.** They are devoid of the sophisticated enzymatic and biosynthetic machinery essential for independent activities of cellular life. This lack of metabolic machinery mandates that they exist as parasites, and they cannot be cultivated outside of a susceptible living cell. Viruses are differentiated from cellular forms of life on the following bases:

1. They are ultramicroscopic and can only be visualized with the electron microscope.
2. They are filterable: They are able to pass through bacteria-retaining filters.
3. They do not increase in size.
4. They must replicate within a susceptible cell.
5. Replication occurs because the viral nucleic acid subverts the synthetic machinery of the host cell (namely, common host cell components and enzyme systems involved in decomposition, synthesis, and bioenergetics) for the purpose of producing new viral components.

6. Viruses are designated either RNA or DNA viruses because they contain one of the nucleic acids but never both.

Much of our knowledge of the mechanism of animal viral infection and replication has been based on our understanding of infection in bacteria by bacterial viruses, called the **bacteriophages,** or **phages.** The bacteriophages were first described in 1915 almost simultaneously by Twort and d'Herelle. The name *bacteriophage*, which in Greek means "to eat bacteria," was coined by d'Herelle because of the destruction through lysis of the infected cell. Bacteriophages exhibit notable variability in their sizes, shapes, and complexities of structure. The T-even (T2, T4, and T6) phages illustrated in Figure VIII.1 demonstrate the greatest morphological complexity.

Phage replication depends on the ability of the phage particle to infect a suitable bacterial host cell. Infection consists of the following sequential events:

1. **Adsorption:** Tail fibers of the phage particle bind to receptor sites on the host's cell wall.
2. **Penetration** (infection): Spiral protein sheath retracts, and an enzyme, early muramidase, perforates the bacterial cell wall,

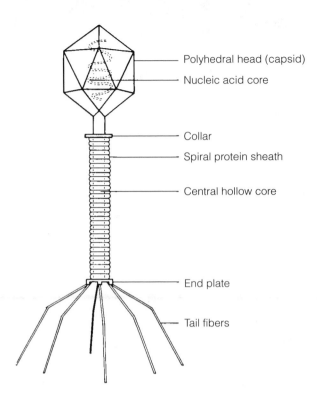

	Polyhedral head (capsid)
	Nucleic acid core
	Collar
	Spiral protein sheath
	Central hollow core
	End plate
	Tail fibers

The functions of these structural components are as follows:

Component	Function
Capsid (protein coat)	Protection of nucleic acid from destruction by DNAses
Nucleic acid core	Phage genome carrying genetic information necessary for replication of new phage particles
Spiral protein sheath	Retracts so that nucleic acid can pass from capsid into host cell's cytoplasm
End plate and **tail fibers**	Attachment of phage to specific receptor sites on a susceptible host's cell wall

FIGURE VIII.1 Bacteriophage: Structural components and their functions

enabling the phage nucleic acid to pass through the hollow core into the host cell's cytoplasm. The empty protein shell remains attached to the cell wall and is called the protein ghost.

3. **Replication:** The phage genome subverts the cell's synthetic machinery, which is then used for the production of new phage components.

4. **Maturation:** The period during which the new phage components are assembled and form complete, mature virulent phage particles.

5. **Release:** Late muramidase (lysozyme) lyses the cell wall, liberating infectious phage particles that are now capable of infecting new susceptible host cells, thereby starting the cycle over again.

Virulent phage particles that infect susceptible host cells always initiate the **lytic cycle** as described above. Other phage particles, called **temperate phages (λ phages),** incorporate their nucleic acid into the host's chromosome. Lysis of the host cell does not occur until it is induced by exogenous physical agents such as ultraviolet, ionizing radiation, or chemical

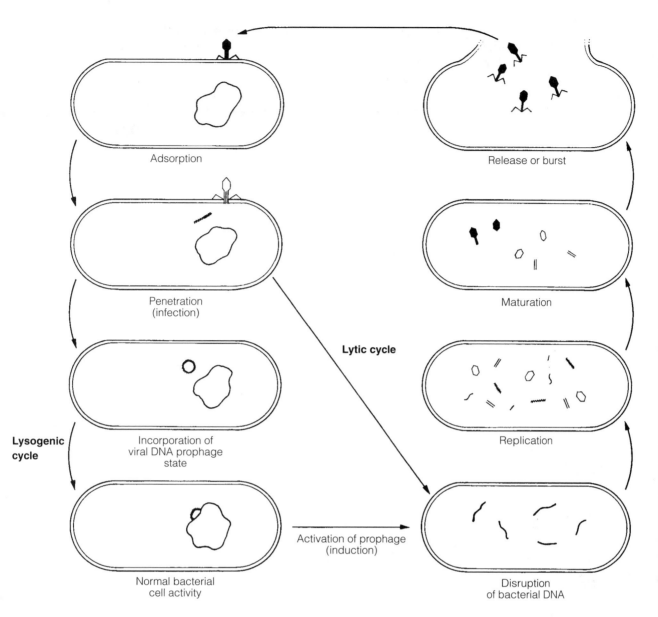

FIGURE VIII.2 The lytic and lysogenic life cycles of a bacteriophage

mutagenic agents. Bacterial cells containing the incorporated phage nucleic acid, the **prophage,** are called **lysogenic cells.** Lysogenic cells appear and function as normal cells, and they reproduce by fission. When induced by physical or chemical agents, these cells will release a virulent prophage from the host's genome, which then initiates the lytic cycle. Figure VIII.2 illustrates the lytic and lysogenic life cycles of a bacteriophage.

Animal viruses differ structurally from bacteriophages in that they lack the spiral protein sheath, base plate, and tail fibers. Their shapes may be helical or cuboidal (icosahedral, containing 20 triangular facets). Some animal viruses are designated as naked viruses because

they are composed solely of nucleocapsids. In others, referred to as enveloped viruses, the nucleocapsid is surrounded by a lipid bilayer that may have glycoproteins associated with it.

The infectious process of the animal virus is very similar to bacteriophage infection. However, there are some notable differences, namely:

1. Adsorption of the virus occurs to receptor sites that are located on the cell membrane of the host cell instead of the cell wall as in the bacterial host.
2. Viral penetration is accomplished by endocytosis, an energy-requiring, receptor-mediated process in which the entire virus enters the host cell.

3. The uncoating of the animal virus, removal of the capsid, occurs within the host cell; with bacteriophage infection, the phage capsid remains on the outside of the host.

4. The latent period, the time between adsorption and the release of virulent viral particles, is considerably longer—hours to days rather than minutes, as in bacteriophage infection.

Cultivation and Enumeration of Bacteriophages

LEARNING OBJECTIVE

Once you have completed this experiment, you should be able to

1. Perform techniques for cultivation and enumeration of bacteriophages.

PRINCIPLE

This exercise demonstrates the ability of viruses to replicate inside a susceptible host cell. For this purpose, you will be provided with a virulent phage and a susceptible host cell culture. This technique also enables you to enumerate phage particles on the basis of plaque formation in a solid agar medium. **Plaques** are clear areas in an agar medium previously seeded with a diluted phage sample and a host cell culture. Each plaque represents the lysis of a phage-infected bacterial cell.

The procedure requires the use of a double-layered culture technique in which the hard agar serves as a base layer, and a mixture of phage and host cells in a soft agar forms the upper overlay. Susceptible *Escherichia coli* cells multiply rapidly and produce a lawn of confluent growth on the medium. When one phage particle adsorbs to a susceptible cell, penetrates the cell, replicates, and goes on to lyse other host cells, the destroyed cells produce a single plaque in the bacterial lawn (Figure 39.1). Each plaque can be designated as a **plaque-forming unit (PFU)** and used to quantitate the number of infective phage particles in the culture.

The number of phage particles contained in the original stock phage culture is determined by counting the number of plaques formed on the seeded agar plate and multiplying this by the dilution factor. For a valid

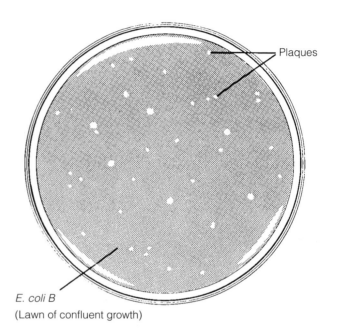

E. coli B
(Lawn of confluent growth)

FIGURE 39.1 Plaque-forming units (PFUs)

phage count, the number of plaques per plate should not exceed 300 nor be less than 30.

Example: 200 PFUs are counted in a 10^{-6} dilution.

$$(200) \times (10^6) = 200 \times 10^6 \text{ or } 2 \times 10^8$$
PFUs per ml of stock phage culture.

Plates showing greater than 300 PFUs are **too numerous to count (TNTC)**; plates showing fewer than 30 PFUs are **too few to count (TFTC).**

MATERIALS

Cultures

24-hour nutrient broth cultures of *Escherichia coli B* and T2 coliphage.

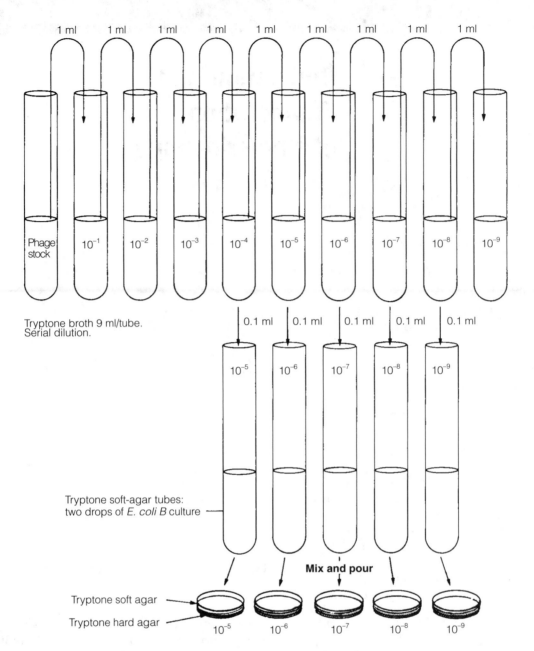

1 ml   **1 ml**   **1 ml**   **1 ml**   **1 ml**   **1 ml**   **1 ml**   **1 ml**   **1 ml**

Phage stock 10^{-1} 10^{-2} 10^{-3} 10^{-4} 10^{-5} 10^{-6} 10^{-7} 10^{-8} 10^{-9}

Tryptone broth 9 ml/tube.
Serial dilution.

0.1 ml   **0.1 ml**   **0.1 ml**   **0.1 ml**   **0.1 ml**

10^{-5} 10^{-6} 10^{-7} 10^{-8} 10^{-9}

Tryptone soft-agar tubes:
two drops of *E. coli B* culture

Mix and pour

Tryptone soft agar
Tryptone hard agar

10^{-5} 10^{-6} 10^{-7} 10^{-8} 10^{-9}

FIGURE 39.2 Dilution procedure for cultivation and enumeration of bacteriophages

Media

Five each of the following per designated student group: tryptone agar plates and tryptone soft agar, 2 ml per tube and nine tryptone broth tubes, 9 ml per tube.

Equipment

Bunsen burner, water bath, thermometer, 1-ml sterile pipettes, sterile Pasteur pipettes, mechanical pipetting device, test tube rack, and glassware marking pencil.

PROCEDURE

To perform the dilution procedure as illustrated in Figure 39.2, do the following:

1. Label all dilution tubes and media as follows:

   a. Five tryptone soft agar tubes: 10^{-5}, 10^{-6}, 10^{-7}, 10^{-8}, 10^{-9}.

   b. Five tryptone hard agar plates: 10^{-5}, 10^{-6}, 10^{-7}, 10^{-8}, 10^{-9}.

   c. Nine tryptone broth tubes: 10^{-1} through 10^{-9}.

2. Place the five labeled soft tryptone agar tubes into a water bath. Water should be of a depth just slightly above that of the agar in the tubes. Bring the water bath to 100°C to melt the agar. Cool and maintain the melted agar at 45°C.

3. With 1-ml pipettes, aseptically perform a ten-fold serial dilution of the provided phage culture using the nine 9-ml tubes of tryptone.

4. To the tryptone soft agar tube labeled 10^{-5}, aseptically add two drops of the *E. coli B* culture with a Pasteur pipette and 0.1 ml of the 10^{-4} tryptone broth phage dilution. Rapidly mix by rotating the tube between the palms of your hands and pour the contents over the hard tryptone agar plate labeled 10^{-5}, thereby forming a double-layered plate culture preparation. Swirl the plate gently and allow to harden.

5. Using separate Pasteur pipettes and 1-ml sterile pipettes, repeat Step 4 for the tryptone broth phage dilution tubes labeled 10^{-5} through 10^{-8} to effect the 10^{-6} through 10^{-9} tryptone soft-agar overlays.

6. Following solidification of the soft agar overlay, incubate all plate cultures in an inverted position for 24 hours at 37°C.

Name _____ Date _____

1. Observe all plates for the presence of plaque-forming units that develop on the bacterial lawn.
2. Count the number of PFUs in the range of 30 to 300 on each plate.
3. Calculate the number of phage particles per ml of the stock phage culture based on your PFU count.
4. Record your results in the chart.

Phage Dilution	Number of PFUs	Calculation: PFUs × Dilution Factor	PFUs/ml of Stock Phage Culture
10^{-5}			
10^{-6}			
10^{-7}			
10^{-8}			
10^{-9}			

REVIEW QUESTIONS

1. Discuss the effects of lytic and lysogenic infections on the life cycle of the host cell.

2. Discuss the factors responsible for the transformation of a lysogenic infection to one that is lytic.

3. Distinguish between the replicative and maturation stages of a lytic phage infection.

4. In this experimental procedure, why is it important to use a hard agar with a soft agar over-lay technique to demonstrate plaque formation?

5. Explain what is meant by plaque-forming units.

6. Determine the number of PFUs per ml in a 10^{-9} dilution of a phage culture that shows 204 PFUs in the agar lawn.

7. The release of phage particles from the host bacterium always occurs by lysis of the cell and results in the death of the host. Animal viruses are released by either the lysis of the host cell or exocytosis, a reverse pinocytosis. Regardless of the mechanism of release, the ani-mal host cell always dies. Explain.

Isolation of Coliphages from Raw Sewage

LEARNING OBJECTIVE

Once you have completed this experiment, you should be able to

1. Isolate virulent coliphages from sewage.

PRINCIPLE

Isolates of bacterial viruses (bacteriophages) can be obtained from a variety of natural sources, including soil, intestinal contents, raw sewage, and some insects such as cockroaches and flies. Their isolation from these environments is not an easy task because the phage particles are usually present in low concentrations. Therefore isolation requires a series of steps:

1. Collection of the phage-containing sample at its source.

2. Addition of an enriched susceptible host cell culture to the sample to increase the number of phage particles for subsequent isolation.

3. Following incubation, centrifugation of the enriched sample for the removal of gross particles.

4. Filtration of the supernatant liquid through a bacteria-retaining membrane filter.

5. Inoculation of the bacteria-free filtrate onto a lawn of susceptible host cells grown on a soft agar plate medium.

6. Incubation and observation of the culture for the presence of phage particles, which is indicated by plaque formation in the bacterial lawn.

In the following experiment, you will use this procedure, as illustrated in Figure 40.1, for the isolation of *Escherichia coli* phage particles from raw sewage. Most bacteriophages that infect *E. coli* (coliphages) are designated by the letter T, indicating types. Seven types have been identified and are labeled T1 through T7. The T-even phages (T2, T4, and T6) differ from the T-odd phages in that the former vary in size, form, and chemical composition. All of the T phages are capable of infecting the susceptible *E. coli B* host cell.

MATERIALS

Cultures

Part 1: 5-ml 24-hour broth cultures of *E. coli B* and 45-ml samples of fresh sewage collected in screw-capped bottles. Part 2: 10-ml 24-hour broth cultures of *E. coli B*.

Media

Per designated student group: Part 1: One 5-ml tube of bacteriophage nutrient broth, 10 times normal concentration. Part 2: Five tryptone agar plates and five 3-ml tubes of tryptone soft agar.

Equipment

Part 1: Sterile 250-ml Erlenmeyer flask and stopper. Part 2: Sterile membrane filter apparatus, sterile 125-ml Erlenmeyer flask and stopper, 125-ml flask, 1000-ml beaker, centrifuge, Bunsen burner, forceps, 1-ml sterile disposable pipettes, sterile Pasteur pipette, mechanical pipetting device, test tube rack, and glassware marking pencil.

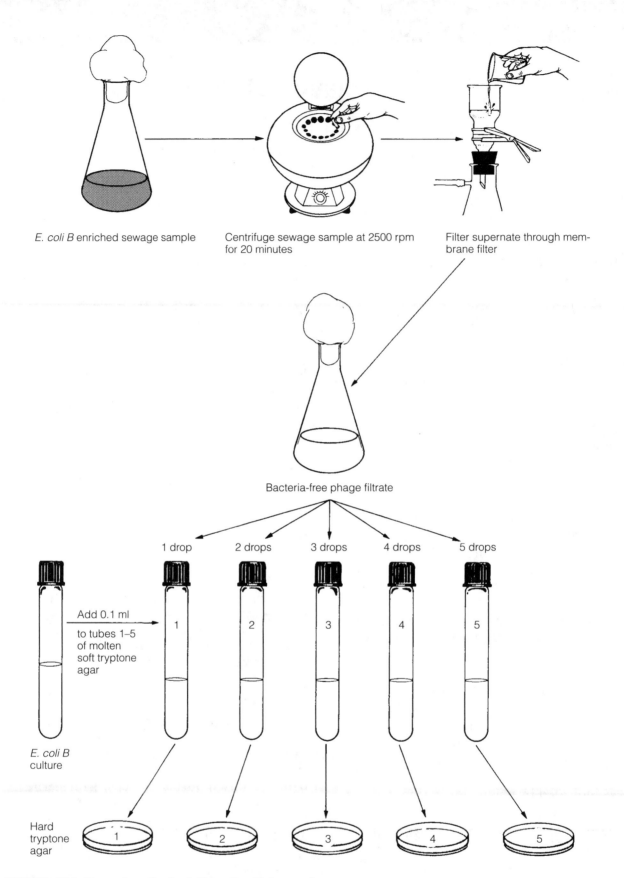

FIGURE 40.1 Procedure for isolation of coliphages from raw sewage

PROCEDURE

> ⚠ Use disposable gloves. It is essential to handle raw sewage with extreme caution because it may serve as vehicle for the transmission of human pathogens.

Part 1: Enrichment of Sewage Sample

1. Aseptically add 5 ml of bacteriophage nutrient broth, 5 ml of the *E. coli B* broth culture, and 45 ml of the raw sewage sample to an appropriately labeled sterile 250-ml Erlenmeyer flask.

2. Incubate the culture for 24 hours at 37°C.

Part 2: Filtration and Seeding

1. Following incubation, pour the phage-infected culture into a 100-ml centrifuge bottle or several centrifuge tubes and centrifuge at 2500 rpm for 20 minutes.

2. Remove the centrifuge bottle or tubes, being careful not to stir up the sediment, and carefully decant the supernate into a 125-ml flask.

3. Pour the supernatant solution through a sterile membrane filter apparatus to collect the bacteria-free, phage-containing filtrate in the vacuum flask below. Refer to Experiment 52 for the procedure in assembling the filter membrane apparatus.

4. Melt the soft tryptone agar by placing the five tubes in a boiling water bath and cool to 45°C.

5. Label the five tryptone agar plates and the five tryptone agar tubes 1, 2, 3, 4, and 5, respectively.

6. Using a sterile 1-ml pipette, aseptically add 0.1 ml of the *E. coli B* culture to all the molten soft-agar tubes.

7. Using a sterile Pasteur pipette, aseptically add 1, 2, 3, 4, and 5 drops of the filtrate to the respectively labeled molten soft-agar tubes. Mix and pour each tube of soft agar into its appropriately labeled agar plate.

8. Allow agar to harden.

9. Incubate all the plates in an inverted position for 24 hours at 37°C.

Name _____   Date _____

1. Examine all the culture plates for plaque formation, which is indicative of the presence of coliphages in the culture.
2. Indicate the presence (+) or absence (−) of plaques in each of the cultures in the chart.

Drops of Phage Filtrate	1	2	3	4	5
Plaque Formation (+) or (−)					

3. Based on your observations, what is the relationship between the number of plaques observed and the number of drops of filtrate in each culture?

REVIEW QUESTIONS

1. Why is enrichment of the sewage sample necessary for the isolation of phage?

2. How is enrichment of the sewage sample accomplished?

3. How are bacteria-free phage particles obtained?

4. Why must you exercise caution when handling raw sewage samples?

Physical and Chemical Agents for the Control of Microbial Growth

---

LEARNING OBJECTIVES

Once you have completed the experiments in this section, you should

1. Know the basic methods for inhibiting microbial growth and the modes of antimicrobial action.

2. Be able to demonstrate the effects of physical agents, moist heat, osmotic pressure, and ultraviolet radiation on selected microbial populations.

3. Be able to demonstrate the effects on selected microbial populations of chemical agents used as disinfectants, antiseptics, and antibiotics.

INTRODUCTION

Control of microorganisms is essential in the home, industry, and medical fields to prevent and treat diseases and to inhibit the spoilage of foods and other industrial products. Common methods of control involve chemical and physical agents that adversely affect microbial structures and functions, thereby producing a microbicidal or microbistatic effect. A **microbicidal effect** is one that kills the microbes immediately; a **microbistatic effect** inhibits the reproductive capacities of the cells and maintains the microbial population at a constant size.

Chemical Methods for Control of Microbial Growth

1. **Antiseptics:** Chemical substances used on living tissue that kill or inhibit the growth of vegetative microbial forms.

2. **Disinfectants:** Chemical substances that kill or inhibit the growth of vegetative microbial forms on nonliving materials.

3. **Chemotherapeutic agents:** Chemical substances that destroy or inhibit the growth of microorganisms in living tissues.

Physical Methods for Control of Microbial Growth

The modes of action of the different chemical and physical agents of control vary, although they all produce damaging effects to one or more essential cellular structures or molecules in order to cause cell death or inhibition of growth. Sites of damage that can result in malfunction are the cell wall, cell membrane, cytoplasm, enzymes, or nucleic acids. The adverse effects manifest themselves in the following ways.

1. **Cell-wall injury:** This can result in one of two ways. First, lysis of the cell wall will leave the protoplast susceptible to osmotic damage, and a hypotonic environment may cause lysis of the vulnerable protoplast. Second, certain agents inhibit cell wall synthesis, which is essential during microbial cell reproduction. Failure to synthesize a missing segment of the cell wall results in an unprotected protoplast.

2. **Cell-membrane damage:** This may be the result of lysis of the membrane, which will cause immediate cell death. Also, the selective nature of the membrane may be affected without causing its complete

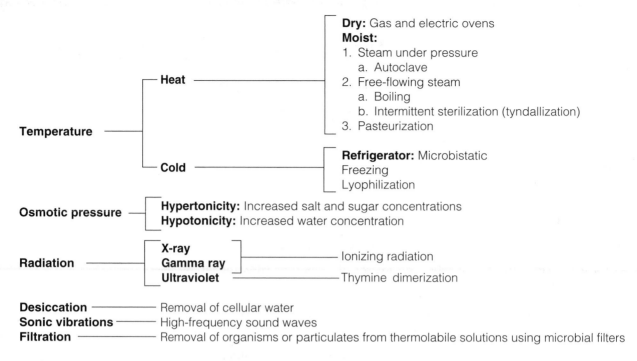

FIGURE IX.1 Physical methods used for the control of microbial growth

disruption. As a result, there may be a loss of essential cellular molecules or interference with the uptake of nutrients. In both cases, metabolic processes will be adversely affected.

3. **Alteration of the colloidal state of cytoplasm:** Certain agents cause denaturing of cytoplasmic proteins. Denaturing processes are responsible for enzyme inactivation and cellular death by irreversibly rupturing the molecular bonds of these proteins and rendering them biologically inactive.

4. **Inactivation of cellular enzymes:** Enzymes may be inactivated competitively or noncompetitively. Noncompetitive inhibition is irreversible and occurs following the application of some physical agent, such as mercuric chloride ($HgCl_2$), that results in the uncoiling of the protein molecule, rendering it biologically inactive. Competitive inhibition occurs when a natural substrate is forced to compete for the active site on an enzyme surface with a chemically similar molecular substrate, which can block the enzyme's ability to create end products. Competitive inhibitions are reversible.

5. **Interference with the structure and function of the DNA molecule:** The DNA molecule is the control center of the cell and may also represent a cellular target area for destruction or inhibition. Some agents have an affinity for DNA and cause breakage or distortion of the molecule, thereby interfering with its replication and role in protein synthesis.

Figure IX.1 illustrates the acceptable physical methods used for the control of microbial growth.

Awareness of the mode of action of the physical and chemical agents is absolutely essential for their proper selection and application in microbial control. The exercises in this section are designed to acquaint you more fully with several commonly employed agents and their uses.

Physical Agents of Control: Moist Heat

LEARNING OBJECTIVE

Once you have completed this experiment, you should understand

1. The susceptibility of microbial species to destruction by the application of moist heat.

PRINCIPLE

Temperature has an effect on cellular enzyme systems and therefore a marked influence on the rate of chemical reactions and thus the life and death of microorganisms. Despite the diversity among microorganisms' temperature requirements for growth, extremes in temperature can be used in microbial growth control. Sufficiently low temperatures will inactivate enzymes and produce a static effect. High temperatures destroy cellular enzymes, which become irreversibly denatured.

The application of heat is a common means of destroying microorganisms. Both dry and moist heat are effective. However, moist heat, which (because of the hydrolyzing effect of water and its greater penetrability) causes coagulation of proteins, kills cells more rapidly and at lower temperatures than does dry heat. **Sterilization,** the destruction of all forms of life, is accomplished in 15 minutes at 121°C with moist heat under pressure; dry heat requires a temperature of 160°C to 180°C for 1½ to 3 hours.

Microbes exhibit differences in their resistance to moist heat. As a general rule, bacterial spores require temperatures above 100°C for destruction, whereas most bacterial vegetative cells are killed at temperatures of 60°C to 70°C in 10 minutes. Fungi can be killed at 50°C to 60°C, and fungal spores require 70°C to 80°C for 10 minutes for destruction. Because of this variability, moist heat can either sterilize or disinfect. Common applications include free-flowing steam under pressure (autoclaving), free-flowing steam at 100°C (tyndallization), and the use of lower temperatures (pasteurization).

Free-flowing steam under pressure requires the use of an autoclave, a double-walled metal vessel that allows steam to be pressurized in the outer jacket (see Figure 41.1). At a designated pressure, the saturated steam is released into the inner chamber, from which all the air has been evacuated. The steam under pressure in the vacuumed inner chamber is now capable of achieving temperatures in excess of 100°C. The temperature is determined by the pounds of pressure applied per square inch, as illustrated:

Pressure (pounds/inch^2)	Temperature (°C)
0 (free-flowing steam)	100
10	115
15	121
20	126
25	130

A pressure of 15 pounds/inch^2 achieves a temperature of 121°C and sterilizes in 15 minutes. This is the usual procedure; however, depending on the heat sensitivity of the material to be sterilized, the operating pressure and time conditions can be adjusted.

Application of **free-flowing steam** requires exposure of the contaminated substance to a temperature of 100°C, which is achieved by boiling water. Exposures to boiling water for 30 minutes will result in disinfection only; all vegetative cells will be killed, but not necessarily the more heat-resistant spores.

Another procedure is **tyndallization,** also referred to as intermittent or fractional

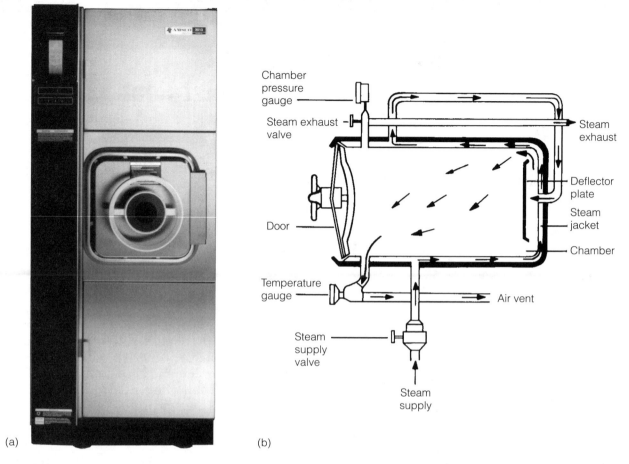

FIGURE 41.1 (a) The autoclave (courtesy of STERIS Corporation), (b) Schematic representation of autoclave

Labels in figure (b): Chamber pressure gauge; Steam exhaust valve; Steam exhaust; Deflector plate; Steam jacket; Chamber; Door; Temperature gauge; Air vent; Steam supply valve; Steam supply.

(a) (b)

sterilization. This procedure requires exposure of the material to free-flowing steam at 100°C for 20 minutes on 3 consecutive days with intermittent incubation at 37°C. The steaming kills all vegetative cells. Any spores that may be present germinate during the period of incubation and are destroyed during subsequent exposure to a temperature of 100°C. Repeating this procedure for 3 days ensures germination of all spores and their destruction in the vegetative form. Because tyndallization requires so much time, it is used only for sterilization of materials that are composed of thermolabile chemicals and that might be subject to decomposition at higher temperatures.

Pasteurization exposes fairly thermolabile products such as milk, wine, and beer for a given period of time to a temperature that is high enough to destroy pathogens and some spoilage-causing microorganisms that may be present, without necessarily destroying all vegetative cells. There are two types of pasteuriza-

tion: the high-temperature, short-time (flash) procedure requires a temperature of 71°C for 15 seconds, and the low-temperature, long-time method requires 63°C for 30 minutes.

MATERIALS

Cultures

48- to 72-hour nutrient broth cultures (50 ml per 250-ml Erlenmeyer flask) of *Staphylococcus aureus* and *Bacillus cereus*; 72- to 96-hour Sabouraud broth cultures (50 ml per 250-ml Erlenmeyer flask) of *Aspergillus niger* and *Saccharomyces cerevisiae*.

Media

Per designated student group (pairs or groups of four): five nutrient agar plates, five Sabouraud agar plates, and one 10-ml tube of nutrient broth.

Equipment

Bunsen burner, 800-ml beaker (waterbath), tripod and wire gauze screen with heat-resistant pad, thermometer, glassware marking pencil, and inoculating loop.

PROCEDURE

1. Label the covers of each of the nutrient agar and Sabouraud agar plates, indicating the experimental heat temperatures to be used: 25 (control), 40, 60, 80, and 100°C.

2. Score the underside of all plates with a glassware marking pencil into two sections. On the nutrient agar plates, label one section *S. aureus* and the other *B. cereus*. On the Sabouraud agar plates, label one section *A. niger* and the second *S. cerevisiae*.

3. Using sterile technique, inoculate the nutrient agar and Sabouraud agar plates labeled 25°C by making a single-line loop inoculation of each test organism in its respective section of the plate.

4. Using a sterile pipette and mechanical pipetter, transfer 10 ml of each culture to four sterile test tubes labeled with the name of the organism and the temperature (40, 60, 80, and 100° C).

5. Set up the waterbath as illustrated below, inserting the thermometer in an uncapped tube of nutrient broth.

6. Slowly heat the water to 40°C; check the thermometer frequently to ensure that it does not exceed the desired temperature. Place the four cultures of the experimental organisms into the beaker and maintain the temperature at 40°C for 10 minutes. Remove the cultures and aseptically inoculate each organism in its appropriate section on the two plates labeled 40°C.

7. Raise the waterbath temperature to 60°C and repeat step 6 for the inoculation of the two plates labeled 60°C.

8. Raise the waterbath temperature to 80°C and repeat step 6 for the inoculation of the two plates labeled 80°C.

9. Raise the waterbath temperature to 100°C and repeat step 6 for the inoculation of the two plates labeled 100°C.

10. Incubate the nutrient agar plate cultures in an inverted position for 24 to 48 hours at 37°C and the Sabouraud agar plate cultures for 4 to 5 days at 25°C in a moist chamber.

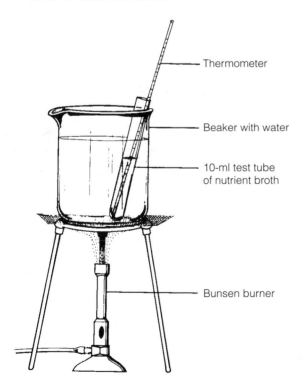

Thermometer

Beaker with water

10-ml test tube of nutrient broth

Bunsen burner

Experiment 41 / Observations and Results

Name _____ Date _____

1. Observe all plates for the amount of growth of the test organisms at each of the temperatures.
2. Record your results in the chart as 0 = none; 1+ = slight; 2+ = moderate; 3+ = abundant.

Microbial Species	Amount of Growth				
	25°C	40°C	60°C	80°C	100°C
B. cereus					
S. aureus					
A. niger					
S. cerevisiae					

3. List the microbial organisms in order of their increasing heat resistance.

_____

_____

_____

_____

REVIEW QUESTIONS

1. Account for the microbistatic effect produced by low temperatures as compared to the microbicidal effect produced by high temperatures.

2. Cite the advantages of each of the modes of sterilization: tyndallization and autoclaving.

3. Discuss the detrimental effects of control agents on the following:
   a. Cytoplasm:

   b. Cell wall:

   c. Nucleic acids:

   d. Cell membrane:

4. Explain why milk is subjected to pasteurization rather than sterilization.

5. A. *niger* and B. *cereus* cultures used in this experiment contained spores. Why is B. *cereus* more heat resistant?

6. Account for the fact that aerobic and anaerobic bacterial spore-formers are more heat re-sistant than the tubercle bacillus, which is also known to tolerate elevated temperatures.

Physical Agents
of Control: Environmental
Osmotic Pressure

LEARNING OBJECTIVE

Once you have completed this experiment, you should understand

1. The possible effect of osmotic pressure environments on microorganisms.

PRINCIPLE

Osmosis is the net movement of water molecules (solvent) across a semipermeable membrane from a solution of their higher concentration to a solution of their lower concentration. The determining factor in the relative water concentrations of the two solutions is their respective solute concentrations. The **hypertonic solution** possesses a higher osmotic pressure and a higher solute concentration, and therefore it has a lower water concentration; it tends to draw in water. The **hypotonic solution** possesses a lower osmotic pressure and has a lower solute concentration, and therefore it has a higher water concentration; it tends to lose water. If two solutions separated by a semipermeable membrane have equal concentrations of solutes and therefore equal water concentrations, there is no osmosis and the solutions are **isotonic.**

The cell and its environment represent two solutions separated by the semipermeable cell membrane. The cell's cytoplasm contains colloidal and solute particles dispersed in water, as does the cell's environment. Water is essential for the transport of materials across the cell membrane. As such, the **osmotic pressure of the environment** in relation to that of the cytoplasm of the cell plays a vital role in the life and death of a cell. The ideal environment of animal cells, which are bounded only by the fragile cell membrane, is completely or closely isotonic so that the cells are not susceptible to damage from osmotic pressure. In a hypertonic, high-pressure environment, all cells lose water by osmosis and become shriveled. This effect is called **plasmolysis;** it inhibits cell reproduction. As water is necessary for the occurrence of many chemical reactions, water loss adversely affects cell metabolism and reproduction. In a hypotonic, low-pressure environment, cells take in water and become swollen. This phenomenon is called **plasmoptysis.** In an environment with sufficiently low osmotic pressure, animal cells undergo **lysis,** which causes their death. Microorganisms and other organisms whose cells possess rigid cell walls are not usually susceptible to lysis in hypotonic environments. Instead, these organisms usually prefer a slightly hypotonic environment to maintain themselves in a turgid (distended) state. The effects of environmental osmotic pressure on cells unprotected by cell walls are illustrated in Figure 42.1.

Osmotic pressure can be used as an effective antimicrobial agent. Microorganisms are not usually adversely affected by low environmental osmotic pressure because of their small sizes and the presence of rigid cell walls that permit rapid adjustment to the hypotonicity. However, hypertonicity is a commonly used method of inhibiting microbial growth.

Because of their varied habitats, microorganisms are generally well-adapted to exist in all types of osmotic pressure environments, such as soil, air, fresh waters, and waters of varying salinity. Different groups of microorganisms require different degrees of salinity for growth, and as a rule, they can adjust to salt concentrations of 0.5% to 3%. Concentrations of 10% to 15% are inhibitory to the growth of most microbes, except for **halophiles,** which are adapted to life in waters of high salinity and require high concentrations for growth. This sensitivity is the basis of food preservation by the process of salting.

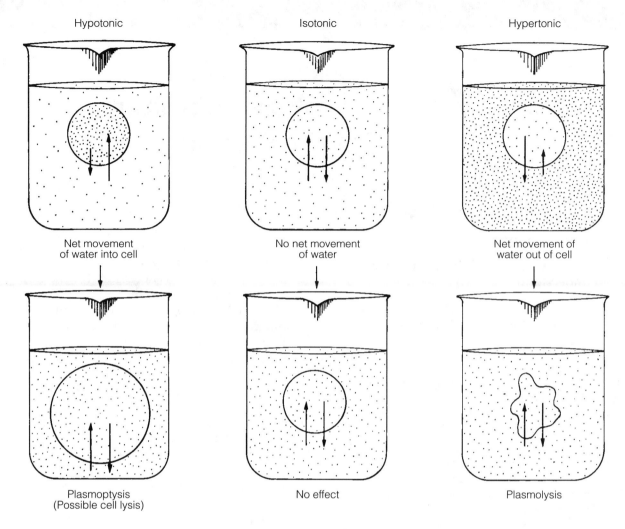

Hypotonic   Isotonic   Hypertonic

Net movement of water into cell   No net movement of water   Net movement of water out of cell

Plasmoptysis (Possible cell lysis)   No effect   Plasmolysis

FIGURE 42.1 Osmotic environments

MATERIALS

Cultures

24- to 48-hour nutrient broth cultures of *Staphylococcus aureus* and *Escherichia coli* and 48-hour *Halobacterium* salt broth culture of *Halobacterium salinarium*.

Media

Per designated student group: one nutrient agar plate of each of the following sodium chloride concentrations: 0.85%, 5%, 10%, 15%, and 25%.

Equipment

Bunsen burner, inoculating loop, and glassware marking pencil.

PROCEDURE

1. Score the underside of each of the five nutrient agar plates into three sections with a glassware marking pencil.

2. Label each of the three sections on each plate with the name of the organism to be inoculated.

3. Using sterile technique, inoculate each of the agar plates with the three experimental organisms by making a single-line loop inoculation of each organism in its appropriately labeled section.

4. Incubate all plates in an inverted position for 4 to 5 days at 25°C.

Name _____   Date _____

1. Observe each of the nutrient agar plate cultures for the amount of growth of each of the experimental species.
2. Record your results in the chart as 0 = no growth; 1+ = scant growth; 2+ = moderate growth; 3+ = abundant growth.

Microbial Species	Salt Concentration of Medium (%)				
	0.85	5	10	15	25
S. aureus					
E. coli					
H. salinarium					

3. For each of the experimental species, indicate (a) the salt concentration range in which growth will occur and (b) the optimal salt concentration.

Microbial Species	Range of Growth	Optimal NaCl Concentration
S. aureus		
E. coli		
H. salinarium		

REVIEW QUESTIONS

1. Compare hypertonic, hypotonic, and isotonic solutions and their effects on cells.

2. Explain how hypertonicity can be used as a means of controlling microbial growth.

3. Early pioneers, traveling to the West, salted their meat products for preservation. Explain the microbiological mechanism by which the salt retarded the spoilage of their meat.

4. How do halophiles survive in their environment?

Physical Agents of Control: Electromagnetic Radiations

LEARNING OBJECTIVE

Once you have completed this experiment, you should understand

1. The microbicidal effect of ultraviolet (UV) radiation on microorganisms.

PRINCIPLE

Certain forms of electromagnetic radiation are capable of producing a lethal effect on cells and therefore can be used for microbial control. Electromagnetic radiations that possess sufficient energy to be microbicidal are the short-wavelength radiations, that is, 300 nm and below. These include UV, gamma rays, and x-rays. The high-wavelength radiations, those above 300 nm, have insufficient energy to destroy cells. The electromagnetic spectrum and its effects on molecules are illustrated in Figure 43.1.

Gamma radiation, originating from unstable atomic nuclei, and **x-radiation,** originating from outside of the atomic nucleus, are representative of **ionizing forms of radiation.** Both transfer their energy through quanta (photons) to the matter through which they pass, causing excitation and the loss of electrons from molecules in their paths. This injurious effect is nonspecific in that any molecule in the path of the radiation will undergo ionization. Essential cell molecules can be directly affected through loss of their chemical structures and activity brought about by the ionization. Also, water, the most abundant chemical constituent of cells, commonly undergoes radiation breakdown, with the ultimate production of highly reactive H^+, OH^-, and, in the presence of oxygen, HO_2 free radicals. These may combine with each other, frequently forming hydrogen peroxide (H_2O_2), which is highly toxic to cells lacking catalase or other peroxidases, or the highly reactive free radicals may combine with any cellular constituents, again resulting in biochemical cell damage.

Because of their high energy content and therefore ability to penetrate matter, x-ray and gamma radiations can be used as means of sterilization, particularly of thermolabile materials. They are not commonly used, however, because of the expense of the equipment and the special facilities necessary for their safe use.

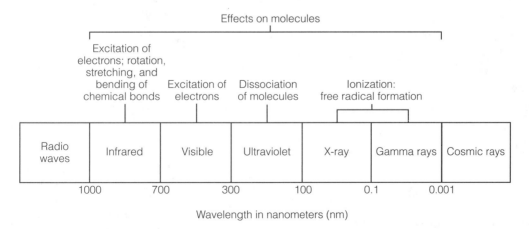

FIGURE 43.1 The electromagnetic spectrum

Ultraviolet light, which has a lower energy content than ionizing radiations, is capable of producing a lethal effect in cells exposed to the low penetrating wavelengths in the range of 210 nm to 300 nm. Cellular components capable of absorbing ultraviolet light are the nucleic acids, with the DNA acting as the primary site of damage. As the pyrimidines especially absorb ultraviolet wavelengths, the major effect of this form of radiation is **thymine dimerization,** which is the covalent bonding of two adjacent thymine molecules on one nucleic acid strand in the DNA molecule. This dimer formation distorts the configuration of the DNA molecule, and the distortion interferes with DNA replication and transcription during protein synthesis.

Some cell types, including some microorganisms, possess enzyme systems for the repair of radiation-induced DNA damage. Two different systems may be operational: (1) The **excision repair system,** which functions in the absence of light; and (2) the **light repair system,** which is made operational by exposure of the irradiated cells to visible light in the wavelength range of 420 nm to 540 nm. The visible light serves to activate an enzyme that splits the dimers and reverses the damage.

Ultraviolet radiation, because of its low penetrability, cannot be used as a means of sterilization, and its practical application is only for surface or air disinfection.

MATERIALS

Cultures

24- to 48-hour nutrient broth cultures of *Serratia marcescens* and *Bacillus cereus;* sterile saline spore suspension of *Aspergillus niger.*

Media

Per designated student group: seven nutrient agar plates.

Equipment

Bunsen burner, inoculating loop, ultraviolet radiation source (254 nm), and glassware marking pencil.

PROCEDURE

⚠ **Wear disposable gloves and do not expose your eyes to the ultraviolet light.**

1. Divide all nutrient agar plates into three sections by scoring the underside of each plate with a glassware marking pencil.
2. Label each of the three sections on each plate with the name of the organism to be inoculated.
3. Using sterile technique, inoculate all the plates by means of a streak inoculation *specifically* as shown in the following illustration:

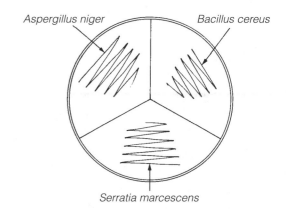

4. Label the cover of each inoculated plate with the exposure time to ultraviolet radiation as 0 (control), 15 seconds, 30 seconds, 1 minute, and 3 minutes. Label two plates as 5 minutes, one of which will serve as the irradiated, covered control.
5. Irradiate all inoculated plates for the designated period of time by placing them 12 inches below the ultraviolet light source, as illustrated below. Make sure first to remove all Petri dish covers except that of the 5-minute irradiated control plate.

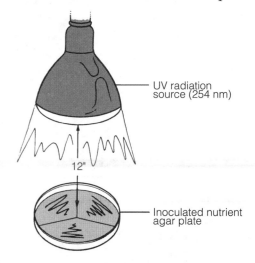

6. Incubate all plates in an inverted position for 4 to 5 days at 25°C.

Name _____ Date _____

1. Observe all nutrient agar plate cultures for the amount of growth of each of the microbial species.
2. Record your observations in the chart as 0 = no growth; 1+ = slight growth; 2+ = moderate growth; 3+ = abundant growth.

Microbial Species	Time of Irradiation						
	Seconds			Minutes			
	0	15	30	1	3	5	5*
B. cereus							
S. marcescens							
A. niger							

*Irradiated, covered plate.

REVIEW QUESTIONS

1. Discuss the effects of ionizing radiation on cellular constituents.

2. Explain why x-rays can be used for sterilization, whereas ultraviolet rays can be used only for surface disinfection of materials.

3. Explain the mechanism of action of ultraviolet radiation on cells.

4. Account for the greater susceptibility of *S. marcescens* than that of *B. cereus* to the effects of ultraviolet radiation.

5. Why is it not essential to shield your hands from ultraviolet light, whereas you must exercise great care to shield your eyes from this type of radiation?

Chemical Agents of Control: Chemotherapeutic Agents

Chemotherapeutic agents are chemical substances used in the treatment of infectious diseases. Their mode of action is to interfere with microbial metabolism, thereby producing a bacteriostatic or bacteriocidal effect on the microorganisms, without producing a like effect in host cells. Chemotherapeutic agents act on a number of cellular targets. Their mechanisms of action include inhibition of cell-wall synthesis, inhibition of protein synthesis, inhibition of nucleic acid synthesis, disruption of the cell membrane, and inhibition of folic acid synthesis. These drugs can be separated into two categories:

1. **Antibiotics** are synthesized and secreted by some true bacteria, actinomycetes, and fungi that destroy or inhibit the growth of other microorganisms. Today, some antibiotics are laboratory synthesized or modified; however, their origins are living cells.

2. **Synthetic drugs** are synthesized in the laboratory.

To determine a therapeutic drug of choice, one must know its mode of action, possible adverse side effects in the host, and the scope of its antimicrobial activity. The specific mechanism of action varies among different drugs, and the short-term or long-term use of many drugs can produce systemic side effects in the host. These vary in severity from mild and temporary upsets to permanent tissue damage (Table 44.1).

SYNTHETIC AGENTS

Sulfadiazine (a sulfonamide) produces a static effect on a wide range of microorganisms by a mechanism of action called **competitive inhibition.** The active component of the drug,

TABLE 44.1   Prototypic Antibiotics

Antibiotic	Mode of Action	Possible Side Effects
Penicillin	Prevents transpeptidation of the N-acetylmuramic acids, producing a weakened peptidoglycan structure	Penicillin resistance; sensitivity (allergic reaction)
Streptomycin	Has an affinity for bacterial ribosomes, causing misreading of codons on mRNA, hereby interfering with protein synthesis	May produce damage to auditory nerve, causing deafness
Chloramphenicol	Has an affinity for bacterial ribosomes, preventing peptide bond formation between amino acids during protein synthesis	May cause aplastic anemia, which is fatal because of destruction of RBC-forming and WBC-forming tissues
Tetracyclines	Have an affinity for bacterial ribosomes; prevent hydrogen bonding between the anticodon on the tRNA–amino acid complex and the codon on mRNA during protein synthesis	Permanent discoloration of teeth in young children
Bacitracin	Inhibits cell-wall synthesis application only	Nephrotoxic if taken internally; used for topical
Polymyxin	Destruction of cell membrane	Toxic if taken internally; used for topical application only

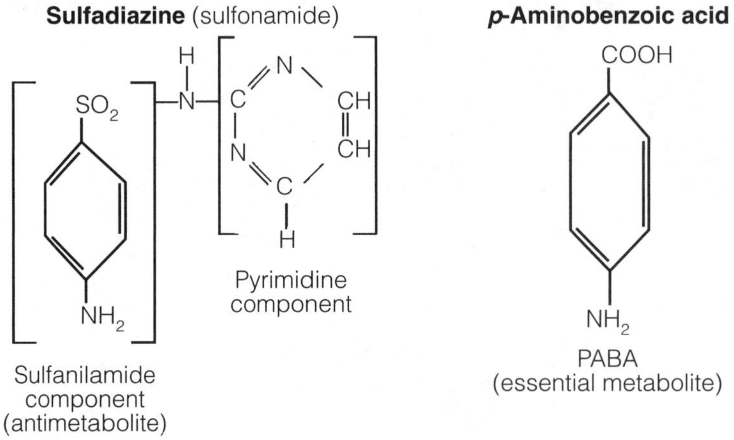

Sulfadiazine (sulfonamide)

Pyrimidine component

Sulfanilamide component (antimetabolite)

p-Aminobenzoic acid

PABA (essential metabolite)

FIGURE 44.1 Chemical similarity of sulfanilamide and PABA

sulfanilamide, acts as an **antimetabolite** that competes with the **essential metabolite,** *p*-aminobenzoic acid (PABA), during the synthesis of folic acid in the microbial cell. Folic acid is an essential cellular coenzyme involved in the synthesis of amino acids and purines. Many microorganisms possess enzymatic pathways for folic acid synthesis and can be adversely affected by sulfonamides. Human cells lack these enzymes, and the essential folic acid enters the cells in a preformed state. Therefore these drugs have no competitive effect on human cells. The similarity between the chemical structure of the antimetabolite sulfanilamide and the structure of the essential metabolite PABA is illustrated in Figure 44.1.

PART A: The Kirby-Bauer Antibiotic Sensitivity Test Procedure

LEARNING OBJECTIVE

Once you have completed this experiment, you should understand

1. The Kirby-Bauer procedure for the evaluation of the antimicrobial activity of chemotherapeutic agents.

PRINCIPLE

The available chemotherapeutic agents vary in their scope of antimicrobial activity. Some have a limited spectrum of activity, being effective against only one group of microorganisms. Others exhibit broad-spectrum activity against a range of microorganisms. The drug susceptibilities of many pathogenic microorganisms are known, but it is sometimes necessary to test several agents to determine the drug of choice.

A standardized filter-paper disc–agar diffusion procedure, known as the Kirby-Bauer method, is frequently used to determine the drug susceptibility of microorganisms isolated from infectious processes (Figure 44.2). This method allows the rapid determination of the efficacy of a drug by measuring the diameter of the zone of inhibition that results from diffusion of the agent into the medium surrounding the disc. In this procedure, filter-paper discs of uniform size are impregnated with specified concentrations of different antibiotics and then placed on the surface of an agar plate that has been seeded with the organism to be tested. The medium of choice is Mueller-Hinton agar, with a pH of 7.2 to 7.4, which is poured into plates to a uniform depth of 5 mm and refrigerated on solidification. Prior to use, the plates are transferred to an incubator at 37°C for 10 to 20 minutes to dry off the moisture that develops on the agar surface. The plates are then heavily inoculated with a standardized inoculum by means of a cotton swab to ensure the confluent growth of the organism. The discs are aseptically applied to the surface of the agar plate at well-spaced intervals. Once applied, each disc is gently touched with a sterile applicator stick to ensure its firm contact with the agar surface.

Following incubation, the plates are examined for the presence of growth inhibition, which is indicated by a clear zone surrounding each disc. The susceptibility of an organism to a drug is determined by the size of this zone, which itself is dependent on variables such as:

1. The ability and rate of diffusion of the antibiotic into the medium and its interaction with the test organism.

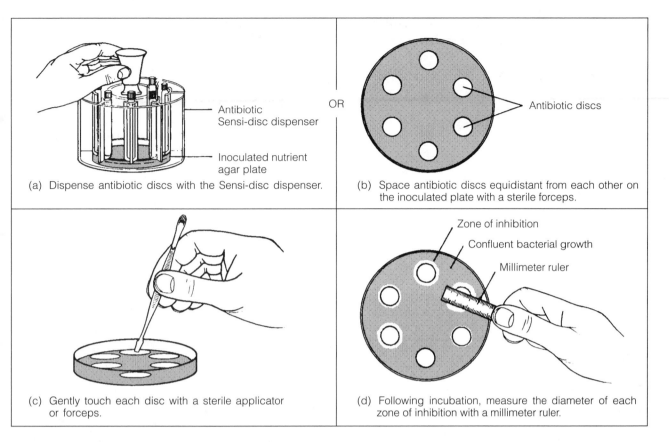

(a) Dispense antibiotic discs with the Sensi-disc dispenser.

Antibiotic
Sensi-disc dispenser

Inoculated nutrient
agar plate

OR

Antibiotic discs

(b) Space antibiotic discs equidistant from each other on the inoculated plate with a sterile forceps.

(c) Gently touch each disc with a sterile applicator or forceps.

Zone of inhibition

Confluent bacterial growth

Millimeter ruler

(d) Following incubation, measure the diameter of each zone of inhibition with a millimeter ruler.

FIGURE 44.2 Kirby-Bauer antibiotic sensitivity procedure

2. The number of organisms inoculated.
3. The growth rate of the organism.
4. The degree of susceptibility of the organism to the antibiotic.

A measurement of the diameter of the zone of inhibition in millimeters is made, and its size is compared to that contained in a standardized chart, which is shown in Table 44.2 on page 284. Based on this comparison, the test organism is determined to be resistant, intermediate, or susceptible to the antibiotic.

MATERIALS

Cultures

0.85% saline suspensions of *Escherichia coli, Staphylococcus aureus, Pseudomonas aeruginosa, Proteus vulgaris, Mycobacterium smegmatis, Bacillus cereus,* and *Enterococcus faecalis* adjusted to an O.D. of 0.1 at 600 nm. *Note: For enhanced growth of* M. smegmatis, *add Tween 80 (1 ml/liter of broth medium) and incubate for 3 to 5 days in a shaking waterbath, if available.*

Media

Per designated student group: seven Mueller-Hinton agar plates.

Antimicrobial-Sensitivity Discs

Penicillin G, 10 μg; streptomycin, 10 μg; tetracycline, 30 μg; chloramphenicol, 30 μg; gentamicin, 10 μg; vancomycin, 30 μg; and sulfanilamide, 300 μg.

Equipment

Sensi-disc™ dispensers or forceps, Bunsen burner, sterile cotton swabs, glassware marking pencil, and millimeter ruler.

PROCEDURE

1. Place agar plates right side up in an incubator heated to 37°C for 10 to 20 minutes with the covers adjusted so that the plates are slightly opened.
2. Label the covers of each of the agar plates with the name of the test organism to be inoculated.

3. Using sterile technique, inoculate all agar plates with their respective test organisms as follows:

   a. Dip a sterile cotton swab into a well-mixed saline test culture and remove excess inoculum by pressing the saturated swab against the inner wall of the culture tube.

   b. Using the swab, streak the entire agar surface horizontally, vertically, and around the outer edge of the plate to ensure a heavy growth over the entire surface.

4. Allow all culture plates to dry for about 5 minutes.

5. Using the Sensi-disc dispenser, apply the antibiotic discs by placing the dispenser over the agar surface and pressing the plunger, depositing the discs simultaneously onto the agar surface. If dispensers are not available, distribute the individual discs at equal distances with forceps dipped in alcohol and flamed.

6. Gently press each disc down with the wooden end of a cotton swab or sterile forceps to ensure that the discs adhere to the surface of the agar. *Note: Do not press the discs into the agar.*

7. Incubate all plate cultures in an inverted position for 24 to 48 hours at 37°C.

PART B: Synergistic Effect of Drug Combinations

LEARNING OBJECTIVE

Once you have completed this experiment, you should be able to

1. Perform the disc–agar diffusion technique for determination of synergistic combinations of chemotherapeutic agents.

PRINCIPLE

Combination chemotherapy, the use of two or more antimicrobial or antineoplastic agents, is being employed in medical practice with ever-increasing frequency. The rationale for using drug combinations is the expectation that effective combinations might lower the incidence of bacterial resistance, reduce host toxicity of the antimicrobial agents (because of decreased dosage requirements), or enhance the agents' bactericidal activity. Enhanced bactericidal activity is known as **synergism.** Synergistic activity is evident when the sum of the effects of the chemotherapeutic agents used in combination is significantly greater than the sum of their effects when used individually. This result is readily differentiated from an **additive (indifferent) effect,** which is evident when the interaction of two drugs produces a combined effect that is no greater than the sum of their separately measured individual effects.

A variety of in vitro methods are available to demonstrate synergistic activity. In this experiment, a disc–agar diffusion technique will be performed to demonstrate this phenomenon. This technique uses the Kirby-Bauer antibiotic susceptibility test procedure, as described in Part A of this experiment, and requires both Mueller-Hinton agar plates previously seeded with the test organisms and commercially prepared, antimicrobial-impregnated discs. The two discs, representing the drug combination, are placed on the inoculated agar plate and separated by a distance (measured in mm) that is equal to or slightly greater than one-half the sum of their individual zones of inhibition when obtained separately. Following the incubation period, an additive effect is exhibited by the presence of two distinctly separate circles of inhibition. If the drug combination is synergistic, the two inhibitory zones merge to form a "bridge" at their juncture, as illustrated in Figure 44.3.

The following drug combinations will be used in this experimental procedure:

1. **Sulfisoxazole, 150 μg,** and **trimethoprim, 5 μg.** Both antimicrobial agents are enzyme inhibitors that act sequentially in the metabolic pathway, leading to folic acid synthesis. The antimicrobial effect of each drug is enhanced when the two drugs are used in combination. The pathway thus exemplifies synergism.

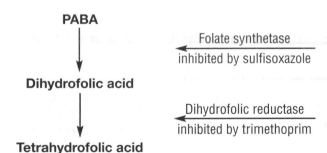

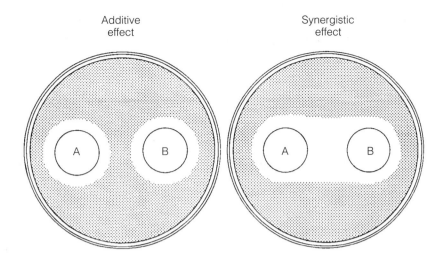

Additive
effect

Synergistic
effect

FIGURE 44.3 Additive and synergistic effects of drug combinations

2. **Trimethoprim, 5 μg, and tetracycline, 30 μg.** The modes of antimicrobial activity of these two chemotherapeutic agents differ; tetracycline acts to interfere with protein synthesis at the ribosomes. Thus, when used in combination, these drugs produce an additive effect.

MATERIALS

Cultures

0.85% saline suspensions of *Escherichia coli* and *Staphylococcus aureus* adjusted to an O.D. of 0.1 at 600 nm.

Media

Per designated student group: four Mueller-Hinton agar plates.

Antimicrobial-Sensitivity Discs

Tetracycline, 30 μg; trimethoprim, 5 μg; and sulfisoxazole, 150 μg.

Equipment

Bunsen burner, forceps, sterile cotton swabs, millimeter ruler, and glassware marking pencil.

PROCEDURE

1. To inoculate the Mueller-Hinton agar plates, follow Steps 1 through 4 as described under the procedure in Part A of this experiment.

2. Using the millimeter ruler, determine the center of the underside of each plate and mark with a glassware marking pencil.

3. Using the glassware marking pencil, mark the underside of each agar plate culture at both sides from the center mark at the distances specified below:

   a. *E. coli*–inoculated plate for trimethoprim and sulfisoxazole combination sensitivity: 12.5 mm on each side of center mark.

   b. *S. aureus*–inoculated plate for trimethoprim and sulfisoxazole combination sensitivity: 14.5 mm on each side of center mark.

   c. *E. coli*– and *S. aureus*–inoculated plates for trimethoprim and tetracycline combination sensitivity: 14.0 mm on each side of center mark.

4. Using sterile forceps, place the antimicrobial discs, in the combinations specified in Step 3, onto the surface of each agar plate culture at the previously marked positions. Gently press each disc down with the sterile forceps to ensure that it adheres to the agar surface.

5. Incubate all plate cultures in an inverted position for 24 to 48 hours at 37°C.

TABLE 44.2   Zone Diameter Interpretive Standards for Organisms Other Than
Haemophilus and *Neisseria gonorrhoeae*

Antimicrobial Agent	Disk Content	Zone Diameter, Nearest Whole mm		
		Resistant	Intermediate[k]	Susceptible
Ampicillin[a]				
when testing gram-negative enteric organisms	10 µg	≤ 13	—	≥ 17
when testing staphylococci[b]	10 µg	≤ 28	—	≥ 29
when testing enterococci[c]	10 µg	≤ 16	—	—
when testing nonenterococcal streptococci[c]	10 µg	≤ 21	—	≥ 30
when testing *Listeria monocytogenes*	10 µg	≤ 19	—	≥ 20
Carbenicillin[b]				
when testing *Pseudomonas*	100 µg	≤ 13	—	≥ 17
when testing other gram-negative organisms	100 µg	≤ 19	—	≥ 23
Cefoxitin[d]	30 µg	≤ 14	—	≥ 18
Cephalothin[d]	30 µg	≤ 14	—	≥ 18
Chloramphenicol	30 µg	≤ 12	13–17	≥ 18
Clindamycin[a]	2 µg	≤ 14	15–20	≥ 21
Erythromycin	15 µg	≤ 13	14–22	≥ 23
Gentamicin[e]	10 µg	≤ 12	13–14	≥ 15
Kanamycin	30 µg	≤ 13	14–17	≥ 18
Methicillin when testing staphylococci[f]	5 µg	≤ 9	10–13	≥ 14
Novobiocin[g]	30 µg	≤ 17	18–21	≥ 22
Penicillin G				
when testing staphylococci[b]	10 units	≤ 28	—	≥ 29
when testing enterococci[c]	10 units	≤ 14	—	—
when testing *L. monocytogenes*	10 units	≤ 19	—	≥ 20
when testing nonenterococcal streptococci	10 units	≤ 19	—	≥ 28
Rifampin	5 µg	≤ 16	17–19	≥ 20
Streptomycin	10 µg	≤ 11	12–14	≥ 15
Tetracycline[h]	30 µg	≤ 14	15–18	≥ 19
Tobramycin[e]	10 µg	≤ 12	13–14	≥ 15
Trimethoprim/sulfamethoxazole[i]	1.25/23.75 µg	≤ 10	—	≥ 16
Vancomycin				
when testing enterococci	30 µg	≤ 9	10–11	—
when testing other gram-positives	30 µg	≤ 9	10–11	≥ 12
Urinary tract specific antimicrobial agents[j]				
Sulfonamides[i]	250 or 300 µg	≤ 12	—	≥ 17
Trimethoprim[i]	5 µg	≤ 10	—	≥ 16

Permission to excerpt portions of M2-A4 (*Performance Standards for Antimicrobial Disk Susceptibility Tests*–Fourth Edition; Approved Standard) has been granted by NCCLS. The interpretive data are valid only if the methodology in M2-A4 is followed. NCCLS frequently updates the interpretive tables through new editions of the standards and supplements. Users should refer to the most-recent edition. The current standard may be obtained from NCCLS, 940 West Valley Road, Suite 1400, Wayne, PA 19087, USA.

[a]The clindamycin disk is used for testing susceptibility to both clindamycin and lincomycin.

[b]Resistant strains of *S. aureus* produce ß-lactamase and the testing of the 10-unit *penicillin G disk is preferred*. Penicillin G should be used to test the susceptibility of all penicillinase-sensitive penicillins, such as ampicillin, amoxicillin, azlocillin, bacampicillin, carbenicillin, hetacillin, mezlocillin, piperacillin, and ticarcillin. Results may also be applied to phenoxymethyl penicillin or phenethicillin.

[c]For enterococci, aerococci, and nonenterococcal streptococci, the designation "moderately susceptible" implies the need for high-dose penicillin or ampicillin for endocarditis and serious invasive tissue infections that may require (always for enterococci) combined therapy with an aminoglycoside (gentamicin) for improved therapeutic response and bactericidal action. Nonenterococcal streptococci should have an MIC determined in case of endocarditis. Urinary isolates should be considered to be susceptible to ampicillin or penicillin alone.

[d]Ampicillin/sulbactam, aztreonam, cefotetan, ceftazidime, ceftriaxone, and imipenem are among the most recently studied ß-lactams having a separate diagnostic disk and a generally wider spectrum of antimicrobial activity, especially against gram-negative bacilli when compared to previously approved cephalosporins such as cephalothin. **The 30-µg Cefazolin test results may not accurately predict susceptibility to other first-generation cephalosporins. Cephalothin should be tested instead to represent cephalothin, cefaclor (except for *Haemophilus*), cephapirin, cephradine, cephalexin, and cefadroxil.** S. aureus *strains exhibiting resistance to one of the penicillinase-resistant penicillins (MRSA) must be reported as resistant to cephalosporins and other newer ß-lactams such as amoxicillin/ clavulanic acid, ampicillin/sulbactam, imipenem, and ticarcillin/ clavulanic acid, regardless of in vitro test results.* This is primarily because in most cases of documented MRSA infection, the patient has responded poorly to the cephalosporin therapy, or convincing clinical data has yet to be derived confirming clinical efficacy (clavulanic acid or sulbactam combinations and imipenem). Methicillin-resistant, coagulase-negative *Staphylococcus* spp. also appear *not* to respond well to the above cited drugs.

[e]The zone sizes obtained with aminoglycosides, particularly when testing *P. aeruginosa*, are very medium-dependent because of variations in divalent cation content. These interpretive standards are to be used only with Mueller-Hinton medium that has yielded zone diameters within the correct range shown when performance tests were done with *P. aeruginosa* ATCC® 27853. Organisms in the intermediate category may be either susceptible or resistant when tested by dilution methods and should therefore more properly be classified as "indeterminate" in their susceptibility.

[f]Of the antistaphylococcal, ß-lactamase-resistant penicillins, either oxacillin or methicillin could be tested, and results can be applied to the other two of these drugs and to cloxacillin and dicloxacillin. *Oxacillin is preferred* due to more resistance to degradation in storage and its application to pneumococcal testing and because it is more likely to detect heteroresistant staphylococcal strains. Do not use nafcillin on blood-containing media. Cloxacillin disks should not be used because they may not detect methicillin-resistant *S. aureus*. When intermediate results are obtained with staphylococci, the strains should be further investigated to determine if they are heteroresistant.

[g]Data on novobiocin **are not** based on the NCCLS standard.

[h]Tetracycline is the class disk for all tetracyclines, and the results can be applied to chlortetracycline, demeclocycline, doxycycline, methacycline, minocycline, and oxytetracycline. However, certain organisms may be more susceptible to doxycycline and minocycline than to tetracycline.

[i]The sulfisoxazole disk can be used for any of the commercially available sulfonamides. Blood-containing media, except media containing lysed horse blood, are not satisfactory for testing sulfonamides. The Mueller-Hinton agar should be as thymidine-free as possible for sulfonamide and/or trimethoprim testing.

[j]Susceptibility data for cinoxacin, nalidixic acid, nitrofurantoin, norfloxacin, sulfonamides, and trimethoprim apply only to organisms isolated from urinary tract infections.

[k]The category "intermediate" should be reported. It generally indicates that the test result be considered equivocal or indeterminate.

Name   Date

PART A: Kirby-Bauer Antibiotic Sensitivity Test Procedure

1. Examine all plate cultures for the presence or absence of a zone of inhibition surrounding each disc.
2. Using a ruler graduated in millimeters, carefully measure each zone of inhibition to the nearest millimeter and record your results in the chart.
3. Compare your results with Table 44.2 and indicate in the chart the susceptibility of each test organism to the chemotherapeutic agent as resistant (R), intermediate (I), or sensitive (S).

Refer to photo number 61 in the color-plate insert for illustration of this reaction.

Chemotherapeutic Agent	Gram-Negative								Acid-Fast	
	E. coli		P. aeruginosa		P. vulgaris				M. smegmatis	
	Zone Size	Susceptibility	Zone Size	Susceptibility	Zone Size	Susceptibility			Zone Size	Susceptibility
Penicillin										
Streptomycin										
Tetracycline										
Chloramphenicol										
Gentamicin										
Vancomycin										
Sulfanilamide										

Chemotherapeutic Agent	Gram-Positive					
	S. aureus		E. faecalis		B. cereus	
	Zone Size	Susceptibility	Zone Size	Susceptibility	Zone Size	Susceptibility
Penicillin						
Streptomycin						
Tetracycline						
Chloramphenicol						
Gentamicin						
Vancomycin						
Sulfanilamide						

4. For each of the chemotherapeutic agents, indicate:
   a. The spectrum of its activity as broad or limited.
   b. The type or types of organisms it is effective against as gram-positive, gram-negative, or acid-fast.

Chemotherapeutic Agent	Spectrum of Activity	Type(s) of Microorganisms
Penicillin		
Streptomycin		
Tetracycline		
Chloramphenicol		
Gentamicin		
Vancomycin		
Sulfanilamide		

PART B: Synergistic Effect of Drug Combinations

Examine all agar plate cultures to determine the zone of inhibition patterns exhibited. Distinctly separate zones of inhibition are indicative of an additive effect, whereas a merging of the inhibitory zones is indicative of synergism. Record your observations and results in the chart.

 Refer to photo number 62 in the color-plate insert for illustration of these reactions.

Cultures	Appearance of Zone Inhibition	Synergistic or Additive Effect
E. coli:		
trimethoprim and sulfisoxazole	_____	_____
trimethoprim and tetracycline	_____	_____
S. aureus:		
trimethoprim and sulfisoxazole	_____	_____
trimethoprim and tetracycline	_____	_____

REVIEW QUESTION

1. Your experimental results indicate that antibiotics such as tetracycline, streptomycin, and chloramphenicol have a broad spectrum of activity against prokaryotic cells. Why do these antibiotics lack inhibitory activity against eukaryotic cells such as fungi?

Determination of Penicillin Activity in the Presence and Absence of Penicillinase

LEARNING OBJECTIVES

Once you have completed this experiment, you should be able to

1. Employ a broth culture system for the determination of the minimal inhibitory concentration (MIC) of penicillin.
2. Demonstrate the reversal of penicillin inhibition against the test organism in the presence of penicillinase (β-lactamase).

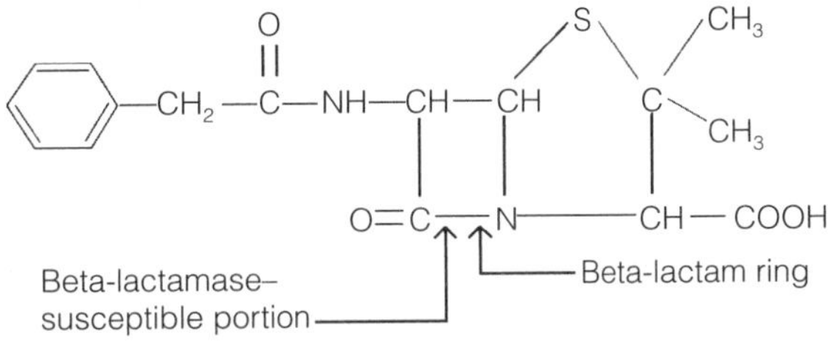

PRINCIPLE

In addition to the Kirby-Bauer paper disc–agar diffusion procedure, the broth tube dilution method may be used to determine the susceptibility of an organism to an antibiotic. The latter procedure, in which dilutions of the antibiotic are prepared in the broth medium, also permits the **minimal inhibitory concentration (MIC)** to be determined for the antibiotic under investigation. The **MIC** is the lowest concentration of an antimicrobial agent that inhibits the growth of the test microorganism. Quantitative data of this nature may be used by a clinician to establish effective antimicrobial regimens for the treatment of a bacterial infection in a host. This data is of particular significance when the toxicity of the antibiotic is known to produce major adverse effects in host tissues.

Penicillin is a potent antibiotic produced by the mold *Penicillium notatum*. Sir Alexander Fleming's discovery of penicillin in 1928 provided the world with the first clinically useful antibiotic in the fight to control human infection. The activity of this antibiotic, as illustrated in Figure 45.1, is associated with the β-lactam ring within its molecular structure. Shortly after the clinical introduction of ben-

FIGURE 45.1 Molecular structure of benzylpenicillin (penicillin G)

zylpenicillin (penicillin G), pathogenic organisms such as *Staphylococcus aureus* were found to be resistant to this "wonder drug." Research revealed that some organisms were genetically capable of producing β-lactamase (penicillinase), an enzyme that breaks a bond in the β-lactam ring portion of the molecule. When the integrity of this ring is compromised, the inhibitory activity of the antibiotic is lost.

In this experiment, the MIC of penicillin will be determined against penicillin-sensitive and penicillinase-producing strains of *Staphylococcus aureus*. The procedure to be followed involves specific concentrations of the penicillin prepared by means of a twofold serial dilution technique in an enriched broth medium. The tubes containing the antibiotic dilutions are then inoculated with a standardized concentration of the test organism. Table 45.1 illustrates the protocol for the antibiotic serial dilution–broth medium setup.

Following incubation, spectrophotometric optical density readings will be used to determine the presence or absence of growth in the cultures. The culture that shows no growth in the presence of the lowest concentration of penicillin represents the minimal inhibitory concentration of this antibiotic against *S. aureus*.

TABLE 45.1 Antibiotic Serial Dilution–Broth Medium Setup

Additions (ml) to:	Tube Number									
	1	2	3	4	5	6	7	8	9	10
Medium	0	2	2	2	2	2	2	2	2	2
Penicillin	2	2	Serial dilution (See protocol)							0
Test culture	2	2	2	2	2	2	2	2	2	2
Total volume	4	4	4	4	4	4	4	4	4*	4
Pencillin (μg/ml)	50	50	25	12.5	6.25	3.12	1.56	0.78	0.39	0
Control	(−)									(+)

*After 2 ml discarded

MATERIALS

Cultures

1:1000 brain–heart infusion (BHI) broth dilutions of 24-hour BHI broth cultures of *Staphylococcus aureus* ATCC™ 27661 (penicillin-sensitive strain) and *Staphylococcus aureus* ATCC 27659 (penicillinase-producing strain).

Media

Per designated student group: 40 ml of brain–heart infusion broth in a 100-ml Erlenmeyer flask and 10 ml of sterile aqueous crystalline penicillin G solution (100 μg/ml).

Equipment

Sterile 13 × 100-mm test tubes, test tube racks, sterile 2-ml and 10-ml pipettes, mechanical pipetting device, Bunsen burner, spectrophotometer, glassware marking pencil, and disinfectant solution in a 500-ml beaker.

PROCEDURE

1. Into each of two test tube racks, place a set of 10 sterile 13 × 100-mm test tubes labeled 1 through 10. Label one rack Set I—penicillin-sensitive and the other rack Set II—penicillin-resistant. *Refer to Table 45.1 for Steps 2 through 7.*

2. Using a sterile 10-ml pipette and mechanical pipetting device, add 2 ml of BHI broth to the tubes labeled 2 through 10 in Sets I and II. *Note: Discard the pipette into the beaker of disinfectant.*

3. With a 2-ml sterile pipette, add 2 ml of the penicillin solution to Tubes 1 and 2 in Sets I and II. Discard the pipette. *Note: Mix the contents of the tubes well.*

4. **Set I Serial Dilution:** Using a sterile 2-ml pipette, transfer 2 ml from Tube 2 to Tube 3. Mix well and transfer 2 ml from Tube 3 to Tube 4. Continue this procedure through Tube 9 into beaker. Discard 2 ml from Tube 9. Tube 10 receives no antibiotic and serves as a positive control. Discard the pipette. *Note: Remember to mix the contents of each tube well between transfers.*

5. **Set II Serial Dilution:** Using a sterile 2-ml pipette, repeat Step 4.

6. Using a sterile 2-ml pipette, add 2 ml of the 1:1000 dilution of the *S. aureus* ATCC 27661 (penicillin-sensitive strain) to all tubes in Set I. Discard the pipette.

7. Repeat Step 6 to inoculate all the tubes in Set II with the 1:1000 dilution of *S. aureus* ATCC 27659 (penicillinase-producing strain). Discard the pipette.

8. Incubate both sets of tubes for 12 to 18 hours at 37°C.

Name   Date

1. Follow the instructions for the use of the spectrophotometer as outlined in Experiment 14 to determine the optical density (O.D.) readings for Tubes 2 through 10 in Sets I and II. Use the Number 1 tubes, the negative controls, as your blanks to adjust the spectrophotometer.

2. Record your O.D. readings in the chart.

Refer to photo number 63 in the color-plate insert for illustration of penicillinase activity.

Optical Density Readings at 600 nm

Tube Number	2	3	4	5	6	7	8	9	10
Penicillin concentration (μg/ml)	50	25	12.5	6.25	3.12	1.56	0.78	0.39	0
Set I									
Set II									

Set I: Minimal inhibitory concentration: _____

Set II: Minimal inhibitory concentration: _____

REVIEW QUESTIONS

1. Was the ability of some microorganisms to produce β-lactamase present prior to their exposure to the antibiotic penicillin? Explain.

2. ⭕ Can the results of an MIC test be used to determine whether an antibiotic is bacteriocidal or bacteriostatic? If not, set up an experimental procedure to determine whether the effect is bacteriocidal or bacteriostatic.

Chemical Agents of Control: Disinfectants and Antiseptics

Antiseptics and disinfectants are chemical substances used to prevent contamination and infection. Many are available commercially for disinfection and asepsis.

Table 46.1 shows the major groups of antimicrobial agents, their modes and ranges of action, and their practical uses.

(text continues on page 296)

TABLE 46.1   Chemical Agents—Disinfectants and Antiseptics

Agent	Mechanism of Action	Use
Phenolic Compounds		
Phenol	1. Germicidal effect caused by alteration of protein structure resulting in protein denaturation. 2. Surface-active agent (surfactant) precipitates cellular proteins and disrupts cell membranes. (Phenol has been replaced by better disinfectants that are less irritating, less toxic to tissues, and better inhibitors of microorganisms.)	1. 5% solution: Disinfection. 2. 0.5% to 1% solutions: Antiseptic effect and relief of itching as it exerts a local anesthetic effect on sensory nerve endings.
Cresols	1. Similar to phenol. 2. Poisonous and must be used externally. 3. 50% solution of cresols in vegetable oil, known as Lysol®.	2% to 5% Lysol solutions used as disinfectants.
Hexachlorophene	Germicidal activity similar to phenol. (This agent is to be used with care, especially on infants, because after absorption it may cause neurotoxic effects.)	1. Reduction of pathogenic organisms on skin; added to detergents, soaps, lotions, and creams. 2. Effective against gram-positive organisms. 3. An antiseptic used topically.
Resorcinol	1. Germicidal activity similar to that of phenol. 2. Acts by precipitating cell protein.	1. Antiseptic. 2. Keratolytic agent for softening or dissolving keratin in epidermis.
Hexylresorcinol	Germicidal activity similar to that of phenol.	1. Treatment of worm infections. 2. Urinary antiseptic.

➡

TABLE 46.1 Chemical Agents—Disinfectants and Antiseptics (continued)

Agent	Mechanism of Action	Use
Thymol	1. Related to the cresols. 2. More effective than phenol.	1. Antifungal activity. 2. Treatment of hookworm infections. 3. Mouthwashes and gargle solutions.
Alcohols Ethyl: CH_3CH_2OH Isopropyl: $(CH_3)_2CHOH$	1. Lipid solvent. 2. Denaturation and coagulation of proteins. 3. Wetting agent used in tinctures to increase the wetting ability of other chemicals. 4. Germicidal activity increases with increasing molecular weight.	Skin antiseptics: Ethyl—50% to 70%. Isopropyl—75%.
Halogens Chlorine compounds: Sodium hypochlorite (Dakin's fluid): NaOCl Chloramine: $CH_3C_6H_4SO_2NNaCl$	1. Germicidal effect resulting from rapid combination with proteins. 2. Chlorine reacts with water to form hypochlorous acid, which is bactericidal. 3. Oxidizing agent. 4. Noncompetitively inhibits enzymes, especially those dealing with glucose metabolism, by reacting with SH and NH_2 groups on the enzyme molecule.	1. Water purification. 2. Sanitation of utensils in dairy and restaurant industries. 3. Chloramine, 0.1% to 2% solutions, for wound irrigation and dressings. 4. Microbicidal.
Iodine compounds: Tincture of iodine Povidone-iodine solution (Betadine®)	1. Mechanism of action is not entirely known, but it is believed that it precipitates proteins. 2. Surface-active agent.	1. Tinctures of iodine are used for skin antisepsis. 2. Treatment of goiter. 3. Effective against spores, fungi, and viruses.
Heavy Metals Mercury compounds: Inorganic: Mercury bichloride Mercurial ointments	1. Mercuric ion brings about precipitation of cellular proteins. 2. Noncompetitive inhibition of specific enzymes caused by reaction with sulfhydryl group (SH) on enzymes of bacterial cells.	1. Inorganic mercurials are irritating to tissues, toxic systemically, adversely affected by organic matter, and incapable of acting on spores. 2. Mercury compounds are mainly used as disinfectants of laboratory materials.
Organic mercurials: Mercurochrome (merbromin) Merthiolate (thimerosal) Metaphen (nitromersol) Merbak (acetomeroctol)	1. Similar to those of inorganic mercurials, but in proper concentrations are useful antiseptics. 2. Much less irritating than inorganic mercurials.	1. Less toxic, less irritating; used mainly for skin asepsis. 2. Do not kill spores.
Silver compounds: Silver nitrate	1. Precipitate cellular proteins. 2. Interfere with metabolic activities of microbial cells. 3. Inorganic salts are germicidal.	Asepsis of mucous membrane of throat and eyes.

TABLE 46.1 (continued)

Agent	Mechanism of Action	Use
Surface-Active Agents Wetting agents: Emulsifiers, soaps, and detergents	1. Lower surface tension and aid in mechanical removal of bacteria and soil. 2. If active portion of the agent carries a negative electric charge, it is called an anionic surface-active agent. If active portion of the agent carries a positive electric charge, it is called a cationic surface-active agent. 3. Exert bactericidal activity by interfering with or by depressing metabolic activities of microorganisms. 4. Disrupt cell membranes. 5. Alter cell permeability.	Weak action against fungi, acid-fast microorganisms, spores, and viruses.

Cationic agents:
  Quaternary ammonium compounds
  Benzalkonium chloride
	1. Lower surface tension because of keratolytic, detergent, and emulsifying properties.
	
	2. Their germicidal activities are reduced by soaps.
		1. Bactericidal, fungicidal; inactive against spores and viruses.
		
		2. Asepsis of intact skin.
		
		3. Disinfectant for operating-room equipment.
		
		4. Dairy and restaurant sanitization.

Anionic agents:
  Tincture of green soap
  Sodium tetradecyl sulfate
	1. Neutral or alkaline salts of high-molecular-weight acids. Common soaps included in this group.
	
	2. Exert their maximum activity in an acid medium and are most effective against gram-positive cells.
	
	3. Same as all surface-active agents.
		1. Cleansing agent.
		
		2. Sclerosing agent in treatment of varicose veins and internal hemorrhoids.

Acids (H^+) **Alkali** (OH^-)	1. Destruction of cell wall and cell membrane. 2. Coagulation of proteins.	Disinfection; however, of little practical value.
Formaldehyde (liquid or gas)	Alkylating agent causes reduction of enzymes.	1. Room disinfection. 2. Alcoholic solution for instrument disinfection. 3. Specimen preservation.
Ethylene Oxide	Alkylating agent causes reduction of enzymes.	Sterilization of heat-labile material.
β-Propiolactone (liquid or gas)	Alkylating agent causes reduction of enzymes.	1. Sterilization of tissue for grafting. 2. Destruction of hepatitis virus. 3. Room disinfection.
Basic Dyes Crystal violet	Affinity for nucleic acids; interfere with reproduction in gram-positive organisms.	1. Skin antiseptic. 2. Laboratory isolation of gram-negative bacteria.

The efficiency of all disinfectants and antiseptics is influenced by a variety of factors, including the following:

1. **Concentration:** The concentration of a chemical substance markedly influences its effect on microorganisms, with higher concentrations producing a more rapid death. Concentration cannot be arbitrarily determined; the toxicity of the chemical to the tissues being treated and the damaging effect on nonliving materials must also be considered.

2. **Length of exposure:** All microbes are not destroyed within the same exposure time. Sensitive forms are destroyed more rapidly than resistant ones. The longer the exposure to the agent, the greater its antimicrobial activity. The toxicity of the chemical and environmental conditions must be considered in assessing the length of time necessary for disinfection or asepsis.

3. **Type of microbial population to be destroyed:** Microorganisms vary in their susceptibility to destruction by chemicals. Bacterial spores are the most resistant forms. Capsulated bacteria are more resistant than noncapsulated forms; acid-fast bacteria are more resistant than non–acid-fast; and older, metabolically less active cells are more resistant than younger cells. Awareness of the types of microorganisms that may be present will influence the choice of agent.

4. **Environmental conditions:** Conditions under which a disinfectant or antiseptic affects the chemical agent are as follows:

    a. **Temperature:** Cells are killed as the result of a chemical reaction between the agent and cellular component. As increasing temperatures increase the rate of chemical reactions, application of heat during disinfection markedly increases the rate at which the microbial population is destroyed.

    b. **pH:** The pH conditions during disinfection may affect not only the microorganisms but also the compound. Extremes in pH are harmful to many microorganisms and may enhance the antimicrobial action of a chemical. Deviation from a neutral pH may cause ionization of the disinfectant; depending on the chemical agent, this may serve to increase or decrease the chemical's microbicidal action.

    c. **Type of material on which the microorganisms exist:** The destructive power of the compound on cells is due to its combination with organic cellular molecules. If the material on which the microorganisms are found is primarily organic, such as blood, pus, or tissue fluids, the agent will combine with these extracellular organic molecules, and its antimicrobial activity will be reduced.

Numerous laboratory procedures are available for evaluating the antimicrobial efficiency of disinfectants or antiseptics. They provide a general rather than an absolute measure of the effectiveness of any agent because test conditions frequently differ considerably from those seen during practical use. Two commonly employed procedures are presented.

PART A: Phenol Coefficient

LEARNING OBJECTIVE

Once you have completed this experiment, you should be able to

1. Compare the effectiveness of disinfectants.

PRINCIPLE

The **phenol coefficient test** compares the antimicrobial activity of a chemical compound to that of phenol under standardized experimental conditions. Equal quantities of a series of dilutions of the chemical being tested and of pure phenol are placed into sterile test tubes. A standardized quantity of a pure culture of the test microorganisms, such as *Staphylococcus aureus* or *Salmonella typhi,* is added to each of the tubes. Subcultures of the test microorganism are made from each dilution of the test chemicals into sterile broth media at intervals of 5, 10, and 15 minutes after introduction of the organisms. All the subcultures are incubated at 37°C for 48 hours and examined for the presence or absence of growth.

The phenol coefficient is determined by dividing the highest dilution of the chemical being tested that destroyed the microorganisms in 10 minutes but not in 5 minutes by the highest dilution of phenol that destroyed the microorganism in 10 minutes but not in 5 minutes. A phenol coefficient no greater than 1 indicates that this agent is as effective as or less effective than phenol. A phenol coefficient

TABLE 46.2 Illustration of Phenol Coefficient Determination

Chemical Agent and Dilution		Presence of Growth in Subcultures (minutes)		
		5	10	15
Phenol	1:80	−	−	−
	1:90*	+	−	−
	1:100	+	+	−
Test chemical	1:400	−	−	−
	1:450†	+	−	−
	1:500	+	+	−

*Phenol dilution of 1:90 showed growth at 5 minutes but no growth at 10 minutes.
†Test chemical dilution of 1:450 showed growth at 5 minutes but no growth at 10 minutes.
Phenol coefficient of test chemical = 450/90 = 5.

greater than 1 suggests that the chemical is more effective than phenol when employed under test conditions. A phenol coefficient of 5 indicates that the chemical agent under evaluation is 5 times as effective as phenol. Table 46.2 illustrates a phenol coefficient determination.

MATERIALS

Cultures

24-hour nutrient broth cultures of *Staphylococcus aureus* dispensed in sterile dropper bottles.

Media

Per designated student group: 19 nutrient broth tubes.

Disinfectants

Per designated student group: phenol dilutions: 1:80, 1:90, 1:100; Lysol dilutions: 1:400, 1:450, 1:500.

Equipment

Bunsen burner, inoculating loop, test tube rack, and glassware marking pencil.

PROCEDURE

1. Label 18 nutrient broth tubes with the name and dilution of the disinfectant and the time interval of subculturing (e.g., phenol 1:80, 5 minutes).

2. In a test tube rack, place one test tube of each of the different phenol and Lysol dilutions.

3. Rapidly introduce one drop of *S. aureus* culture into each of the test tubes of disinfectant. Note the time when you start introducing the microorganisms into the disinfectants.

4. Agitate all the test tubes to ensure contact between the disinfectant and the microbes.

5. Using sterile technique, at intervals of 5, 10, and 15 minutes transfer one loopful from each of the test tubes containing the disinfectant and microorganisms into the appropriately labeled sterile tube of nutrient broth.

6. Incubate all nutrient broth cultures for 48 hours at 37°C.

PART B: Agar Plate–Sensitivity Method

LEARNING OBJECTIVE

Once you have completed this experiment, you should be able to

1. Evaluate the effectiveness of antiseptic agents against selected test organisms.

PRINCIPLE

This procedure requires the heavy inoculation of an agar plate with the test organism. Sterile, color-coded filter-paper discs are impregnated with a different antiseptic and equally spaced on the inoculated agar plate. Following incubation, the agar plate is examined for zones of inhibition (areas of no microbial growth) surrounding the discs. A zone of inhibition is indicative of microbicidal activity against the organism. Absence of a zone of inhibition indicates that the chemical was ineffective against the test organism. *Note: The size of the zone of inhibition is not indicative of the degree of effectiveness of the chemical agent.*

MATERIALS

Cultures

24- to 48-hour trypticase soy broth cultures of *Escherichia coli, Bacillus cereus, Staphylococcus aureus,* and *Mycobacterium smegmatis,* and a 7-day-old trypticase soy broth culture of *Bacillus cereus.*

Media

Per designated student group: five trypticase soy agar plates.

Antiseptics/Disinfectants

Per designated student group: 10 ml of each of the following dispensed in 25-ml beakers: tincture of iodine, 3% hydrogen peroxide, 70% isopropyl alcohol, and 5% chlorine bleach.

Equipment

Four different-colored, sterile Sensi-discs™; forceps; sterile cotton swabs; Bunsen burner; and glassware marking pencil.

PROCEDURE

1. Aseptically inoculate the appropriately labeled agar plates with their respective test organisms by streaking each plate in horizontal and vertical directions and around the edge with a sterile swab.

2. Color-code the Sensi-discs according to the chemical agents to be used (e.g., red = chlorine bleach).

3. Using forceps dipped in alcohol and flamed, expose five discs of the same color by placing them into the solution of one of the chemical agents. Drain the saturated discs on absorbent paper immediately prior to placing one on each of the inoculated agar plates. Place each disc approximately 2 cm in from the edge of the plate. Gently press the discs down with the forceps so that they adhere to the surface of the agar.

4. Impregnate the remaining discs as described in Step 3. Place one of each of the three remaining colored discs on the surface of each of the five inoculated agar plates equidistant from each other around the periphery of the plate.

5. Incubate all plate cultures in an inverted position for 24 to 48 hours at 37°C.

Name _____ Date _____

PART A: Phenol Coefficient

1. Observe all nutrient broth cultures for the presence of growth. Use the sterile test tube of nutrient broth as a control in determining the presence or absence of growth.
2. Record your observations in the chart as growth (+) or no growth (−).

Disinfectant	Dilution	Growth in Subcultures (minutes)		
		5	10	15
Phenol	1:80			
	1:90			
	1:100			
Lysol	1:400			
	1:450			
	1:500			

3. Calculate the phenol coefficient of Lysol.

PART B: Agar Plate–Sensitivity Method

1. Observe all the plates for the presence of a zone of inhibition surrounding each of the impregnated discs.
2. Record your observations in the chart on the following page as absence of a zone of inhibition (0), or presence of a zone of inhibition (+).

📷 *Refer to photo number 64 in the color-plate insert for illustration of the reaction.*

Bacterial Species	Antimicrobial Agent			
	Tincture of Iodine	3% Hydrogen Peroxide	70% Isopropyl Alcohol	5% Chlorine Bleach
E. coli gram-negative				
S. aureus gram-positive				
M. smegmatis acid-fast				
B. cereus spore-former gram-positive				
B. cereus spore-former gram-positive 7-day-old				

3. Indicate which of the antiseptics exhibited microbicidal activity against each of the following groups of microorganisms.

Bacterial Group	Tincture of Iodine	3% Hydrogen Peroxide	70% Isoprophl Alcohol	5% Chlorine Bleach
Gram-negative				
Gram-positive				
Acid-fast				
Spore-former				

4. Which of the experimental chemical compounds appears to have the broadest range of microbicidal activity? The narrowest range of microbicidal activity?

REVIEW QUESTIONS

1. Evaluate the effectiveness of a disinfectant with a phenol coefficient of 40.

2. Can the disinfection period (exposure time) be arbitrarily increased? Explain.

3. A household cleanser is labeled germicidal. Explain what this means to you.

Microbiology of Food

LEARNING OBJECTIVES

Once you have completed the experiments in this section, you should be familiar with

1. The endogenous and exogenous organisms that may be found in food products.

2. The analysis of food products as a means of determining their quality from the public health point of view.

3. The microbiological production of wine and sauerkraut.

INTRODUCTION

Microbiologists have always been aware that foods, especially milk, have served as important inanimate vectors in the transmission of disease. Foods contain the organic nutrients that provide an excellent medium to support the growth and multiplication of microorganisms under suitable temperatures.

Food and dairy products may be contaminated in a variety of ways and from a variety of sources:

1. **Soil and water:** Food-borne organisms that may be found in soil and water and that may contaminate food are members of the genera *Alcaligenes, Bacillus, Citrobacter, Clostridium, Pseudomonas, Serratia, Proteus, Enterobacter,* and *Micrococcus.* The common soil and water molds include *Rhizopus, Penicillium, Botrytis, Fusarium,* and *Trichothecium.*

2. **Food utensils:** The type of microorganisms found on utensils depends on the type of food and the manner in which the utensils were handled.

3. **Enteric microorganisms of humans and animals:** The major members of this group are *Bacteroides, Lactobacillus,*

Clostridium, Escherichia, Salmonella, Proteus, Shigella, Staphylococcus, and *Streptococcus.* These organisms find their way into the soil and water, from which they contaminate plants and are carried by wind currents onto utensils or prepared and exposed foods.

4. **Food handlers:** People who handle foods are especially likely to contaminate them because microorganisms on hands and clothing are easily transmitted. A major offending organism is *Staphylococcus,* which is generally found on hands and skin, and in the upper respiratory tract. Food handlers with poor personal hygiene and unsanitary habits are most likely to contaminate foods with enteric organisms.

5. **Animal hides and feeds:** Microorganisms found in water, soil, feed, dust, and fecal debris can be found on animal hides. Infected hides may serve as a source of infection for workers, or the microorganisms may migrate into the musculature of the animal and remain viable following its slaughter.

By enumerating microorganisms in milk and foods, the quality of a particular sample can be determined. Although the microorganisms cannot be identified, the presence of a high number suggests a good possibility that pathogens may be present. Even if a sample contains a low microbial count, it can still transmit infection.

In the laboratory procedures that follow, you will have an opportunity directly and indirectly to enumerate the number of microorganisms present in milk and other food products and to thereby determine the quality of the samples.

Methylene Blue Reductase Test

LEARNING OBJECTIVE

Once you have completed this experiment, you should be familiar with

1. An enzymatic test to determine the quality of a milk sample.

PRINCIPLE

A milk sample that contains a large population of actively metabolizing microorganisms will contain a markedly decreased concentration of dissolved oxygen because of the vigorous growth of the organisms. In other words, the oxidation–reduction potential of the sample is greatly lowered. The dye methylene blue (MB), a redox indicator, loses its color in an anaerobic environment and is then said to be reduced.

The methylene blue reductase test is designed to screen the quality of raw milk, which may contain large populations of enteric organisms and *Lactococcus lactis*, which are potent reducers of the dye. The speed at which reduction occurs following addition of MB to a sample of milk indicates the milk's quality. This determination is made as follows:

1. Reduction within 30 minutes is indicative of very poor quality.

2. Reduction occurring between 30 minutes and 2 hours is indicative of poor quality.

3. Reduction occurring between 2 and 6 hours is indicative of fair quality.

4. Reduction occurring between 6 and 8 hours is indicative of good quality.

MATERIALS

Cultures

One raw milk sample and one pasteurized milk sample stored at room temperature for 48 hours.

Reagent

Methylene blue solution (1:250,000).

Equipment

Sterile screw-cap test tubes, test tube rack, sterile 10-ml and 1-ml pipettes, mechanical pipetting device, 37°C waterbath, Bunsen burner, and glassware marking pencil.

PROCEDURE

1. Label the test tubes as raw milk and pasteurized milk.

2. Using a different 10-ml pipette each time, transfer 10 ml of each type of milk into its appropriately labeled test tube.

3. Add 1 ml of methylene blue dye to each test tube.

4. Insert stoppers, invert the test tubes gently about four times, and place in waterbath. Record the time of incubation, that is, record the amount of time elapsed for the color to turn white.

5. Allow the tubes to stabilize for 5 minutes, remove them from the waterbath, invert them gently once, and replace them in the waterbath.

Name _____ Date _____

1. Observe the milk samples for methylene blue reduction every 30 minutes for 3 to 6 hours. Reduction is demonstrated by a change in the color of the sample to white.
2. Record in the chart the time required for reduction in both milk samples.
3. Based on your observations, determine and record in the chart the quality of each sample as very poor, poor, fair, or good.

	Raw Milk	Pasteurized Milk
Reduction time		
Quality of milk sample		

REVIEW QUESTIONS

1. Explain the function of the methylene blue dye in this experiment.

2. Explain why milk sours in the absence of refrigeration.

3. Milk also sours, although at a slower rate, when refrigerated. Why?

4. If a milk sample is judged to be of good quality, can it still serve as a source of human infection? Explain.

5. If milk is a sterile body fluid, explain how it may become contaminated during the milking process.

Microbiological Analysis of Food Products: Bacterial Count

LEARNING OBJECTIVES

Once you have completed this experiment, you should be able to determine

1. The total number of microorganisms present in food products.
2. The number of coliform bacteria in the selected food products.

PRINCIPLE

The presence of microorganisms in food may be considered harmful in some cases, while in others it is definitely beneficial. Certain microorganisms are necessary in preparation of foods such as cheese, pickles, sauerkraut, yogurt, and sausage. The presence of other microorganisms, however, is responsible for serious and sometimes fatal food poisoning and toxicity as well as spoilage. As with milk or water, the presence and number of coliform bacteria and other enteric organisms in food is indicative of fecal contamination and may suggest the presence of pathogens.

Figure 48.1 illustrates the procedure for enumerating microorganisms in foods and determining the presence of coliform bacteria.

MATERIALS
Cultures

Samples of thawed frozen vegetables, ground beef, and dried fruit.

Media

Per designated student group: nine brain–heart infusion agar deep tubes, three eosin–methylene blue (EMB) agar plates, three 99-ml sterile water blanks, and three 180-ml sterile water blanks.

Equipment

Bunsen burner, waterbath, Quebec® or electronic colony counter, balance, sterile glassine weighing paper, blender with three sterile jars, sterile Petri dishes, 1-ml pipettes, mechanical pipetting device, inoculation loop, and glassware marking pencil.

PROCEDURE

1. Label three sets of three Petri dishes for each of the food samples to be tested and their dilutions (10^{-2}, 10^{-3}, 10^{-4}). Label the three EMB agar plates with the names of the food.

2. Melt the brain–heart infusion agar deep tubes in a waterbath, cool, and maintain at 45°C.

3. Place 20 gm of each food sample, weighed on sterile glassine paper, into its labeled blender jar. Add 180 ml of sterile water to each of the blender jars and blend each mixture for 5 minutes. You will have made a 1:10 (10^{-1}) dilution of each food sample.

4. Transfer 1 ml of the 10^{-1} ground beef suspension into its labeled 99-ml sterile water blank, thereby effecting a 10^{-3} dilution, and 0.1 ml to the appropriately labeled 10^{-2} Petri dish. Shake the 10^{-3} sample dilution, and using a different pipette, transfer 1 ml to the plate labeled 10^{-3} and 0.1 ml to the plate labeled 10^{-4}. Add a 15-ml aliquot of the molten and cooled agar to each of the three plates. Swirl the plates gently to obtain a uniform distribution and allow the plates to solidify.

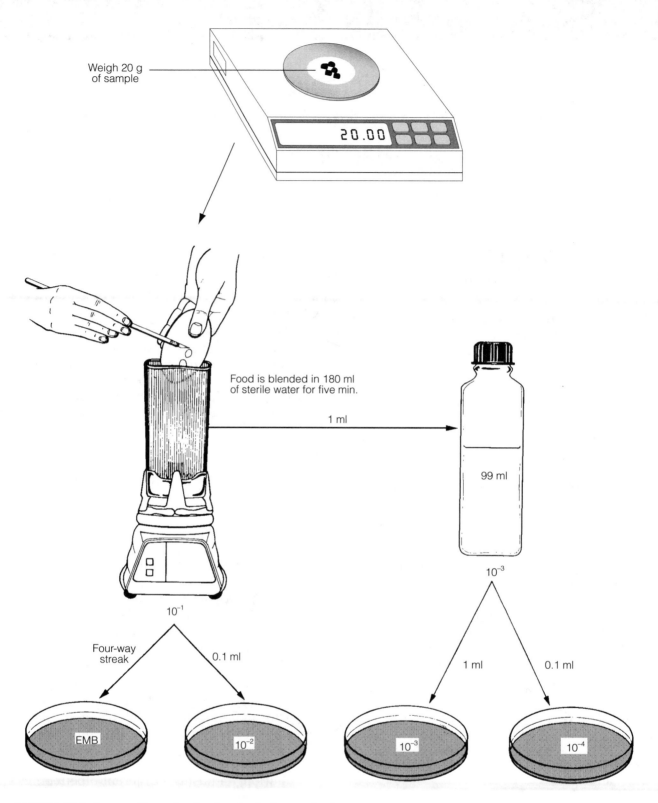

Weigh 20 g of sample

20.00

Food is blended in 180 ml of sterile water for five min.

1 ml

99 ml

10^{-3}

10^{-1}

Four-way streak

0.1 ml

1 ml

0.1 ml

EMB

10^{-2}

10^{-3}

10^{-4}

FIGURE 48.1 Preparation of a food sample for analysis

5. Repeat Step 4 for the remaining two 10^{-1} test food sample dilutions.

6. Aseptically prepare a four-way streak plate, as described in Experiment 2, and inocu-late each 10^{-1} food sample dilution on its appropriately labeled EMB agar plate.

7. Incubate all plates in an inverted position for 24 to 48 hours at 37°C.

Name _____   Date _____

1. Using either the Quebec or electronic colony counter, count the number of colonies on each plate. Count only statistically valid plates that contain between 30 and 300 colonies. Designate plates with fewer than 30 colonies as **too few to count (TFTC)** and plates with more than 300 colonies as **too numerous to count (TNTC).**

2. Determine the number of organisms per milliliter of each food sample on plates not designated as TFTC or TNTC by multiplying the number of colonies counted by the dilution factor.

3. Record in the chart the number of colonies per plate and the number of organisms per milliliter of each food sample.

Type of Food	Dilution	Number of Colonies per Plate	Number of Organisms per ml
Ground beef	10^{-2}		
	10^{-3}		
	10^{-4}		
Frozen vegetables	10^{-2}		
	10^{-3}		
	10^{-4}		
Dried fruits	10^{-2}		
	10^{-3}		
	10^{-4}		

4. Examine the eosin–methylene blue agar plate cultures for colonies with a metallic green sheen on their surfaces, which is indicative of *E. coli*. Indicate in the chart the presence or absence of *E. coli* growth and the possibility of fecal contamination of the food.

Sample	*E. coli* (+) or (−)	Fecal Contamination (+) or (−)
Ground beef		
Frozen vegetables		
Dried fruit		

REVIEW QUESTIONS

1. Indicate some possible ways in which foods may become contaminated with enteric organisms.

2. Explain why it is not advisable to thaw and then refreeze food products without having cooked them.

3. Would detection of *E. coli* in meat be indicative of contamination or spoilage of the product? Explain.

4. Following a Fourth of July picnic lunch of ham, sour pickles, potato salad, and cream puffs, a group of students was admitted to the hospital with severe gastrointestinal distress. A diagnosis of staphylococcal food poisoning was made. Explain how the staphylococci can multiply in these foods and produce severe abdominal distress.

Wine Production

Once you have completed this experiment, you should understand

1. Wine production by the fermentative activities of yeast cells.

PRINCIPLE

Wine is a product of the natural fermentation of the juices of grapes and other fruits, such as peaches, pears, plums, and apples, by the action of yeast cells. This biochemical conversion of juice to wine occurs when the yeast cells enzymatically degrade the fruit sugars fructose and glucose first to acetaldehyde and then to alcohol, as illustrated in Figure 49.1.

Grapes containing 20% to 30% sugar concentration will yield wines with an alcohol content of approximately 10% to 15%. Also present in grapes are acids and minerals whose concentrations are increased in the finished product and that are responsible for the characteristic tastes and bouquets of different wines. For red wine, the crushed grapes must be fermented with their skins to allow extraction of their color into the juice. White wine is produced from the juice of white grapes.

The commercial production of wine is a long and exacting process. First, the grapes are crushed or pressed to express the juice, which is called **must.** Potassium metabisulfite is added to the must to retard the growth of acetic acid bacteria, molds, and wild yeast that are endogenous to grapes in the vineyard. A wine-producing strain of yeast, *Saccharomyces cerevisiae* var. *ellipsoideus,* is used to inoculate the must, which is then incubated for 3 to 5 days under aerobic conditions at 21°C to 32°C. This is followed by an anaerobic incubation period. The wine is then aged for a period of 1 to 5 years in aging tanks or wooden barrels. During this time, the wine is clarified of any turbidity, thereby producing volatile esters that are responsible for characteristic flavors. The clarified product is then filtered, pasteurized at 60°C for 30 minutes, and bottled.

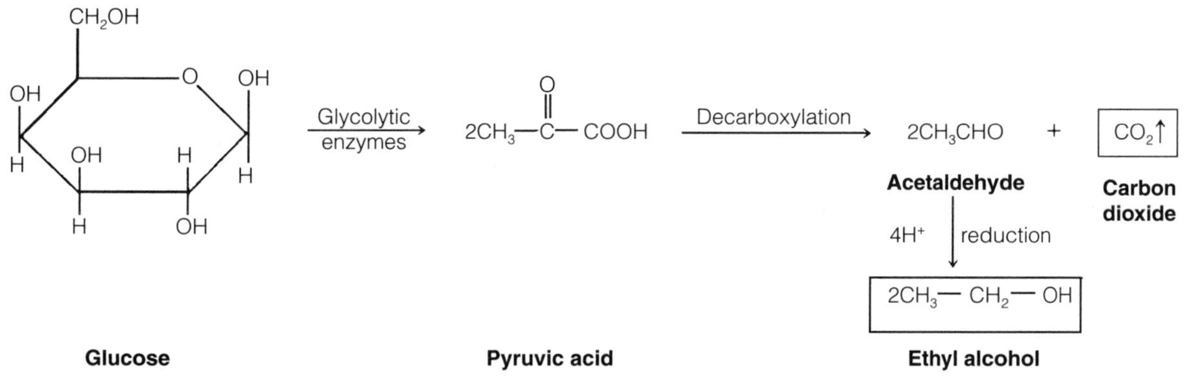

FIGURE 49.1 Biochemical pathway for alcohol production

This experiment is a modified method in which white wine is produced from white grape juice. You will examine the fermenting wine at 1-week intervals during the incubation period for:

1. Total acidity (expressed as % tartaric acid): To a 10-ml aliquot of the fermenting wine, add 10 ml of distilled water and 5 drops of 1% phenolphthalein solution. Mix and titrate to the first persistent pink color with 0.1N sodium hydroxide. Calculate total acidity using the following formula:

$$\% \text{ tartaric acid} = \frac{\text{ml alkali} \times \text{normality of alkali} \times 7.5}{\text{weight of sample in gm*}}$$

*1 ml = 1 gm

2. Volatile acidity (expressed as % acetic acid): Following titration, calculate volatile acidity using the following formula:

$$\% \text{ acetic acid} = \frac{\text{ml alkali} \times \text{normality of alkali} \times 6.0}{\text{weight of sample in gm*}}$$

*1 ml = 1 gm

3. Alcohol (expressed as volume %): Optional; can be determined by means of an ebulliometer.

4. Aroma: Fruity, yeastlike, sweet, none.

5. Clarity: Clear, turbid.

MATERIALS
Cultures

50 ml of white grape juice broth culture of *Saccharomyces cerevisiae* var. *ellipsoideus* incubated for 48 hours at 25°C.

Media

Per designated student group: 500 ml of pasteurized Welch's® commercial white grape juice.

Reagents

1% phenolphthalein solution, 0.1N sodium hydroxide, and sucrose.

Equipment

1-liter Erlenmeyer flask, one-holed rubber stopper containing a 2-inch glass tube plugged with cotton, pan balance, spatula, glassine paper, 10-ml graduated cylinder, ebulliometer (optional), and burette or pipette for titration.

PROCEDURE

1. Pour 500 ml of the white grape juice into the 1-liter Erlenmeyer flask. Add 20 gm of sucrose and the 50 ml of *S. cerevisiae* grape juice broth culture (10% starter culture). Close the flask with the stopper containing a cotton plugged air vent.

2. After 2 days and 4 days of incubation, add 20 gm of sucrose to the fermenting wine.

3. Incubate the fermenting wine for 21 days at 25°C.

ame   Date

. Using uninoculated white grape juice:

   a. Perform a titration to determine total acidity and volatile acidity.

   b. Note aroma and clarity.

   c. Determine volume % alcohol (optional).

. Record your results in the chart.

. At 7-day intervals, using samples of the fermenting wine, repeat Steps 1a though 1c and record your results in the chart.

	Grape Juide	Fermenting Wine		
		7 days	14 days	21 days
% Tartaric acid				
% Acetic acid				
Volume % alcohol				
Aroma				
Clarity				

REVIEW QUESTIONS

. What is the purpose of adding sulfite to the must?

. Explain what occurs during the aging process in the commercial preparation of wine.

3. What are the chemical end products of fermentation?

4. How are white and red wines produced?

5. Why is wine pasteurized? Would it be preferable to sterilize the wine? Explain.

Sauerkraut Production

LEARNING OBJECTIVE

Once you have completed this experiment, you should understand

1. The microbiological production of sauerkraut.

PRINCIPLE

Sauerkraut is a classic example of a food of plant origin produced by microbial fermentation. Its preparation requires the fermentative activities of a mixed microbial flora, including *Leuconostoc mesenteroides, Lactobacillus plantarum, Lactobacillus brevis,* and *Enterococcus faecalis.*

In the production of sauerkraut, shredded cabbage is treated with sodium chloride, which creates an osmotic environment in which plasmolysis occurs, thereby extracting the juice from the cabbage tissue. The resultant brine solution favors the growth of lactic acid–producing microorganisms and inhibits the growth of other microorganisms. The lactic acid is responsible for the characteristic kraut flavor and also acts as a preservative by inhibiting the growth of microorganisms that cause food spoilage.

Production of the lactic acid is initiated by *L. mesenteroides,* which are cocci, and sustained by *L. plantarum,* which are bacilli. When the acid concentration reaches a level of 0.7% to 1%, the fermentative activities of *L. mesenteroides* cease and the final stages of the process are carried out by *L. plantarum, L. brevis,* and *E. faecalis.* The finished product contains a total acidity of 1.5% to 2%, of which lactic acid represents 1% to 1.5%.

In this experiment, you will prepare two samples of sauerkraut, one for sampling and testing of the final product and the other for testing at specific intervals during incubation for:

1. Odor: Acid, earthy, spicy, or putrid.
2. Color: Brown, pink, straw yellow, pale yellow, or colorless.
3. Texture:
   a. Soft: Fermentation initiated by *L. plantarum* rather than *L. mesenteroides.*
   b. Slimy: Rapid growth of *Lactobacillus cucumeris* at elevated temperatures.
   c. Rotted: Spoilage by bacteria, yeast, or molds.
4. pH: The pH of the finished product should be in the range of 3.1 to 3.7.
5. Total acidity expressed as % lactic acid:
   a. Place 10 ml of the fermentation juice and 10 ml distilled water into an Erlenmeyer flask. Boil to drive off the CO_2.
   b. Cool and add 5 drops of 1% phenolphthalein to the diluted juice.
   c. Titrate to the first persistent sample with pink color with 0.1N NaOH.
   d. Calculate % lactic acid as follows:

   $$\% \text{ lactic acid} = \frac{\text{ml of alkali} \times \text{normality of alkali} \times 9}{\text{weight of sample in gm*}}$$

   *1 ml = 1 gm

6. Microscopic appearance of the microbial flora.

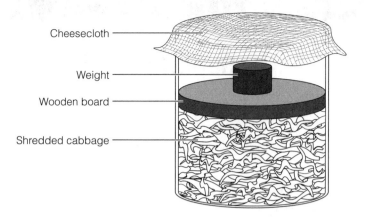

Cheesecloth

Weight

Wooden board

Shredded cabbage

FIGURE 50.1 Setup for sauerkraut preparation

MATERIALS

Media

Per designated student group: two heads of cabbage.

Reagents

1% phenolphthalein, 0.1N NaOH, methylene blue, and uniodized table salt.

Equipment

Two wide-mouthed jars with covers, two wooden boards to fit into jars, two heavy weights, cheesecloth, pH meter or indicator paper, pan balance, microscope, Bunsen burner, inoculating loop, glass slides, coverslips, 10-ml disposable pipettes, mechanical pipetting device, knife, and Erlenmeyer flask.

PROCEDURE

1. Remove the outer leaves and all bruised tissues from each of the cabbage heads.
2. Halve, core, and wash the heads in tap water.
3. Shred the cabbage.
4. Weigh the shredded cabbage on a pan balance and separate into two equal portions.
5. Weigh out the table salt in amounts equal to 3% of the weight of each of the portions of shredded cabbage.
6. Place the shredded cabbage and salt in alternating layers in the two wide-mouthed jars.
7. Place a wooden board over each of the mixtures and press gently to squeeze out a layer of juice from the cabbage.
8. Place a weight on each of the boards and cover the jars with cheesecloth (Figure 50.1).
9. Incubate the jars for 14 days at 30°C.

Name   Date

Examine the sauerkraut preparation on Days 2, 7, 14, and 21 of incubation as follows:

1. Examine the fermenting cabbage for aroma, texture, and color. Record your results in the chart.
2. With a 10-ml pipette, remove 10 ml of the fermentation juice.
   a. Using methylene blue, prepare a stained slide preparation for microscopic examination.
   b. Using the pH meter or indicator paper, determine the pH of the juice.
   c. Perform a titration of the juice to determine the % lactic acid present.
3. Record your results in the chart.

Result	Sauerkraut Preparation			
	2 days	7 days	14 days	21 days
Odor				
Color				
Texture				
% lactic acid				
pH				
Draw microbial flora if present.				

4. Based on your observations, was there any indication of microbial spoilage of your sauerkraut? Explain.

REVIEW QUESTIONS

1. Discuss the importance of the specific sequential activity of the microbial flora responsible for sauerkraut production.

2. What is the function of the salt used in preparing sauerkraut?

3. Why is uniodized salt used in this procedure?

4. Explain the production of slimy or rotten kraut.

5. How does the process of fermentation aid in food preservation?

Bacterial Structure and Staining Techniques

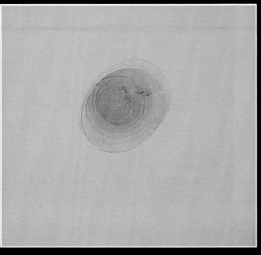

1 A bacterial smear following fixation and staining. Experiment 8.

Diplobacilli

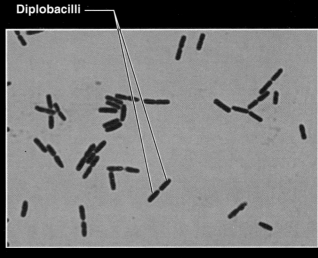

2 Rod-shaped bacilli and diplobacilli. (4000×) Experiment 9.

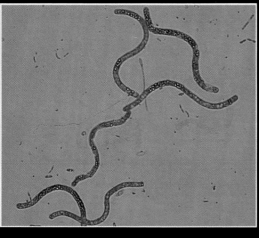

3 Spirilli (spiral-shaped) bacteria. (2500×) Experiment 9.

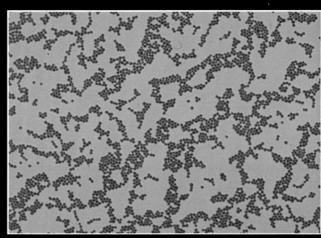

4 Cocci (spherical-shaped) bacteria: *Staphylococcus*. (3600×) Experiment 9.

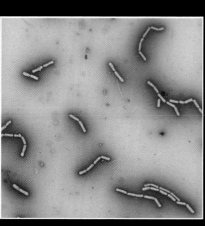

5 Negative staining: Bacilli. (3600×) Experiment 10.

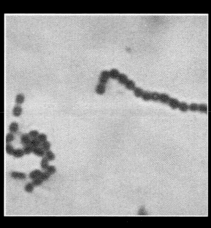

6 Gram stain: Gram-positive streptococci. (enlarged view) Experiments 11 and 65.

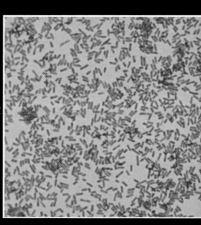

7 Gram stain: Gram-negative *E. coli*. (3600×) Experiment 11.

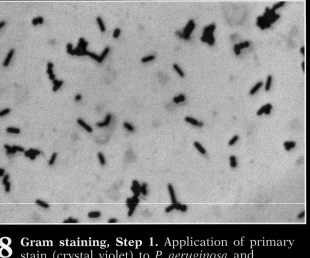

8 **Gram staining, Step 1.** Application of primary stain (crystal violet) to *P. aeruginosa* and *S. aureus*. Experiment 11.

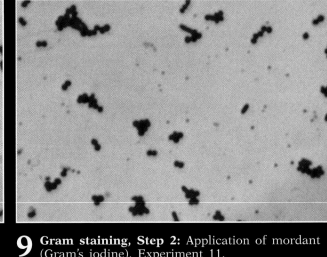

9 **Gram staining, Step 2:** Application of mordant (Gram's iodine). Experiment 11.

10 **Gram staining, Step 3:** Application of decolorizing agent (ethyl alcohol). Experiment 11.

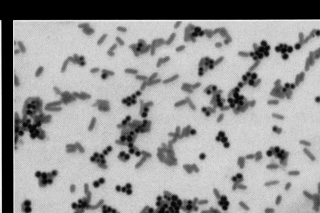

11 **Gram staining, Step 4:** Application of counterstain (safranin). Gram-positive *S. aureus* appears purple and gram-negative *P. aeruginosa* appears red. Experiment 11.

─ Vegetative cells ─ Free spores

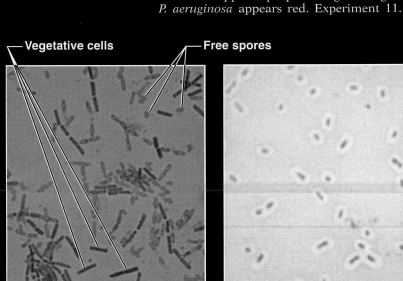

12 **Acid-fast stain:** Acid-fast mycobacteria. (3600×) Experiment 12.

13 **Spore stain showing free spores and vegetative bacilli.** (4250×)

14 **Capsule stain:** Capsulated diplococci. Experiment 13.

15 **Nutrient agar plate:** Four-way streak-plate inoculation with *Serratia marcescens.* Experiment 2.

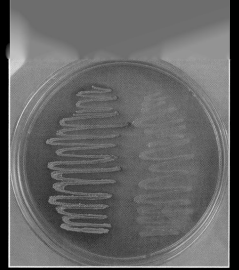

16 **MacConkey agar plate:** Lactose fermenter on left; lactose nonfermenter on right. Experiments 15 and 70.

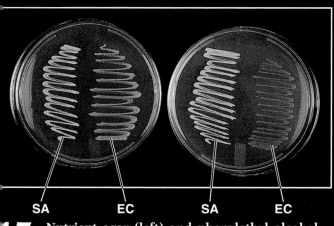

SA EC SA EC

17 **Nutrient agar (left) and phenylethyl alcohol agar (PEA, right):** Each plate inoculated with *Staphylococcus aureus* (SA) and *Escherichia coli* (EC). Gram-negative *E. coli* exhibits reduced growth on PEA. Experiment 15.

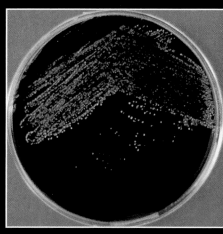

18 **Eosin-methylene blue agar plate:** *E. coli* exhibiting a green metallic sheen. Experiments 15 and 51.

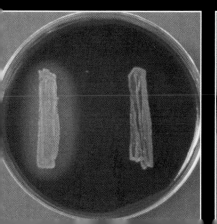
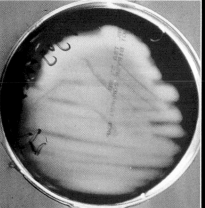
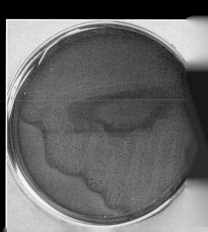

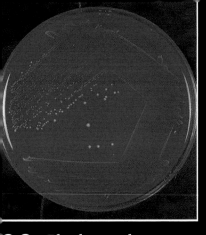

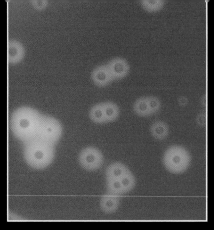

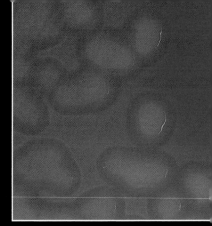

22 **Blood agar plate:** Gamma hemolysis. Experiments 15, 63, 65, and 70.

23 **Blood agar plate:** Beta hemolysis. Experiments 15 and 63.

24 **Blood agar plate:** Alpha hemolysis. Experiments 15 and 63.

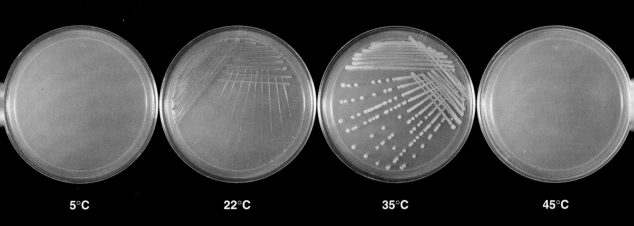

5°C   22°C   35°C   45°C

25 **Effect of temperature on bacterial growth and pigmentation.** Experiment 16.

Bubbles   **Uniform growth**   **Bottom growth**

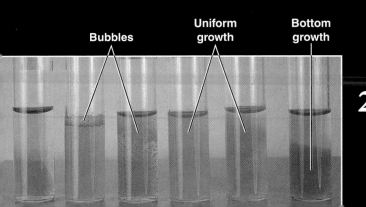

26 **Bacterial growth patterns in thioglycollate broth tubes.** Bubbles (tubes 2 and 3) are indicative of gas-producing bacteria; uniform growth (tubes 4 and 5) is indicative of facultative anaerobic bacteria; and bottom growth is indicative of anaerobic bacteria. Tube 1 is the uninoculated control. Experiment 19

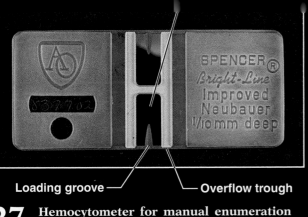

Loading groove ———   ——— Overflow trough

27 **Hemocytometer for manual enumeration of bacterial cells.** Experiment 20.

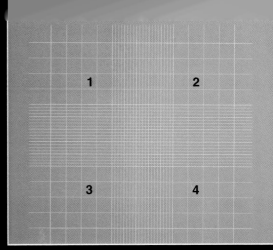

28 **Microscopic view of hemocytometer chambers for bacterial cell counts (chambers 1–4).** Experiment 20.

1×10^{-1} 1×10^{-2} 1×10^{-3} 1×10^{-4} 1×10^{-5} 1×10^{-6} 1×10^{-7}

29 **Serial dilution of bacterial culture for quantitation of viable cell numbers.** Experiment 20.

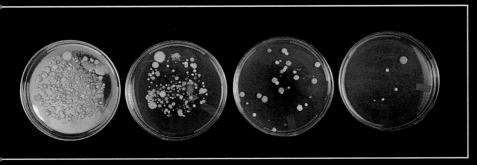

30 **Agar plating procedure for viable cell counts using dilutions (left to right) 1×10^3, 1×10^4, 1×10^5, 1×10^6.** Experiment 20.

31 **Counter for the enumeration of** ~~b~~ ct ~~i~~ l c ~~l~~ i Experiment 20.

Biochemical Tests for the Identification of Microorganisms

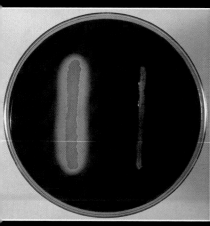

32 **Starch agar plate:** Starch hydrolysis on left; no starch hydrolysis on right. Experiment 22.

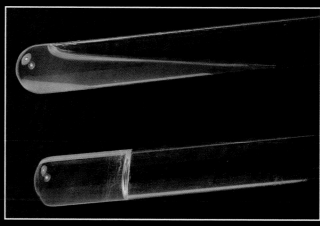

33 **Nutrient gelatin tubes.** Top tube shows gelatin liquefaction; bottom tube is negative for gelatin liquefaction. Experiment 22.

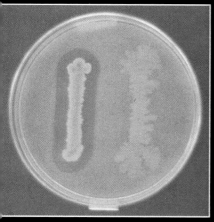

34 **Tributyrin agar plate:** Lipid hydrolysis on left; no lipid hydrolysis on right. Experiment 22.

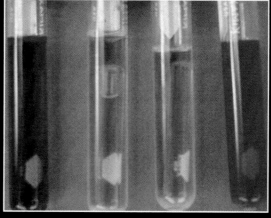

35 **Carbohydrate fermentation test:** (from left to right) uninoculated, acid and gas, acid, negative. Experiment 23.

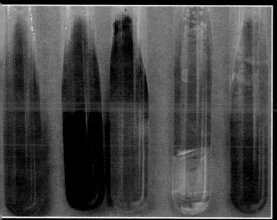

36 **Triple sugar–iron agar test:** (from left to right) uninoculated; alkaline slant/acid butt, H₂S; alkaline slant/

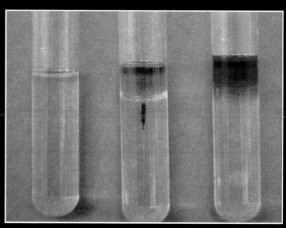

37 **Indole production test:** (from left to right) uninoculated, negative, positive. Experiment 25.

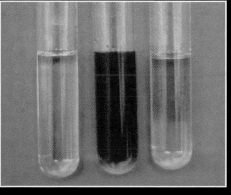

38 Methyl red test: (from left to right) uninoculated, positive, negative. Experiment 25.

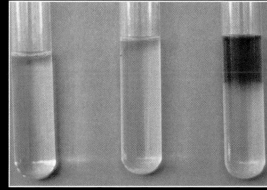

39 Voges-Proskauer test: (from left to right) uninoculated, negative, positive. Experiment 25.

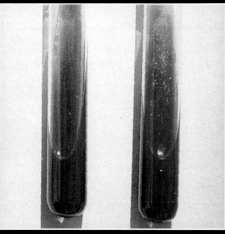

40 Citrate utilization test: Negative on left, no growth on slant surface; positive on right, growth on slant surface. Experiment 25.

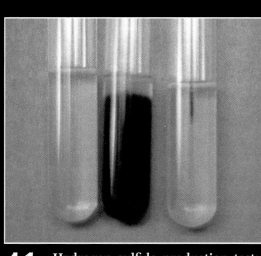

41 Hydrogen sulfide production test: (from left to right) negative, positive with motility, positive with no motility. Experiment 26.

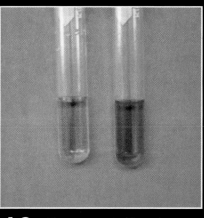

42 Urease test: Negative on left; positive on right. Experiment 27.

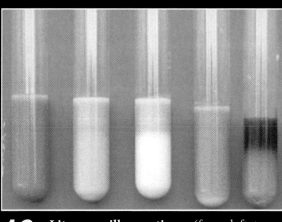

43 Litmus milk reactions: (from left to right) uninoculated, acid, acid with reduction and curd, alkaline, proteolysis. Experiment 28.

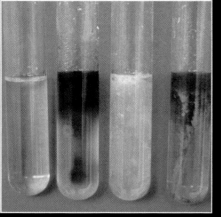

44 Nitrate reduction test: (from left to right) uninoculated, positive with Solutions A + B, positive with Solutions A + B + zinc powder, negative with Solutions A + B + zinc powder. Experiment 29.

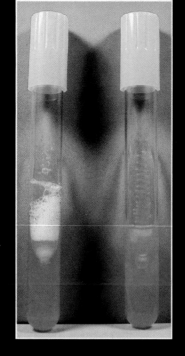

45 Catalase test, tube method: Positive on the left, as evidenced by the evolution of O₂ bubbles; negative on the right. Experiment 30.

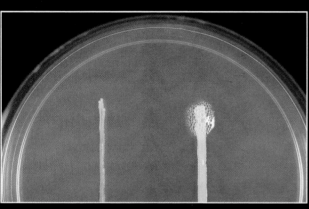

46 Catalase test, plate method: Negative on the left; positive on the right. Experiment 30.

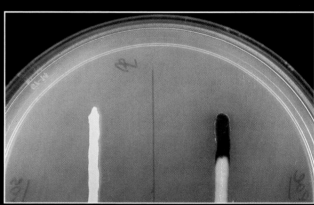

47 Oxidase test: Negative on the left; positive on the right. Experiments 31 and 70.

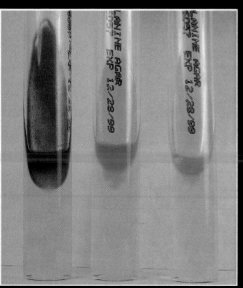

48 Phenylalanine deaminase test: (from left to right) positive, negative, uninoculated. Experiment 32.

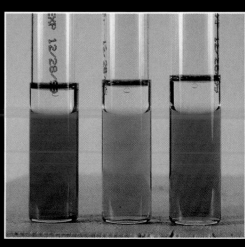

49 Decarboxylase test: (from left to right) positive, negative, uninoculated. Experiment 32.

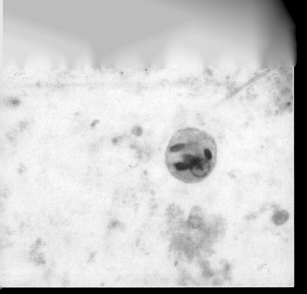

50 *Entamoeba histolytica:* Causative of amebic dysentery. (3600×) Experiment 35.

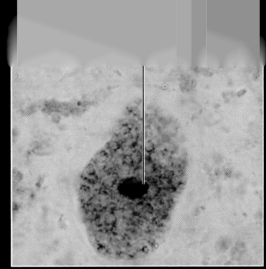

51 A trophozoite of the amoeboid protozoan *Entamoeba histolytica.* (3600×) Experiment 35.

Single ring

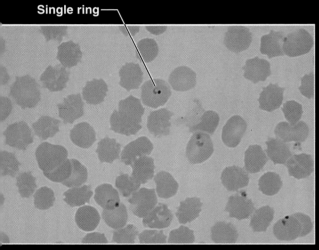

52 Red blood cells infected by the ring stage (signet ring) of *Plasmodium vivax*, **the causative agent of malaria.** (3600×) Experiment 35.

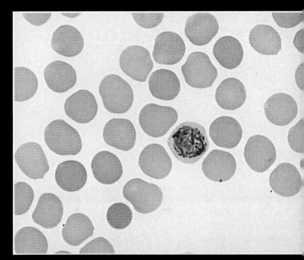

53 *Plasmodium vivax:* Malarial parasite. (3600×) Experiment 35.

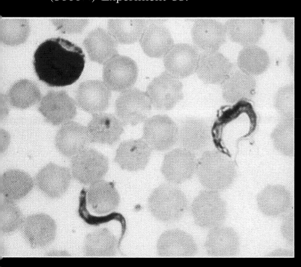

Trypanosoma gambiense: Causative agent

Macronucleus

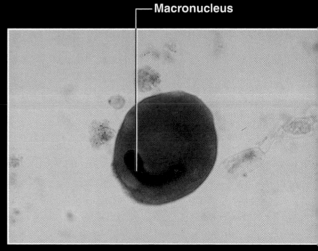

A cyst of the ciliated protozoan *Balantidiu*

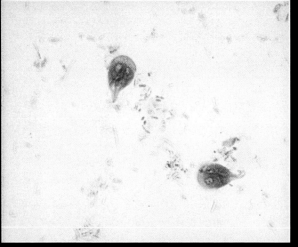

56 *Giardia lamblia:* Causative agent of gastrointestinal diarrhea. (3600×) Experiment 35.

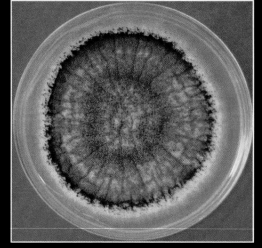

57 **Sabouraud agar plate:** Colony of *Aspergillus niger*. Experiments 36, 63, and 70.

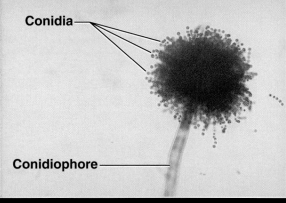

Conidia

Conidiophore

58 **Conidiophore and conidia of mold** *Aspergillus niger.* (550×) Experiment 36.

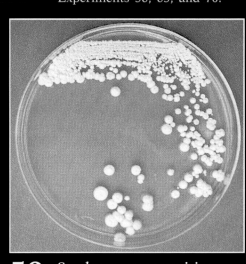

59 *Saccharomyces cerevisiae:* Colonies of yeast cells. Experiments 37, 63, and 70.

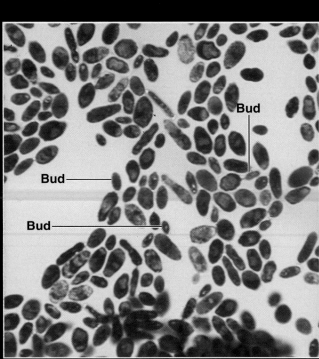

Bud

Bud

Bud

60 *Saccharomyces cerevisiae* **microcospic cell structure.** (3600×) Experiment 37.

Antibiotic and Disinfectant Activities

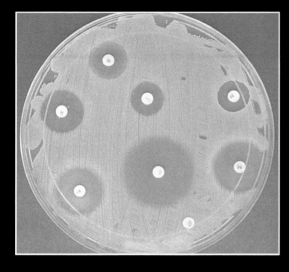

61 Kirby-Bauer antibiotic sensitivity test: Experiment 44.

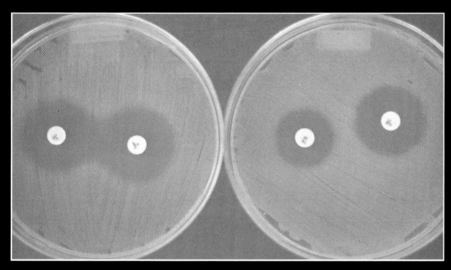

62 **Synergistic effects of drug combinations:** Synergism on the left; no synergism on the right. Experiment 44.

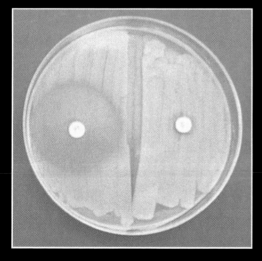

63 **Penicillinase activity:** Penicillin sensitivity on left; penicillin resistance on right. Experiment 45.

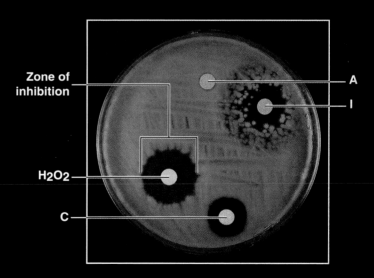

Zone of inhibition

H2O2

C

A

I

64 Antiseptic susceptibility test with discs saturated with chlorine bleach (C), hydrogen peroxide (H$_2$O$_2$), isopropyl alcohol (A), and tincture of iodine (I). Experiment 46.

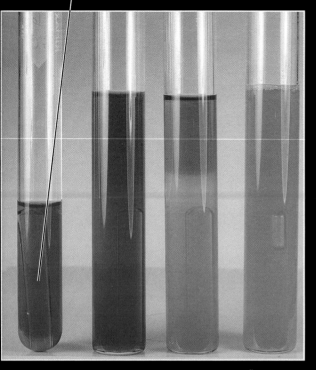

Durham tube

UC   NC(−)   A(−)   A/G(+)

65 **Possible MPN presumptive test results:**
(from left to right) uninoculated control
(UC); inoculated tube, no change (NC);
inoculated tube, acid production only (A);
inoculated tube, acid and gas production
(A/G). Only the tube on the right (A/G) is
considered positive; the other inoculated
tubes (NC and A) are considered negative.
Experiment 51.

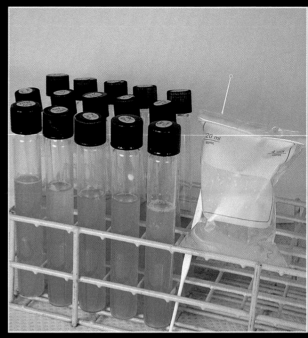

66 **The MPN presumptive test results for a
water sample:** There were five positives
(acid and gas) for the 10-ml tubes in the
front row, five positives for the 1-ml tubes
in the middle row, and five negatives (acid
only) in the back row. The score (5, 5, 0)
indicates 240 coliforms per 100 ml of
water (see Table 51.1). This represents a
positive presumptive test for the presence
of coliforms in the water sample under
test. Experiment 51.

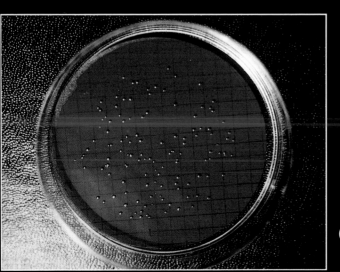

67 **Membrane filter method for
quantitative water analysis for
enumeration of coliform bacteria.**
Experiment 52.

Epithelial cell — G(+) rods — G(+) streptococci

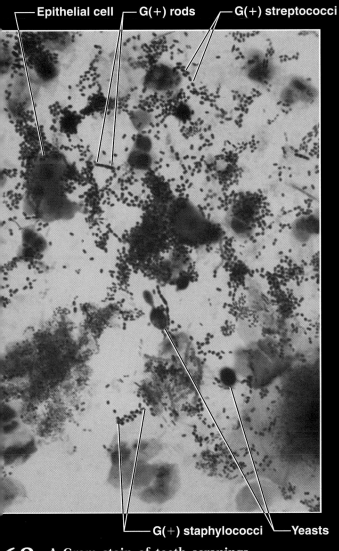

G(+) staphylococci — Yeasts

68 **A Gram stain of teeth scrapings.** Experiment 62.

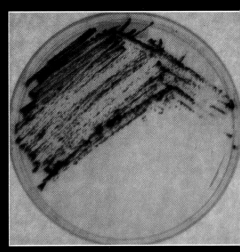

69 **Tinsdale or Mueller-Hinton tellurite agar plate:** Positive for the presence of diphtheroids. Experiment 63.

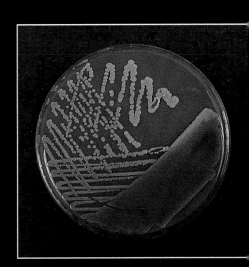

70 **Chocolate agar plate:** Four-

Identification of Staphylococcal Pathogens

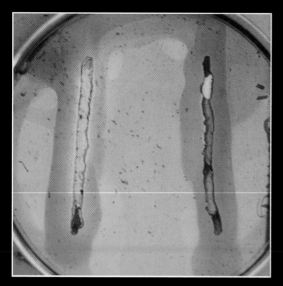

71 **DNase agar plate:** Positive on the left; negative on the right. Experiment 64.

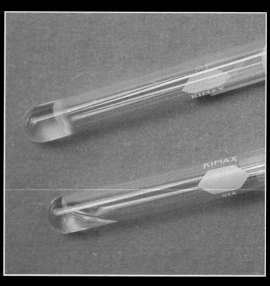

72 **Coagulase test:** Negative on the bottom; positive on the top. Experiment 64.

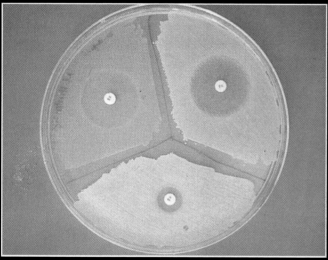

73 **Novobiocin test:** Top: *Staphylococcus aureus, Staphylococcus epidermidis*—sensitive. Bottom: *Staphylococcus saprophyticus*—resistant. Experiment 64.

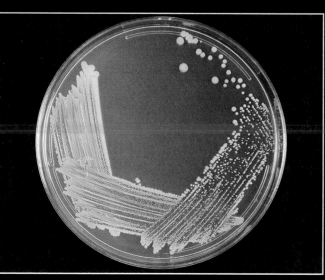

74 **Streak-plate of *Staphylococcus aureus:*** Note the circular, convex, smooth, and cream-colored to golden-yellow appearance. Experiment 64.

75 **6.5% sodium chloride test:** Group D enterococci on left; Group D nonenterococci on right. Experiment 65.

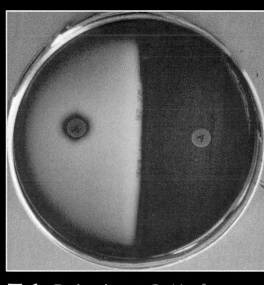

76 **Bacitracin test:** Positive for beta-hemolytic group A streptococci on the left; negative on the right. Experiment 65.

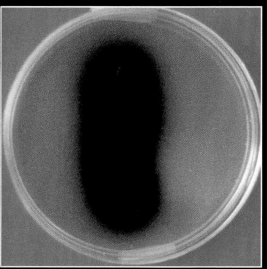

77 **Bile esculin agar plate:** Positive for Group D streptococci. Experiment 65.

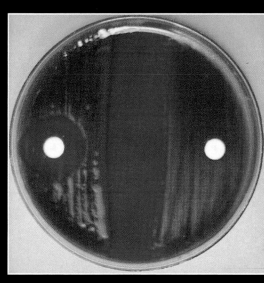

78 **Optochin test:** Alpha-hemolytic *S. pneumoniae* on the left; other alpha-hemolytic streptococcal species on right. Experiment 66.

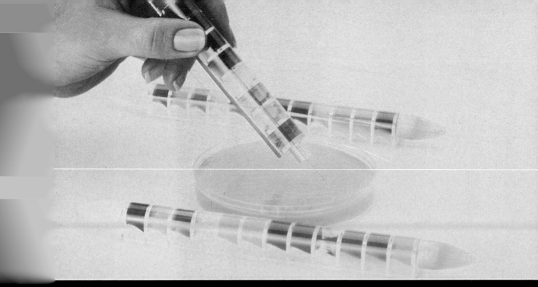

79 **Enterotube Multitest System:** Rear tube is uninoculated control. The center tube demonstrates the inoculation procedure, and the front tube is inoculated with the test organism. Experiment 67.

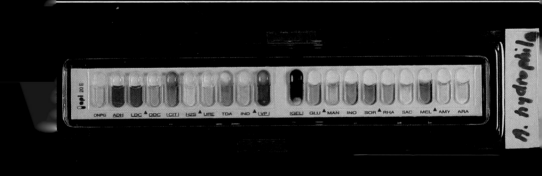

80 **API Multitest System.** Experiment 67.

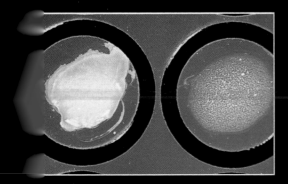

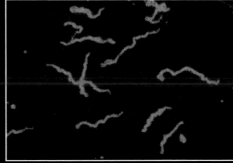

81 Agglutination reaction:

82 Bacterial immunofluorescence

Microbiology of Water

LEARNING OBJECTIVES

Once you have completed the experiments in this section, you should be familiar with

1. The types of microorganisms present in water.
2. The methods to determine the potability of water using standard qualitative and quantitative procedures.

INTRODUCTION

The importance of potable (drinking) water supplies cannot be overemphasized. With increasing industrialization, water sources available for consumption and recreation have been adulterated with industrial as well as animal and human wastes. As a result, water has become a formidable factor in disease transmission. Polluted waters contain vast amounts of organic matter that serve as excellent nutritional sources for the growth and multiplication of microorganisms. The presence of nonpathogenic organisms is not of major concern, but intestinal contaminants of fecal origin are important. These pathogens are responsible for intestinal infections such as **bacillary dysentery, typhoid fever, cholera,** and **paratyphoid fever.**

The World Health Organization (WHO) estimates that 1.7 million deaths per year result from unsafe water supplies. Most of these are from diarrheal diseases, and 90% of these deaths occur in children living in developing countries where sanitary facilities and potable water are at a minimum. The WHO indicates that there are about 3.4 million deaths annually caused by dangerous waterborne enteric bacterial pathogens such as *Shigella dysenteriae, Campylobacter jejuni, Salmonella thyphosa,* and *Vibrio cholerae.*

In addition to bacterial infections, unsafe water supplies are responsible for numerous parasitological infections, including helminth diseases such as schistosomiasis and especially guinea worm *(Dracunculus medinensis),* which infects about 200 million people worldwide each year. Intestinal, hepatic, and pulmonary flukes such as *Fasolopsis buski, Clonorchis sinensis,* and *Paragonimus westermani* are responsible for human infection and are all associated with unsafe water and sanitation. The parasitic protozoa *Entamoeba histolytica, Giardia lamblia,* and *Balantidium coli* are just a few of the protozoa responsible for major diarrheal disease in humans.

Although waterborne infections occur in the United States, their incidence in comparison to the rest of the world is much lower, and they occur sporadically. This can be attributed to the diligent attention given to our water supplies and sewage disposal systems.

Analysis of water samples on a routine basis would not be possible if each pathogen required detection. Therefore water is examined to detect *Escherichia coli*, the bacterium that indicates fecal pollution. Since *E. coli* is always present in the human intestine, its presence in water alerts public health officials to the possible presence of other human or animal intestinal pathogens. Both qualitative and quantitative methods are used to determine the sanitary condition of water.

Standard Qualitative Analysis of Water

The three basic tests to detect coliform bacteria in water are presumptive, confirmed, and completed (Figure 51.1). The tests are performed sequentially on each sample under analysis. They detect the presence of coliform bacteria (indicators of fecal contamination), the gram-negative, nonspore-forming bacilli that ferment lactose with the production of acid and gas that is detectable following a 24-hour incubation period at 37°C.

PART A: Presumptive Test: Determination of the Most Probable Number of Coliform Bacteria

LEARNING OBJECTIVES

Once you have completed this experiment, you should be able to

1. Determine the presence of coliform bacteria in a water sample.
2. Obtain an index indicating the possible number of organisms present in the sample under analysis.

PRINCIPLE

The **presumptive test** is specific for detection of coliform bacteria. Measured aliquots of the water to be tested are added to a lactose fermentation broth containing an inverted gas vial. Because these bacteria are capable of using lactose as a carbon source (the other enteric organisms are not), their detection is facilitated by use of this medium. In this experiment, you will use lactose fermentation broth containing an inverted Durham tube for gas collection.

Tubes of this lactose medium are inoculated with 10-ml, 1-ml, and 0.1-ml aliquots of the water sample. The series consists of at least three groups, each composed of five tubes of the specified medium. The tubes in each group are then inoculated with the designated volume of the water sample as described under "Procedure." The greater the number of tubes per group, the greater the sensitivity of the test. Development of gas in any of the tubes is *presumptive* evidence of the presence of coliform bacteria in the sample. The presumptive test also enables the microbiologist to obtain some idea of the number of coliform organisms present by means of the **most probable number (MPN) test.** The MPN is estimated by determining the number of tubes in each group that show gas following the incubation period (Table 51.1).

MATERIALS
Cultures

Water samples from sewage plant, pond, and tap.

Media

Per designated student group: 15 double-strength lactose fermentation broths (LB2X) and 30 single-strength lactose fermentation broths (LB1X).

Equipment

Bunsen burner, 45 test tubes, test tube rack, sterile 10-ml pipettes, sterile 1-ml pipettes, sterile 0.1-ml pipettes, mechanical pipetting device, and glassware marking pencil.

(text continues on page 326)

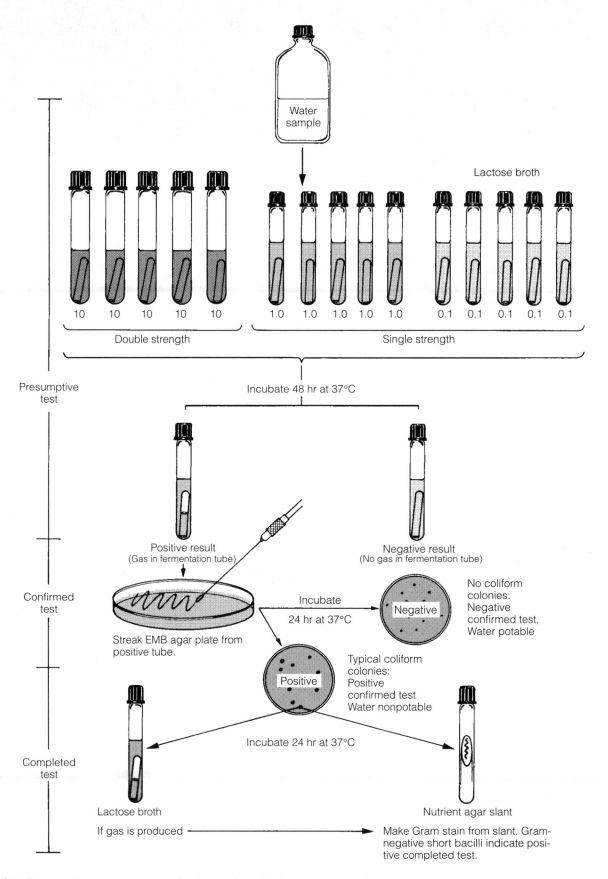

FIGURE 51.1 Standard method for bacteriological water analysis

TABLE 51.1 The MPN Index per 100 ml for Combinations of Positive and Negative Presumptive Test Results when Five 10-ml, Five 1-ml, and Five 0.1 ml Portions of Sample Are Used.

Number of Tubes with Positive Results						**Number of Tubes with Positive Results**					
Five of 10 ml each	Five of 1 ml each	Five of 0.1 ml each	MPN index per 100 ml	95% Confidence Limits		Five of 10 ml each	Five of 1 ml each	Five of 0.1 ml each	MPN index per 100 ml	95% Confidence Limits	
				Lower	Upper					Lower	Upper
0	0	0	<2	0	6	4	2	1	26	7	67
0	0	1	2	<0.5	7	4	3	0	27	9	78
0	1	0	2	<0.5	7	4	3	1	33	9	78
0	2	0	4	<0.5	11	4	4	0	34	11	93
1	0	0	2	0.1	10	5	0	0	23	7	70
1	0	1	4	0.7	10	5	0	1	31	11	89
1	1	0	4	0.7	12	5	0	2	43	14	100
1	1	1	6	1.8	15	5	1	0	33	10	100
1	2	0	6	1.8	15	5	1	1	46	14	120
						5	1	2	63	22	150
2	0	0	5	<0.5	13						
2	0	1	7	1	17	5	2	0	49	15	150
2	1	0	7	1	17	5	2	1	70	22	170
2	1	1	9	2	21	5	2	2	94	34	230
2	2	0	9	2	21	5	3	0	79	22	220
2	3	0	12	3	28	5	3	1	110	34	250
						5	3	2	140	52	400
3	0	0	8	2	22						
3	0	1	11	4	23	5	3	3	180	70	400
3	1	0	11	5	35	5	4	0	130	36	400
3	1	1	14	6	36	5	4	1	170	58	400
3	2	0	14	6	36	5	4	2	220	70	440
3	2	1	17	7	40	5	4	3	280	100	710
3	3	0	17	7	40	5	4	4	350	100	710
4	0	0	13	4	35	5	5	0	240	70	710
4	0	1	17	6	36	5	5	1	350	100	1100
4	1	0	17	6	40	5	5	2	540	150	1700
4	1	1	21	7	42	5	5	3	920	220	2600
4	1	2	26	10	70	5	5	4	1600	400	4600
4	2	0	22	7	50	5	5	5	≥2400	700	---

Sources: pp 9–51, *Standard Methods for the Examination of Water and Wastewater*, 20th Edition (1998). M. J. Taras, A. E. Greenberg, R. D. Hoak, and M. C. Rand, eds. American Public Health Association, Washington, D.C. Copyright 1998, American Public Health Association, and *Bacteriological Analytical Manual (BAM)*, 8th Edition, Food and Drug Administration, 1998.

PROCEDURE

⚠ **Exercise care in handling sewage waste water sample because enteric pathogens may be present.**

1. Set up three separate series consisting of three groups, a total of 15 tubes per series, in a test tube rack; for each tube, label the water source and volume of sample inoculated as illustrated.

Series 1: Sewage water	5 tubes of LB2X-10 ml 5 tubes of LB1X-1 ml 5 tubes of LB1X-0.1 ml
Series 2: Pond water	5 tubes of LB2X-10 ml 5 tubes of LB1X-1 ml 5 tubes of LB1X-0.1 ml
Series 3: Tap water	5 tubes of LB2X-10 ml 5 tubes of LB1X-1 ml 5 tubes of LB1X-0.1 ml

2. Mix sewage plant water sample by shaking thoroughly.
3. Flame bottle and then, using a 10-ml pipette, transfer 10-ml aliquots of water sample to the five tubes labeled LB2X-10 ml.
4. Flame bottle and then, using a 1-ml pipette, transfer 1-ml aliquots of water sample to the five tubes labeled LB1X-1 ml.
5. Flame bottle and then, using a 0.1-ml pipette, transfer 0.1-ml aliquots of water sample to the five tubes labeled LB1X-0.1 ml.
6. Repeat Steps 2 through 5 for the tap and pond water samples.
7. Incubate all tubes for 48 hours at 37°C.

PART B: Confirmed Test

LEARNING OBJECTIVE

Once you have completed this experiment, you should be able to

1. Confirm the presence of coliform bacteria in a water sample for which the presumptive test was positive.

PRINCIPLE

The presence of a positive or doubtful presumptive test immediately suggests that the water sample is nonpotable. Confirmation of these results is necessary, since positive presumptive tests may be the result of organisms of noncoliform origin that are not recognized as indicators of fecal pollution.

The **confirmed test** requires that selective and differential media such as eosin–methylene blue (EMB) or Endo agar be streaked from a positive lactose broth tube obtained from the presumptive test. The nature of differential and selective media was discussed in Experiment 15 but is reviewed briefly here. Eosin–methylene blue contains the dye methylene blue, which inhibits the growth of gram-positive organisms. In the presence of an acid environment, EMB forms a complex that precipitates out onto the coliform colonies, producing dark centers and a green metallic sheen. This reaction is characteristic for *Escherichia coli*, the major indicator of fecal pollution. Endo agar is a nutrient medium containing the dye fuchsin, which is present in the decolorized state. In the presence of acid produced by the coliform bacteria, fuchsin forms a dark pink complex that turns the *E. coli* colonies and the surrounding medium pink.

MATERIALS

Cultures

One 24-hour-old positive lactose broth culture from each of the three series from the presumptive test.

Media

Three each per designated student group: eosin–methylene blue agar plates or Endo agar plates.

Equipment

Bunsen burner, glassware marking pencil, and inoculating loop.

PROCEDURE

1. Label the covers of the three EMB plates or the three Endo agar plates with the source of the water sample (sewage, pond, and tap).
2. Using a positive 24-hour lactose broth culture from the sewage water series from the presumptive test, streak the surface of one EMB or one Endo agar plate, as described in Experiment 2, to obtain discrete colonies.

3. Repeat Step 2 using the positive lactose broth cultures from the pond and tap water series to inoculate the remaining plates.

4. Incubate all plate cultures in an inverted position for 24 hours at 37°C.

PART C: Completed Test

LEARNING OBJECTIVE

Once you have completed this experiment, you should be able to

1. Confirm the presence of coliform bacteria in a water sample, or if necessary, confirm a suspicious or doubtful result from the previous test.

PRINCIPLE

The **completed test** is the final analysis of the water sample. It is used to examine the coliform colonies that appeared on the EMB or Endo agar plates used in the confirmed test. An isolated colony is picked from the confirmatory test plate and inoculated into a tube of lactose broth and streaked on a nutrient agar slant to perform a Gram stain. Following inoculation and incubation, tubes showing acid and gas in the lactose broth and the presence of gram-negative bacilli on microscopic examination are further confirmation of the presence of *E. coli*, and they are indicative of a positive completed test.

MATERIALS

Cultures

One 24-hour coliform-positive EMB or Endo agar culture from each of the three series of the confirmed test.

Media

Three each per designated student group: nutrient agar slants and lactose fermentation broths.

Reagents

Crystal violet, Gram's iodine, 95% ethyl alcohol, and safranin.

Equipment

Bunsen burner, staining tray, inoculating loop, lens paper, bibulous paper, microscope, and glassware marking pencil.

PROCEDURE

1. Label each tube with the source of its water sample.

2. Inoculate one lactose broth and one nutrient agar slant from the same isolated *E. coli* colony obtained from an EMB or an Endo agar plate from each of the experimental water samples.

3. Incubate all tubes for 24 hours at 37°C.

Name _____ Date _____

PART A: Presumptive Test

1. Examine all tubes after 24 and 48 hours of incubation. Record your results in the chart as follows:

    a. Positive: 10% or more of gas appears in a tube in 24 hours.

    b. Doubtful: Gas develops in a tube after 48 hours.

    c. Negative: There is no gas in a tube after 48 hours.

2. Determine and record the MPN using Table 51.1.

📷 *Refer to photo numbers 65 and 66 in the color-plate insert for illustration of these reactions.*

Example: If gas appeared in all five tubes labeled LB2X-10, in two of the tubes labeled LB1X-1, and in one labeled LB1X-0.1, the series would be read as 5-2-1. From the MPN table, such a reading would indicate that there would be approximately 70 microorganisms per 100 ml of water, with a 95% probability that there are between 22 and 170 microorganisms present.

Water Sample	Gas															Reading	MPN	Range 95% Probability
	LB2X-10					LB1X-1					LB1X-0.1							
	Tube					Tube					Tube							
	1	2	3	4	5	1	2	3	4	5	1	2	3	4	5			
Sewage																		
Pond																		
Tap																		

PART B: Confirmed Test

1. Examine all the plates for the presence or absence of *E. coli* colonies. Record your results in the chart.

📷 *Refer to photo number 18 in the color-plate insert for an illustration of* E. coli *growth on an EMB plate.*

2. Based on your results, determine and record whether each of the samples is potable or non-potable. The presence of *E. coli* is a positive confirmed test, indicating that the water is non-potable. The absence of *E. coli* is a negative test, indicating that the water is not contaminated with fecal wastes and is therefore potable.

Water Sample	Coliforms		Potable	Nonpotable
	EMB Plate	Endo Agar Plate		
Sewage				
Pond				
Tap				

PART C: Completed Test

1. Examine all lactose fermentation broth cultures for the presence or absence of acid and gas. Record your results in the chart.
2. Prepare a Gram stain, using the nutrient agar slant cultures of the organisms that showed a positive result in the lactose fermentation broth (refer to Experiment 11 for the staining procedure).
3. Examine the slides microscopically for the presence of gram-negative short bacilli, which are indicative of *E. coli* and thus nonpotable water. Record your results for Gram stain reaction and morphology of the cells in the chart.

Water Source	Lactose Broth A/G (+) or (−)	Gram Stain Reaction/Morphology	Potability Potable	Nonpotable
Sewage				
Pond				
Tap				

REVIEW QUESTIONS

1. What is the rationale for selecting *E. coli* as the indicator of water potability?

2. Why is this procedure qualitative rather than quantitative?

3. Explain why it is of prime importance to analyze water supplies that serve industrialized communities.

4. Account for the presence of microorganisms in natural bodies of water and sewage systems. What is their function? Explain.

Quantitative Analysis of Water: Membrane Filter Method

LEARNING OBJECTIVE

Once you have completed this experiment, you should be able to

1. Determine the quality of water samples using the membrane filter method.

PRINCIPLE

Bacteria-tight **membrane filters** capable of retaining microorganisms larger than 0.45 micrometer (μm) are frequently used for analysis of water. These filters offer several advantages over the conventional, multiple-tube method of water analysis: (1) results are available in a shorter period of time, (2) larger volumes of sample can be processed, and (3) because of the high accuracy of this method, the results are readily reproducible. A disadvantage involves the processing of turbid specimens that contain large quantities of suspended materials; particulate matter clogs the pores and inhibits passage of the specific volume of water.

A water sample is passed through a sterile membrane filter that is housed in a special filter apparatus contained in a suction flask. Following filtration, the filter disc that contains the trapped microorganisms is aseptically transferred to a sterile Petri dish containing an absorbent pad saturated with a selective, differential liquid medium. Following incubation, the colonies present on the filter are counted with the aid of a microscope. This procedure is illustrated in Figure 52.1.

This experiment is used to analyze a series of dilutions of water samples collected upstream and downstream from an outlet of a sewage treatment plant. A total count of coliform bacteria determines the potability of the water sources. Also, the types of fecal pollution, if any, are established by means of a fecal coliform count, indicative of human pollution, and a fecal streptococcal count, indicative of pollution from other animal origins. The ratio of the fecal coliforms to fecal streptococci per milliliter of sample is interpreted as follows: Between 2 and 4 indicates human and animal pollution; >4 indicates human pollution; and <0.7 indicates poultry and livestock pollution.

MATERIALS

Cultures

Water samples collected upstream (labeled U) and downstream (labeled D) from an outlet of a sewage treatment plant.

Media

Per designated student group for analysis of one water sample: one 20-ml tube of m-Endo broth, one 20-ml tube of m-FC broth, one 20-ml tube of KF broth, four 90-ml sterile water blanks, and one 300-ml flask of sterile water.

Equipment

Sterile membrane filtration apparatus (i.e., Millipore; Pall® Gelman; sterile, plastic, disposable membrane filters), 1-liter suction flask, 15 sterile membrane filters and absorbent pads, 15 sterile 50-mm Petri dishes, 12 10-ml pipettes, mechanical pipetting device, small beaker of 95% alcohol, membrane forceps, waterproof tape, watertight plastic bags, 44.5°C waterbath, dissecting microscope, and glassware marking pencil.

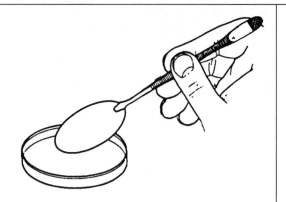

(a) Aseptically place an absorbent pad in a 50-mm Petri dish.

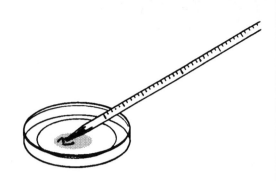

(b) Saturate the absorbent pad with the specified selective broth medium.

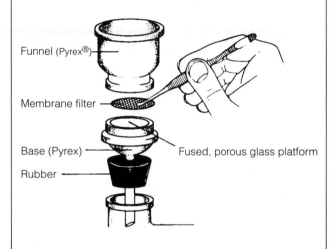

Funnel (Pyrex®)

Membrane filter

Base (Pyrex)

Rubber

Fused, porous glass platform

(c) Assemble the filter apparatus and insert membrane filter.

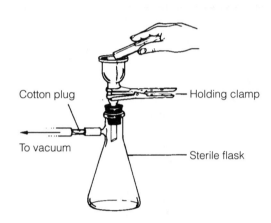

Cotton plug

Holding clamp

To vacuum

Sterile flask

(d) Pour test sample into funnel, filter under vacuum, and rinse with sterile water.

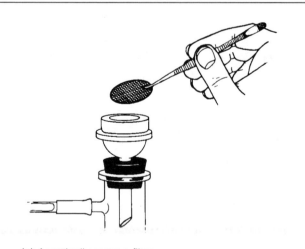

(e) Aseptically remove filter.

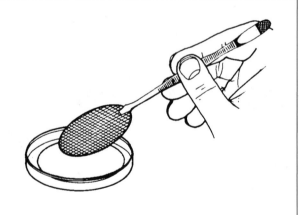

(f) Place filter in Petri dish on top of medium-saturated pad and incubate.

FIGURE 52.1 Membrane filter technique

PROCEDURE

The following instructions are for analysis of one of the provided water samples using the Millipore system. Different samples may be assigned to individual groups.

> ⚠ **Use disposable gloves when handling the water samples in this experiment.**

1. Label the four 90-ml water blanks with the source of the water sample and dilution (10^{-1}, 10^{-2}, 10^{-3}, and 10^{-4}).

2. Using 10-ml pipettes, aseptically perform a ten-fold serial dilution of the assigned undiluted water sample, using the four 90-ml water blanks to effect the 10^{-1}, 10^{-2}, 10^{-3}, and 10^{-4} dilutions.

3. Arrange the 15 Petri dishes into three sets of five plates. Label each set as follows:

   a. For total coliform count (TCC) and dilutions (undiluted, 10^{-1}, 10^{-2}, 10^{-3}, and 10^{-4}).

   b. For fecal coliform count (FCC) and dilutions as in Step 3a.

   c. For fecal streptococcal count (FSC) and dilutions as in Step 3a.

4. Using sterile forceps dipped in 95% alcohol and flamed, add a sterile absorbent pad to all Petri dishes.

5. With sterile 10-ml pipettes, aseptically add:

   a. Two ml of m-Endo broth to each pad in the plates labeled TCC.

   b. Two ml of m-FC broth to each pad in the plates labeled FCC.

   c. Two ml of KF broth to each pad in the plates labeled FSC.

6. Aseptically assemble the sterile paper-wrapped membrane filter unit as follows:

   a. Unwrap and insert the sintered glass filter base into the neck of a 1-liter side-arm suction flask.

   b. With sterile forceps, place a sterile membrane filter disc, grid side up, on the sintered glass platform.

   c. Unwrap and carefully place the funnel section of the apparatus on top of the filter disc. Using the filter clamp, secure the funnel to the filter base.

   d. Attach a rubber hose from the side-arm on the vacuum flask to a vacuum source.

7. Using the highest sample dilution (10^{-4}) and a pipette, place 20 ml of the dilution into the funnel and start the vacuum.

8. When the entire sample has been filtered, wash the inner surface of the funnel with 20 ml of sterile water.

9. Disconnect the vacuum, unclamp the filter assembly, and with sterile forceps, remove the membrane filter and place it on the medium-saturated pad in the Petri dish labeled TCC, 10^{-4}.

10. Aseptically place a new membrane on the platform, reassemble the filtration apparatus, and repeat Steps 7 through 9 twice, adding the filter discs to the 10^{-4} dilution plates labeled FCC and FSC.

11. Repeat Steps 8 through 10, using 20 ml of the 10^{-3}, 10^{-2}, and 10^{-1} dilutions and the undiluted samples.

12. Incubate the plates in an inverted position as follows:

    a. TCC and FSC plates for 24 hours at 37°C.

    b. FCC plates sealed with waterproof tape and placed in a weighted watertight plastic bag, which is then submerged in a 44.5°C waterbath for 24 hours.

Name _____ Date _____

1. Using sterile membrane forceps, remove the filter discs from the Petri dishes and allow to dry on absorbent paper for 1 hour.
2. Using membrane forceps, place each dry filter disc into its Petri dish cover. *Keep the discs within the covers at all times for further observation.*
3. Examine all filter discs under a dissecting microscope and perform colony counts on each set of discs as follows:
   a. TCC: Count colonies on m-Endo agar that present a golden metallic sheen (performed on a disc showing 20 to 80 of these colonies).
   b. FCC: Count colonies on m-FC agar that are blue (performed on a disc showing 20 to 60 of these colonies).
   c. FSC: Count colonies on KF agar that are pink to red (performed on a disc showing 20 to 100 of these colonies).

   Dilution samples that show fewer colonies than indicated are designated as TFTC, and those showing a greater number of colonies are designated as TNTC.
4. For each of the three counts, determine the number of fecal organisms present in 100 ml of the water sample, using the following formula:

$$\frac{\text{colony count} \times \text{dilution factor}}{\text{ml of sample used}} \times 100$$

5. Record your results in the chart.

 Refer to photo number 67 in the color-plate insert for illustration of this reaction.

Dilution	Upstream Water			Downstream Water		
	TCC	FCC	FSC	TCC	FCC	FSC
Undiluted						
10^{-1}						
10^{-2}						
10^{-3}						
10^{-4}						

6. Determine the fecal coliform to fecal streptococcal (FC:FS) ratio. Record your results in the chart.

Dilution	Upstream Water			Downstream Water		
	cells/ml*			cells/ml*		
	FCC	FSC	FC:FS Ratio	FCC	FSC	FC:FS Ratio
Undiluted						
10^{-1}						
10^{-2}						
10^{-3}						
10^{-4}						

*Cells/ml $= \dfrac{\text{Cells/100 ml}}{100}$

7. Based on your FC:FS ratio, indicate the type of fecal pollution, if any, in:
   a. Upstream water sample:

   b. Downstream water sample:

REVIEW QUESTIONS

1. What are the advantages of the membrane filter method in the analysis of water samples?

2. What are the disadvantages of the membrane filter method?

3. What is the purpose of determining the FC:FS ratio?

4. Cite some other microbiological applications of the membrane filter technique in environmental studies.

Microbiology of Soil

LEARNING OBJECTIVES

Once you have completed the experiments in this section, you should be able to

1. Understand the characteristics and activities of soil microorganisms.
2. Demonstrate the role of soil microorganisms in the biotransformation of nitrogenous compounds via the nitrogen cycle.
3. Enumerate soil microorganisms.
4. Demonstrate the ability of some soil microorganisms to produce antibiotics.
5. Demonstrate the use of enrichment cultures for the isolation of specific soil microorganisms.

INTRODUCTION

Soil is often thought of as an inert substance by the average layperson. However, contrary to this belief, it serves as a repository for many life forms, including a huge and diverse microbial population. The beneficial activities of these soil inhabitants far outweigh their detrimental effects.

Life on this planet could not be sustained in the absence of microorganisms that inhabit the soil. This flora is essential for degradation of organic matter deposited in the soil, such as dead plant and animal tissues and animal wastes. Hydrolysis of these macromolecules by microbial enzymes supplies and replenishes the soil with basic elemental nutrients. By means of enzymatic transformations, plants assimilate these nutrients into organic compounds essential for their growth and reproduction. In turn, these plants serve as a source of nutrition for animals. Thus, many soil microorganisms play a vital role in a number of elemental cycles. Included among these are the nitrogen cycle, the carbon cycle, and the sulfur cycle.

Nitrogen Cycle

The nitrogen cycle involves the biochemical transformation of nitrogen-containing organic compounds and atmospheric nitrogen into inorganic forms that can be utilized by plants. This cycle is presented in detail in Experiment 53.

Carbon Cycle

Carbon dioxide is the major carbon source for the synthesis of organic compounds. The carbon cycle is basically represented by the following two steps:

1. Oxidation of organic compounds to carbon dioxide with the production of energy and heat by heterotrophs.
2. Fixation of carbon dioxide into organic compounds by green plants and some bacteria, the autotrophic soil flora.

Sulfur Cycle

Elemental sulfur and proteins cannot be utilized by plants for growth. They must first undergo enzymatic conversions into inorganic sulfur-containing compounds. The basic steps in the sulfur cycle are:

1. Degradation of proteins into hydrogen sulfide (H_2S) by many heterotrophic microorganisms.
2. Oxidation of H_2S to sulfur (S) by a number of bacterial genera, such as *Beggiatoa*.
3. Oxidation of sulfur to utilizable sulfate (SO_4^{2-}) by several chemoautotrophic genera, such as *Thiobacillus*.

Some soil microorganisms also play a role in the enzymatic transformation of other elements, such as phosphorus, iron, potassium, zinc, manganese, and selenium. These biochemical changes make these minerals available to plants in a soluble form.

Many members of the soil flora, because of their fermentative and synthetic capabilities, play an important role in the synthesis of a variety of industrial products as cited below:

1. **Food.** *Penicillium* spp. are used in the production of such cheeses as Camembert, Roquefort, and Brie.
2. **Beverages.** *Saccharomyces* spp. are utilized in the wine, beer, and ale industries.
3. **Vitamins.** *Eremothecium ashbyi* and *Pseudomonas denitrificans*, respectively, synthesize riboflavin and vitamin B_{12}.
4. **Enzymes.** Amylases, pectinases, and proteases are produced by *Aspergillus* spp.
5. **Antibiotics.** *Penicillium* spp. (penicillin), *Streptomyces* spp. (kanamycins and tetracyclines), and *Bacillus* spp. (bacitracin).
6. **Steroids.** *Rhizopus, Streptomyces,* and *Curvularia* are microorganisms that are used to carry out specific reactions, bioconversions, to aid in the manufacture of these lipid compounds.
7. **Industrial chemicals.** *Clostridium acetobutylicum* is used in the production of acetone and butanol, and *Aspergillus niger* is used in the synthesis of citric acid.

The major adverse effect of some soil organisms is their ability to produce disease. Included in this flora are both plant and animal pathogens. The ability to produce infections in humans is associated with some bacteria, such as members of the spore-forming genera *Clostridium and Bacillus,* and with some fungal genera, such as *Cryptococcus* and *Coccidioides*.

Nitrogen Cycle

The **nitrogen cycle** is concerned with the enzymatic conversion of nitrogenous compounds found in the soil and gaseous nitrogen from the atmosphere into inorganic nitrogen compounds that are used by plants for synthesis of essential macromolecules such as nucleic acids and proteins. The four distinct phases of this cycle are shown in Figure 53.1.

1. **Ammonification:** Sequential degradation of nitrogenous organic compounds with the release of ammonia.

2. **Nitrification:** Step-by-step oxidation of ammonia to nitrite (NO_2^-) and then to nitrate (NO_3^-), a nutritional form of nitrogen that is assimilated by plants.

3. **Denitrification:** Reduction of nitrates not used by plants to gaseous nitrogen ($N_2\uparrow$).

4. **Nitrogen fixation:** Chemical combination of free nitrogen ($N_2\uparrow$) with other elements to form fixed nitrogen (nitrogen-containing compounds).

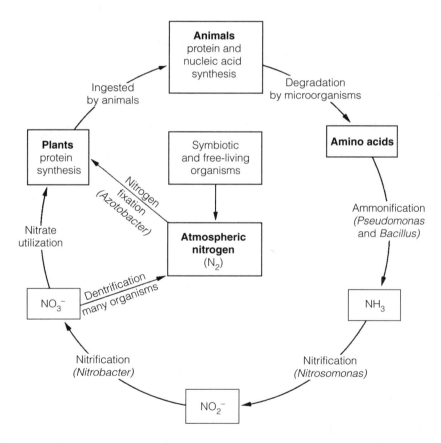

FIGURE 53.1 Nitrogen cycle

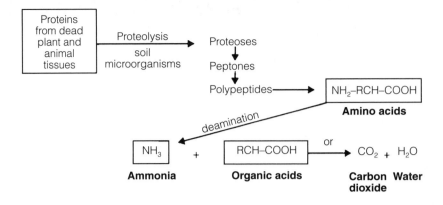

FIGURE 53.2 Ammonification

PART A: Ammonification

LEARNING OBJECTIVE

Once you have completed this experiment, you should be able to

1. Demonstrate the liberation of ammonia from nitrogenous organic compounds.

PRINCIPLE

Ammonification, one of the essential phases of the nitrogen cycle, involves degradation of nitrogenous biopolymers and subsequent release of ammonia. This process is initiated by excretion of extracellular proteolytic enzymes that are commonly produced by such soil organisms as members of the genera *Bacillus*, *Clostridium*, and *Streptomyces*. These enzymes act sequentially to hydrolyze the proteins of plant and animal origins into their constituent amino acids. The amino acids are subsequently enzymatically deaminated, with the release of ammonia, as shown in Figure 53.2.

In this experiment, peptone broth, which contains an organic nitrogen substrate, is used to demonstrate the ability of some microorganisms to degrade proteins, with the resultant formation of ammonia. Following incubation, the presence of ammonia, indicative of ammonification, is detectable by the yellow color when Nessler's reagent is added to samples of the test cultures. The relative amount of ammonia can be determined by differences in the degree of yellow coloration.

MATERIALS

Cultures

24-hour nutrient broth cultures of *Bacillus cereus*, *Pseudomonas fluorescens*, and *Proteus vulgaris*.

Soils

0.1-g samples each of rich and poor garden soil.

Reagent

Nessler's reagent.

Media

Per designated student group: six 4% peptone broth tubes.

Equipment

Bunsen burner, inoculating loop, spot plates, Pasteur pipettes, and glassware marking pencil.

PROCEDURE

1. Label each of the peptone broth tubes with the name of the organism and soil sample to be inoculated. Label the last tube of medium as a control.
2. Using sterile inoculating technique, inoculate each experimental organism and add each soil sample to its appropriately labeled tube.
3. Incubate the culture tubes for 7 days at 25°C.

PART B: Nitrification

LEARNING OBJECTIVE

Once you have completed this experiment, you should be able to

1. Demonstrate the enzymatic conversion of ammonia to nitrates by soil micro-organisms.

PRINCIPLE

Ammonia that is liberated during the ammonification process seldom accumulates to any appreciable extent in the soil. In an aerobic environment, it undergoes bacterial oxidation to nitrates by a two-step process primarily involving the chemoautotrophic soil organisms *Nitrosomonas* and *Nitrobacter*. This process is called **nitrification,** and the chemical transformations are illustrated as follows:

Step 1. **Nitrite formation**

$$NH_4^+ + 1\frac{1}{2}O_2 \xrightarrow{Nitrosomonas} NO_2^- + H_2O + 2H^+$$

Ammonium ion   **Nitrite ion**

Step 2. **Nitrate formation**

$$NO_2^- + \frac{1}{2}O_2 \xrightarrow{Nitrobacter} NO_3^-$$

Nitrite ion   **Nitrate ion**

The nitrates released in the soil are highly soluble and are assimilated by terrestrial plants and some microorganisms for the biosynthesis of cellular proteins.

 In this procedure ammonium sulfate broth is used to demonstrate the ability of some microorganisms to oxidize ammonia to nitrite. The nitrite broth illustrates the further oxidation of nitrite to nitrate. Following incubation, the presence of nitrite and nitrate is determined as follows:

Determination of nitrite production in ammonium sulfate broth soil cultures:

1. Test for the presence of ammonia by use of Nessler's reagent. A yellow color indicates that ammonia was not oxidized to nitrite. No color change indicates the absence of ammonia, and, therefore, nitrite should be present.

2. Test for the presence of nitrite by use of Trommsdorf's reagent and sulfuric acid.

The presence of a blue-black color is indicative of the presence of nitrite; no color change is indicative of the absence of nitrite.

Determination of nitrate production in nitrite broth soil cultures:

1. Test for the presence of nitrite by use of Trommsdorf's reagent and sulfuric acid. The appearance of a blue-black color indicates that nitrites are present and that nitrates have not been formed.

2. Test for the presence of nitrate by use of diphenylamine reagent and sulfuric acid if the absence of nitrites (no color change with Trommsdorf's reagent) was ascertained. The appearance of a deep-blue color indicates that nitrates are present.

MATERIALS

Soil

0.1-g samples of slightly alkaline and acidic garden soil.

Media

Per designated student group: two 25-ml flasks each of ammonium sulfate broth and nitrite broth.

Reagents

Nessler's reagent, diphenylamine reagent, Trommsdorf's reagent, concentrated sulfuric acid, and dilute sulfuric acid (one part concentrated H_2SO_4 plus three parts H_2O).

Equipment

Bunsen burner, Pasteur pipettes, spot plate, wooden applicator sticks, and glassware marking pencil.

PROCEDURE

1. Label each of the flasks of ammonium sulfate broth and nitrite broth with the type of soil sample to be inoculated.

2. Add the provided soil samples to the appropriately labeled flasks. Shake vigorously.

3. Incubate all of the flasks for 3 weeks at 25°C.

PART C: Denitrification

LEARNING OBJECTIVE

Once you have completed this experiment, you should be able to

1. Demonstrate the reduction of nitrates to nitrogen gas.

PRINCIPLE

Nitrates produced in the soil by microbial nitrification or deposited by the use of nitrite fertilizers must be used rapidly or they are lost by leaching or by microbial reduction. In the latter, nitrates are reduced to nitrites, ammonia, nitrous oxide, and finally to elemental nitrogen in the form of nitrogen gas. This process is called **denitrification.**

Some of the soil organisms involved in denitrification are members of the genera *Pseudomonas, Bacillus,* and *Micrococcus.* Under anaerobic conditions, these organisms have the enzyme profile enabling them to use nitrate rather than gaseous oxygen as a final electron acceptor in their energy metabolism, thereby reducing the nitrate to gaseous nitrogen:

$$NO_3^- \rightarrow NO_2^- \rightarrow NO \rightarrow N_2$$

Anaerobiosis

The following experiment uses a medium containing nitrate, the substrate for nitrogen gas formation, and a Durham tube for detection of the evolution of nitrogen gas. Following incubation, the evolution of gaseous nitrogen, the end product of denitrification, can be detected by the presence of an air bubble in the Durham tube. If denitrification does not occur, there will not be an air bubble in the tube.

MATERIALS

Cultures

24-hour nutrient broth cultures of *Pseudomonas aeruginosa* and *Proteus vulgaris.*

Soil

0.1-g samples of rich and poor soil.

Media

Per designated student group: four nitrate broth tubes containing Durham tubes.

Equipment

Bunsen burner, inoculating loop, test tube rack, and glassware marking pencil.

PROCEDURE

1. Label each of the nitrate broth tubes with the name of the organisms and soil samples to be inoculated.

2. Using sterile inoculating technique, inoculate each test organism and soil sample in its appropriately labeled tube. *Note: Do not shake the culture tubes during inoculation.*

3. Incubate the culture tubes for 2 weeks at 25°C.

PART D: Nitrogen Fixation

LEARNING OBJECTIVE

Once you have completed this experiment, you should be able to

1. Demonstrate the fixation of nitrogen by symbiotic and nonsymbiotic microorganisms.

PRINCIPLE

Nitrogen fixation is the phase of the nitrogen cycle during which enzymatically competent microorganisms convert atmospheric nitrogen into nitrogenous compounds, thereby fixing the nitrogen. This vital process replenishes the soil with usable nitrogen that is rapidly removed by plant use, denitrification, and leaching.

Nitrogen fixation is mediated by two microbial systems. One consists of nonsymbiotic, free-living microorganisms such as members of the genera *Azotobacter, Beijerinckia, Clostridium,* and the cyanobacteria, which are capable of using nitrogen gas as their nitrogen source. A distinguishing characteristic of these organisms is their ability to form thick walls, thereby producing a dormant cyst that is resistant to drying and ultraviolet radiation; however, it is sensitive to heat. The second system involves symbiotic microbial forms, such as members of the genus *Rhizobium,* which, following infection, grow in a tumorlike nodule in the roots of leguminous plants. A mutually beneficial association is thus established in which the microorganisms use nutrients in

TABLE 53.1 Distinguishing Characteristics of *Azotobacter* Species

Species	Production of Water-Insoluble Pigment	Color of Agar Colonies	Production of Water-Soluble Pigment	Fluorescence Under Ultraviolet Light	Microscopic Appearance
A. chroococcum	+	Brown to black with age	−	None	Gram-negative; large ovoid rods, often in pairs; cysts may be present.
A. vinelandii	−	Colorless	+	Green	Gram-negative; large ovoid rods, often in pairs; cysts may be present
A. macrocytogenes*	−	Colorless	−	None	Gram-variable; large ovoid to coccoid cells, occurring singly or in pairs; no cysts

*Azomonas and *Azotobacter* are closely related genera; therefore, your isolate may be a member of this genus.

the plant sap and act to fix atmospheric nitrogen as ammonia for its subsequent assimilation into plant proteins.

This procedure consists of the following two parts designed to isolate symbiotic and nonsymbiotic nitrogen fixers:

1. *Rhizobium:* A crushed leguminous root nodule is used as the source for a stained slide preparation. The *Rhizobium* organisms, called bacteroids when present in plant cells, are distinguishable by their pleomorphism; they appear in a variety of X, Y, T, V, and stellate shapes.

2. *Azotobacter:* Colonies of these species will be isolated from an alkaline soil sample on a nitrogen-free mannitol agar medium. The differentiation between the three mannitol-utilizing *Azotobacter* species is made on the basis of the distinguishing characteristics shown in Table 53.1.

MATERIALS
Cultures

1-g samples of slightly alkaline soil. Freshly picked leguminous plants (peas, clover, or alfalfa) with root nodules.

Media

Per designated student group: one 50-ml nitrogen-free mannitol broth in screw-cap bottle and one nitrogen-free mannitol agar plate.

Reagents

Methylene blue, crystal violet, Gram's iodine, ethyl alcohol, and safranin.

Equipment

Bunsen burner, inoculating loop, glass slides, staining tray, ultraviolet lamp, and glassware marking pencil.

PROCEDURE
Azotobacter Isolation

1. Add a 1-g soil sample to a bottle of appropriately labeled, nitrogen-free mannitol broth. Tighten the cap and shake vigorously until a uniform mixture is established.

2. Incubate the culture for 4 to 7 days at room temperature, 25°C.

3. At the end of the incubation period, examine the surface of the culture for the presence of a thin film. *Note: Do not shake the flask or disturb the film.*

4. Using sterile inoculating technique, transfer a loopful of the surface film to an appropriately labeled nitrogen-free mannitol agar plate. Perform a four-way streak inoculation for the isolation of colonies (see Experiment 2 to review the procedure).

5. Incubate the plate in an inverted position for 4 to 6 days at room temperature, 25°C.

Rhizobium Isolation

1. Thoroughly rinse a nodule obtained from the roots of a leguminous plant.
2. Crush the nodule between two slides.
3. Spread a loopful of the material from the nodule in a thin layer on the surface of a clean slide.
4. Air-dry, heat fix, and stain the smear preparation with methylene blue for 1 minute. Rinse the slide under running water and blot dry with bibulous paper.

Name   Date

PART A: Ammonification

1. Test each culture as described below for the presence of ammonia on days 3, 5, and 7 following inoculation:

   Using a Pasteur pipette, place one drop of Nessler's reagent in six separate depressions on the spot plate. Add a loopful of each experimental culture and the control to a separate depression containing the Nessler's reagent. Mix by rotating the plate.

2. Using the chart, determine whether ammonia is present and, if so, its relative amount.

Color		Result
No change	0	No ammonia
Pale yellow	1+	Small amount of ammonia
Deep yellow	2+	Moderate amount of ammonia
Brown precipitate	3+	Large amount of ammonia

3. Record the following in the chart:
   a. The observed color of each Nessler's reagent culture mixture.
   b. The absence or presence, as 0, 1+, 2+, or 3+, of ammonia in each test culture.

Test Samples of Organisms	Day 3		Day 5		Day 7	
	Color	Result	Color	Result	Color	Result
Rich garden soil						
Poor garden soil						
B. cereus						
P. fluorescens						
P. vulgaris						
Control						

4. Based on the results you obtained in the soil sample cultures:
   a. Was there a difference in the amount of ammonia present in the two cultures? Explain.

   b. Was the amount of ammonia present in each soil culture different from one day of observation to the next?

5. Describe any differences observed in each of the following samples:

   a. Poor garden soil sample:

   b. Rich garden soil sample:

PART B: Nitrification

As described below, examine all flasks at 7-day intervals for nitrite and nitrate production.

1. Nitrite production in ammonium sulfate soil cultures

    a. Determine the presence or absence of ammonia with Nessler's reagent. Refer to the Principle section of Part B for the procedure to be followed.

    b. Determine the presence or absence of nitrite by mixing three drops of Trommsdorf's reagent with one drop of dilute sulfuric acid in two separate depressions on the spot plate. With a Pasteur pipette, add one drop of each soil culture to a separate reagent mixture. Mix gently with a wooden applicator. Observe for a color change.

2. Nitrate production in nitrite broth soil cultures

    a. Repeat Step 1b to determine the presence or absence of nitrite. Check for nitrate only if nitrite is absent, since the nitrate test will give a positive result in the presence of both nitrate and nitrite.

    b. Determine the presence or absence of nitrate by placing one drop of diphenylamine and two drops of concentrated H_2SO_4 into separate depressions on the spot plate. Add one drop of each soil culture to a separate reagent mixture. Mix and observe for color change.

3. Record in the chart the observed color change, if any, and the presence (+) or absence (−) of the compound being tested for the following:

	Alkaline Soil Sample			Acidic Soil Sample		
	Week			Week		
	1	2	3	1	2	3
Ammonium Sulfate Cultures Nessler's reaction for ammonia						
Trommsdorf's reaction for nitrite						
Nitrite Cultures Trommsdorf's reaction for nitrite						
Diphenylamine reaction for nitrate						

4. Cite any observable differences in the rate at which nitrite and nitrate are formed in the two experimental soil cultures.

Do you feel that your experimental result is the same as the expected result? Explain.

PART C: Denitrification

1. At 1-week intervals, examine all the test cultures for the presence or absence of an air bubble in the Durham tube. Record your observations as (+) or (−) in the chart.
2. Based on your observations, determine and record whether denitrification has taken place.

Organisms and Soils	7th Day		14th Day	
	Gas Production	Denitrification	Gas Production	Denitrification
P. aeruginosa				
P. vulgaris				
Rich soil				
Poor soil				

PART D: Nitrogen Fixation

Azotobacter Identification

⚠ **Do not look into the UV source. Use of safety glasses is recommended.**

1. Select two or three isolated colonies whose appearances differ on the agar plates.
2. Observe the selected colonies for the presence or absence of pigmentation.
3. Place two loopfuls of each colony on a separate slide. Place each slide under ultraviolet light to determine the presence or absence of green fluorescence.
4. Prepare a Gram stain of each isolate. Examine all slides microscopically to determine the Gram reaction and the size, shape, and arrangement of the cells.

5. Based on your observations, record your results in the chart.

	Colony 1	Colony 2	Colony 3
Draw a representative microscopic field.			
Microscopic morphology			
Color of colony on agar culture			
Fluorescence under ultraviolet light			
Identification of isolate			

Rhizobium Identification

1. Examine your methylene blue slide preparation under the oil-immersion objective for the presence of pleomorphic cell forms.
2. Draw a representative field of your observations.

Representative Field of *Rhizobium*	Microscopic Cell Morphology

REVIEW QUESTIONS

1. Is ammonification restricted to solely aerobic or anaerobic organisms? Explain.

2. Does proteolysis occur extracellularly or intracellularly? Explain.

3. Distinguish among ammonification, nitrification, denitrification, and nitrogen fixation.

4. How would you explain to a farmer what denitrification does to farmland soil and what measures could be taken to prevent this from occurring?

5. Distinguish between symbiotic and nonsymbiotic nitrogen fixation.

6. 🔍 Explain why only some species of microorganisms are capable of ammonification.

7. 🔍 Are anaerobic organisms capable of nitrification? Explain.

8. 🔍 Why is the medium used for isolation of *Azotobacter* devoid of nitrogen-containing compounds?

9. 🔍 What are the methods by which soil is depleted of usable forms of nitrogen?

Microbial Populations in Soil: Enumeration

LEARNING OBJECTIVES

Once you have completed this experiment, you should be

1. Familiar with the microbial soil flora.
2. Able to determine the number of bacteria and fungi present in a soil sample.

PRINCIPLE

Soil contains myriads of microorganisms, including bacteria, fungi, protozoa, algae, and viruses. Despite this diversity, bacteria, including the moldlike actinomycetes, and fungi are the most prevalent, as shown below.

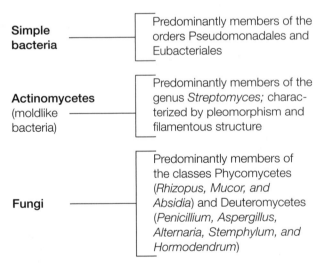

Simple bacteria	Predominantly members of the orders Pseudomonadales and Eubacteriales
Actinomycetes (moldlike bacteria)	Predominantly members of the genus *Streptomyces*; characterized by pleomorphism and filamentous structure
Fungi	Predominantly members of the classes Phycomycetes (*Rhizopus, Mucor, and Absidia*) and Deuteromycetes (*Penicillium, Aspergillus, Alternaria, Stemphylum, and Hormodendrum*)

It is essential to bear in mind that the soil environment differs from one location to another and from one period of time to another. Therefore factors such as moisture, pH, temperature, gaseous oxygen content, and organic and inorganic composition of soil are crucial in determining the specific microbial flora of a particular sample.

Just as the soil differs, microbiological methods used to analyze soil also vary. A single technique cannot be used to count all the different types of microorganisms present in a given soil sample because there is no one laboratory cultivation procedure that can provide all the physical and nutritional requirements necessary for the growth of a greatly diverse microbial population. In this experiment, only the relative numbers of bacteria, actinomycetes, and fungi are determined. The method used is the serial dilution–agar plate procedure described in Experiment 20 and illustrated in Figure 54.1. Different media are employed to support the growth of these three types of microorganisms: glycerol yeast agar for the isolation of actinomycetes, Sabouraud agar for the isolation of fungi, and nutrient agar for the isolation of bacteria. The glycerol yeast agar and Sabouraud agar are supplemented with 10 µg of chlortetracycline (Aureomycin) per milliliter of medium to inhibit the growth of bacteria.

MATERIALS

Soil

1-g sample of finely pulverized, rich garden soil in a flask containing 99 ml of sterile water; flask labeled 1:100 dilution (10^{-2}).

Media

Per designated student group: four glycerol yeast agar deep tubes, four Sabouraud agar deep tubes, four nutrient agar deep tubes, and two 99-ml flasks of sterile water.

Equipment

Bunsen burner, 12 Petri dishes, Quebec® colony counter, mechanical hand counter, sterile 1-ml pipettes, mechanical pipetting device, turntable (optional), 95% alcohol in a 500-ml beaker, and glassware marking pencil.

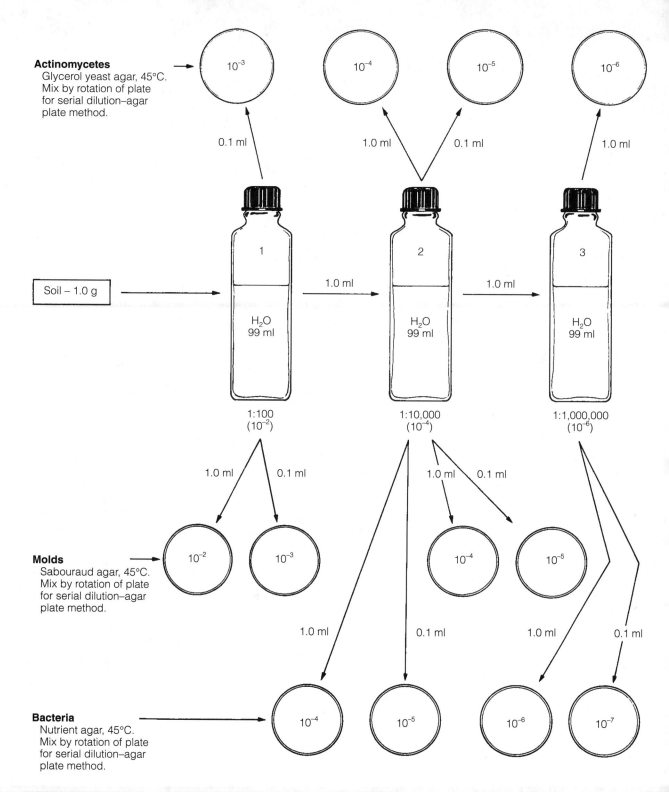

FIGURE 54.1 Procedure for enumeration of soil microorganisms

The figure contains the following labels:

Actinomycetes
Glycerol yeast agar, 45°C.
Mix by rotation of plate
for serial dilution–agar
plate method.

10^{-3} 10^{-4} 10^{-5} 10^{-6}

0.1 ml   1.0 ml   0.1 ml   1.0 ml

Soil – 1.0 g

1.0 ml   1.0 ml

Bottle 1: H_2O 99 ml — 1:100 (10^{-2})
Bottle 2: H_2O 99 ml — 1:10,000 (10^{-4})
Bottle 3: H_2O 99 ml — 1:1,000,000 (10^{-6})

Molds
Sabouraud agar, 45°C.
Mix by rotation of plate
for serial dilution–agar
plate method.

1.0 ml   0.1 ml   1.0 ml   0.1 ml

10^{-2} 10^{-3} 10^{-4} 10^{-5}

1.0 ml   0.1 ml   1.0 ml   0.1 ml

Bacteria
Nutrient agar, 45°C.
Mix by rotation of plate
for serial dilution–agar
plate method.

10^{-4} 10^{-5} 10^{-6} 10^{-7}

PROCEDURE

1. Liquefy the glycerol yeast, Sabouraud, and nutrient agar deep tubes in an autoclave or by boiling. Cool the molten agar tubes and maintain in a waterbath at 45°C.

2. Divide the Petri dishes into three groups of four; using a glassware marking pencil, label the groups as nutrient agar, glycerol yeast extract agar, and Sabouraud agar. Then, label each set of Petri dishes as follows:

   Nutrient agar: 10^{-4}, 10^{-5}, 10^{-6}, and 10^{-7} (to be used for enumeration of bacteria).

   Glycerol yeast extract agar: 10^{-3}, 10^{-4}, 10^{-5}, and 10^{-6} (to be used for enumeration of actinomycetes).

   Sabouraud agar: 10^{-2}, 10^{-3}, 10^{-4}, and 10^{-5} (to be used for enumeration of fungi).

3. With a glassware marking pencil, label the 99-ml sterile water Flasks 2 and 3.

4. Vigorously shake the provided soil sample dilution of 1:100 (10^{-2}) approximately 30 times, with your elbow resting on the table.

5. With a sterile 1-ml pipette, transfer 1 ml of the provided soil sample dilution to Flask 2 and shake vigorously as before. The final dilution is 1:10,000 (10^{-4}).

6. Using another sterile 1-ml pipette, transfer 1 ml of Dilution 2 to Flask 3 and shake vigorously as before. The final dilution is 1:1,000,000 (10^{-6}).

7. Using sterile 1-ml pipettes and antiseptic technique, add the proper amount of each dilution into each Petri dish as indicated in a–c and shown in Figure 54.1.

   a. **For molds**—in plates labeled Sabouraud agar:
   1 ml of Dilution 1 in plate to effect a 10^{-2} dilution.
   0.1 ml of Dilution 1 in plate to effect a 10^{-3} dilution.
   1 ml of Dilution 2 in plate to effect a 10^{-4} dilution.
   0.1 ml of Dilution 2 in plate to effect a 10^{-5} dilution.

   b. **For actinomycetes**—in plates labeled glycerol yeast extract agar:
   0.1 ml of Dilution 1 in plate to effect a 10^{-3} dilution.
   1 ml of Dilution 2 in plate to effect a 10^{-4} dilution.
   0.1 ml of Dilution 2 in plate to effect a 10^{-5} dilution.
   1 ml of Dilution 3 in plate to effect a 10^{-6} dilution.

   c. **For bacteria**—in plates labeled nutrient agar:
   1 ml of Dilution 2 in plate to effect a 10^{-4} dilution.
   0.1 ml of Dilution 2 in plate to effect a 10^{-5} dilution.
   1 ml of Dilution 3 in plate to effect a 10^{-6} dilution.
   0.1 ml of Dilution 3 in plate to effect a 10^{-7} dilution.

8. Check the temperature of the molten agar medium to be sure that the temperature is 45°C. Remove the tubes from the waterbath and wipe the outside surface dry with a paper towel. Using the pour-plate technique, pour the liquefied agar into the plates as shown in Figure 20.2 on page 130 and rotate gently to ensure uniform distribution of the cells in the medium.

9. Incubate the plates in an inverted position at 25°C. Perform colony counts on nutrient agar plate cultures in 2 to 3 days and on the remaining agar plate cultures in 4 to 7 days.

Name _____   Date _____

1. Using an electronic colony counter or a Quebec colony counter and a mechanical hand counter, observe all the colonies on each plate. Plates with more than 300 colonies cannot be counted and should be designated as **too numerous to count (TNTC);** plates with fewer than 30 colonies should be designated as **too few to count (TFTC).** Count only plates with between 30 and 300 colonies.

2. Determine the number of organisms per milliliter of original culture on all plates other than those designated as TFTC or TNTC by multiplying the number of colonies counted by the dilution factor. Refer to the Observations and Results section of Experiment 20 for examples of the calculation of cell counts.

3. Record your observations and calculated cell counts per gram of sample in the chart.

Organism	Dilution	Number of Colonies	Organisms per Gram of Soil
Bacteria	10^{-4}		
	10^{-5}		
	10^{-6}		
	10^{-7}		
Actinomycetes	10^{-3}		
	10^{-4}		
	10^{-5}		
	10^{-6}		
Molds	10^{-2}		
	10^{-3}		
	10^{-4}		
	10^{-5}		

4. Based on your results, which of the three types of soil organisms were most abundant in your sample? Least abundant?

REVIEW QUESTIONS

1. Would you expect to be able to duplicate your results if a soil sample were taken from the same location at a different time of the year? Explain.

2. In the experiment performed, why wasn't the same medium used for enumeration of all three types of soil organisms?

3. Would you expect to be able to isolate an anaerobic organism from any of your cultures? Explain.

4. Explain why most microorganisms are present in the upper layers of the soil.

5. Following the nuclear disaster at Chernobyl, the regional microbial flora was destroyed. What impact did this have on higher forms of plant and animal life in this area?

Isolation of Antibiotic-Producing Microorganisms and Determination of Antimicrobial Spectrum of Isolates

LEARNING OBJECTIVES

Once you have completed this experiment, you should be able to

1. Isolate antibiotic-producing microorganisms.
2. Determine the spectrum of antimicrobial activity of the isolated antibiotic.

PRINCIPLE

Soil is the major repository of microorganisms that produce **antibiotics** capable of inhibiting the growth of other microorganisms. Clinically useful antibiotics have been isolated from four groups of soil microorganisms—*Streptomyces, Bacillus, Penicillium*, and *Cephalosporium*—that represent three microbial types, namely, actinomycetes, true bacteria, and molds.

Although soils from all parts of the world are continually screened in industrial laboratories for the isolation of new antibiotic-producing microorganisms, industrial microbiology is directing its energies toward chemical modification of existing antibiotic substances. This is accomplished by adding or replacing chemical side chains, reorganizing intramolecular bonding, or producing mutant microbial strains capable of excreting a more potent form of the antibiotic. The establishment of chemical congeners has been responsible for the circumvention of antibiotic resistance, minimizing adverse side effects in the host and increasing the effective spectrum of a given antibiotic.

In Part A of this experiment, you will use the **crowded-plate technique** for isolation of antibiotic-producing microorganisms from two soil samples, one of which is seeded with *Streptomyces griseus* to serve as a positive control. Figure 55.1 illustrates the procedure to be followed. In Part B, isolates exhibiting antibiotic activity will be screened against several different microorganisms to establish their effectiveness.

MATERIALS

Cultures

For Part B: 24-hour trypticase soy broth cultures of *Escherichia coli, Staphylococcus aureus, Mycobacterium smegmatis*, and *Pseudomonas aeruginosa*.

Soil Suspensions

For Part A: 1:500 dilution of soil sample suspension (0.1 g of soil per 50 ml of tap water) to serve as an unknown; 1:500 dilution of soil sample seeded with *S. griseus* (0.1 g of soil per 50 ml of tap water) to serve as a positive control.

Media

Per designated student group: Part A: Six 15-ml trypticase soy agar deep tubes, and two trypticase soy agar slants. Part B: Two trypticase soy agar slants.

Equipment

Part A: 500-ml beaker, test tubes, test tube rack, sterile Petri dishes, inoculating needle, hot plate, thermometer, 1-ml and 5-ml pipettes, mechanical pipetting device, and magnifying hand lens. Part B: Bunsen burner, inoculating loop, and glassware marking pencil.

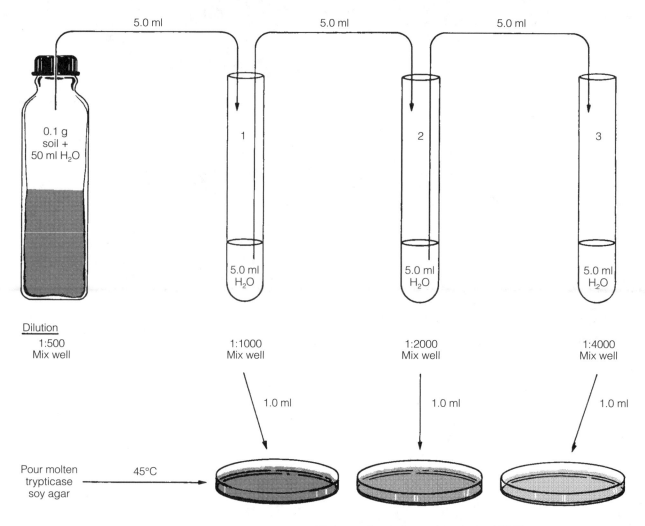

FIGURE 55.1 Crowded-plate technique for isololation of antibiotic-producing microorganisms

PART A: Isolation of Antibiotic-Producing Microorganisms

PROCEDURE

1. Label two sets of three sterile Petri dishes with the types of soil samples being used and dilutions (1:1000, 1:2000, and 1:4000).

2. Place six trypticase soy agar deep tubes into a beaker of water and bring to 100°C on a hot plate. Once agar is liquefied, add cool water to the waterbath. Cool to 45°C, checking the temperature with a thermometer.

3. Prepare a serial dilution of the unknown and positive control 1:500 soil samples as follows (refer to Figure 55.1):

    a. Label three test tubes 1, 2, and 3. With a pipette, add 5 ml of tap water to each tube.

    b. Shake the provided 1:500 soil sample thoroughly for 5 minutes to effect a uniform soil–water suspension.

    c. Using a 5-ml pipette, transfer 5 ml from the 1:500 dilution to Tube 1 and mix. The final dilution is 1:1000.

    d. Using another pipette, transfer 5 ml from Tube 1 to Tube 2 and mix. The final dilution is 1:2000.

e. Using another pipette, transfer 5 ml from Tube 2 to Tube 3 and mix. The final dilution is 1:4000.

f. Using separate 1-ml pipettes, transfer 1 ml of the 1:1000, 1:2000, and 1:4000 dilutions to their appropriately labeled Petri dishes.

g. Pour one tube of molten trypticase soy agar, cooled to 45°C, into each plate and mix by gentle rotation.

h. Allow all plates to solidify.

4. Incubate all plates in an inverted position for 2 to 4 days at 25°C.

5. After incubation, aseptically isolate one colony showing a zone of growth inhibition from each soil culture with an inoculating needle and streak onto trypticase soy agar slants labeled with the soil sample from which the isolate was obtained.

6. Incubate the slants for 2 to 4 days at 25°C. These will serve as stock cultures of antibiotic-producing isolates to be used in Part B.

PART B: Determination of Antimicrobial Spectrum of Isolates

PROCEDURE

1. Label the trypticase soy agar plates with the soil sample source of the isolate.

2. Using sterile technique, make a single-line streak inoculation of each isolate on the surface of an agar plate so as to divide the plate in half as shown:

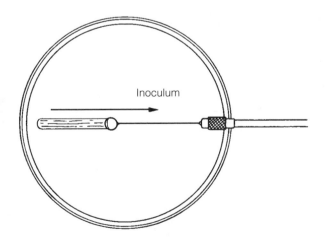

3. Incubate the plates in an inverted position for 3 to 5 days at 25°C.

4. Following incubation, on the bottom of each plate draw four lines perpendicular to the growth of the antibiotic-producing isolate as shown:

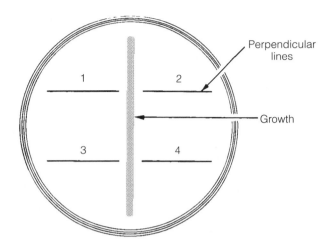

5. Aseptically make a single-line streak inoculation of each of the four test cultures following the inoculation template on each plate. Start close to, but not touching, the growth of the antibiotic-producing isolate and streak toward the edge of the plate.

6. Incubate the plates in an inverted position for 24 hours at 37°C.

Name _____   Date _____

PART A: Isolation of Antibiotic-Producing Microorganisms

1. Examine all crowded-plate dilutions for colonies exhibiting zones of growth inhibition. Use a hand magnifying lens if necessary.
2. Record in the chart the number of colonies showing zones of inhibition.

	Number of Colonies		
	Dilutions		
Soil Sample	1:1000	1:2000	1:4000
Unknown			
Positive control			

PART B: Determination of Antimicrobial Spectrum of Isolates

1. Examine all plates for inhibition of test organisms.
2. Draw a representation of your observed antibiotic activity against the test organisms.

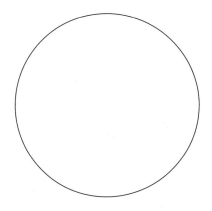

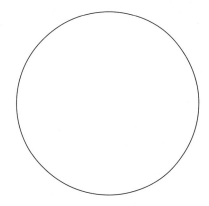

Antibiotic-Producing Isolate 1   Antibiotic-Producing Isolate 2

3. Based on your observations, record in the chart the presence (+) or absence (−) of antibiotic activity against each of the test organisms and the spectrum of antimicrobial activity (broad or narrow).

	Test Organisms				
Soil Sample	E. coli Gram-Negative	S. aureus Gram-Positive	P. aeruginosa Gram-Negative	M. smegmatis Acid-Fast	Spectrum
Unknown					
Positive control					

REVIEW QUESTIONS

1. Why is it frequently advantageous to modify antibiotics in industrial laboratories?

2. Is the ability to produce antibiotics limited only to bacterial species? Explain.

3. Do you feel that sufficient test organisms were used in Part B to determine fully the spectrum of activity of each isolated antibiotic? Explain.

Isolation of *Pseudomonas* Species by Means of the Enrichment Culture Technique

LEARNING OBJECTIVE

Once you have completed this experiment, you should

1. Understand the enrichment culture technique for the isolation of a specific microbial cell type.

PRINCIPLE

The enrichment culture technique is used for the isolation of a specific type of microorganism from an environment that is replete with different types of microbes. In such an environment, the desired organism may be present only in very small numbers because of the competitive activities of this diverse microbial population. Under these circumstances, the use of conventional enriched media is not suitable for the selection of a specific cell type. These special-purpose media are supplemented with a variety of enriching nutrients capable of supporting the growth of many organisms rather than a single cell type in the test sample. Enrichment broths, on the other hand, are designed to contain a limited number of specific substrates that will preferentially promote the growth of the desired microorganisms.

The enrichment culture technique employs such a specifically designed enrichment broth for the initial inoculation of the test sample. Once growth occurs in the primary culture, it is sequentially transferred into a fresh medium of the same composition until the desired microorganisms are predominant in the culture. These organisms are capable of exponential growth because of their ability to adapt to the medium and to enzymatically use the incorporated substrate(s) as an energy source. Most of the competitors, however, are incapable of utilizing the substrate(s) and

therefore remain in the lag phase of the growth curve. In some instances the organisms to be isolated do not grow more rapidly than their competitors. Instead, they produce a growth inhibitor that greatly suppresses the growth of the competing population. After the serial transfer through the broth medium, the culture is streaked on an agar plate of the same composition as the enrichment broth for the isolation and subsequent identification of the discrete colonies.

The use of the enrichment culture technique has a wide range of applications in clinical, industrial, and environmental microbiology. Medically, the procedure is routinely used for the isolation of intestinal pathogens from fecal samples where these organisms may be present only in low concentrations during the infectious process. Enrichment methods may be used to isolate and cultivate specific soil microorganisms for the production of industrial products such as steroids, enzymes, and vitamins. Likewise, a beneficial environmental application may involve the isolation by enrichment of petroleum-utilizing microorganisms such as *Pseudomonas* that would be capable of degrading environmentally destructive oil spills in waterways.

In this experimental procedure, a compost or a rich garden soil sample will be used to isolate *Pseudomonas* species by means of the enrichment culture procedure. Members of the genus *Pseudomonas* can utilize mandelic acid aerobically as their sole carbon and energy source. Therefore, this compound is the most important factor in the enrichment broth, which also contains a number of inorganic salts. The pseudomonads are gram-negative, motile organisms that generally produce a diffusible yellow-green pigment. In addition, they commonly reduce nitrates (NO_3^-) and produce

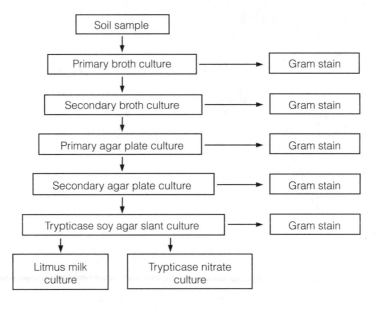

FIGURE 56.1 Enrichment culture procedure schema

an alkaline or proteolytic reaction in litmus milk. The schema for the experimental procedure to be followed is illustrated in Figure 56.1.

MATERIALS
Cultures
Rich garden soil or compost sample.

Media
Per designated student group: Two Erlenmeyer flasks containing 20 ml of basal salts broth supplemented with 2 ml of 2.5% mandelic acid, two agar plates of the same composition as the broth, one trypticase nitrate broth, one litmus milk, and one trypticase soy agar slant.

Reagents
Crystal violet, Gram's iodine, 95% ethanol, safranin, Solution A (sulfanilic acid), Solution B (alpha-naphthylamine), and zinc powder. *Note: Solutions A and B are not Barritt's reagent.*

Equipment
Sterile 10-ml, 5-ml, and 1-ml pipettes, mechanical pipetting device, microspatula, Bunsen burner, staining tray, glass slides, lens paper, bibulous paper, inoculating loop, and glassware marking pencil.

PROCEDURE
Session 1: Primary Broth Culture Preparation

1. Inoculate an appropriately labeled Erlenmeyer flask containing the enrichment broth by adding an amount of the soil sample equivalent to the size of a pea with a microspatula. Gently swirl the flask to mix the culture.

2. Incubate the primary broth culture for 24 hours at 30°C.

Session 2: Secondary Broth Culture Preparation

1. Examine the primary culture for presence of growth. If growth is not present, return the flask to the incubator for an additional 24 hours.

2. If growth is present, aseptically transfer 1 ml of the primary culture to an appropriately labeled Erlenmeyer flask containing fresh enrichment medium. Swirl the flask.

3. Incubate the secondary broth culture for 24 hours at 30°C.

4. Prepare and examine a Gram-stained smear from the primary culture.

5. Refrigerate the primary broth culture.

Session 3: Primary Agar Plate Preparation

1. If growth is present in the secondary broth culture, aseptically perform a four-way streak inoculation on the appropriately labeled agar plate of the enrichment medium (refer to Experiment 2).
2. Incubate the agar plate culture in an inverted position for 24 hours at 30°C.
3. Prepare and examine a Gram-stained smear of the secondary broth culture.
4. Refrigerate the secondary broth culture.

Session 4: Secondary Agar Plate Preparation

1. Examine the primary plate culture for the presence of discrete colonies. Using a discrete colony:
   a. Aseptically prepare and examine a Gram-stained smear.
   b. Aseptically perform a four-way streak inoculation on an appropriately labeled agar plate of the enrichment medium.
2. Incubate the secondary agar plate culture in an inverted position for 24 hours at 30°C.
3. Refrigerate the primary agar plate culture.

Session 5: Pure Culture Isolation

1. Examine the secondary agar plate culture. If the cultural characteristics of discrete colonies appear to be similar:
   a. Prepare and examine a Gram-stained smear from a discrete colony.
   b. Aseptically pick a discrete colony and inoculate a trypticase soy agar slant by means of a streak inoculation.
2. Incubate the agar slant culture for 24 to 48 hours at 30°C.
3. Refrigerate the secondary agar plate culture.

Session 6: Genus Identification of Isolate

1. Prepare and examine a Gram-stained smear from the trypticase agar slant culture.
2. Using the trypticase agar slant cultures, aseptically inoculate the appropriately labeled tubes of trypticase nitrate broth and litmus milk by means of a loop inoculation.
3. Incubate the litmus milk and trypticase nitrate broth cultures for 24 to 48 hours at 30°C.

Experiment 56 / Observations and Results

Name Date

1. Examine the Gram-stained smear preparations from the enrichment broth, agar plate, and trypticase soy agar cultures. Record your observations of cellular morphology and Gram reaction in the chart below.
2. Examine the enrichment agar plate cultures for the presence of discrete colonies. Record your observations of the cultural characteristics of these colonies in the chart below (refer to Experiment 3).

Culture	Gram Stain	Cultural Characteristics
Primary broth culture		
Secondary broth culture		
Primary plate culture		
Secondary plate culture		
Trypticase soy agar slant culture		

3. Observe the litmus milk culture. Below, record the type of reaction that has taken place (refer to Experiment 28). Litmus milk reaction:

4. Perform the nitrate-reduction test on the trypticase nitrate broth culture. Record your result below (refer to Experiment 29). Nitrate reduction (+ or −):

REVIEW QUESTIONS

1. A child with a severe gastroenteritis that is suspected to be food poisoning caused by a *Salmonella* species is hospitalized. Explain why the laboratory supervisor uses an enrichment broth technique rather than selective media to confirm her suspicions.

2. A patient is afflicted with a disease that generates a large volume of gelatinous abdominal ascites. Drainage by surgical means is not successful. The use of a microbial enzyme capable of degrading this viscous ascites is suggested. Explain how you would go about isolating an organism that is enzymatically competent to act on this unusual substrate.

Bacterial Genetics

LEARNING OBJECTIVES

Once you have completed the experiments in this section, you should be able to demonstrate the applicability of bacterial test systems in genetic-related studies. The procedures include

1. Enzyme induction.
2. Transfer of genetic material by means of conjugation.
3. Isolation of a streptomycin-resistant mutant.
4. Detection of potential chemical carcinogens.
5. Transformation of an antibiotic-sensitive bacterium to one that is antibiotic resistant by the insertion of a free DNA plasmid.

INTRODUCTION

In recent years, bacteria have proved to be essential organisms in research into the structure and function of DNA, the universal genetic material. Their use is predicated on the following:

1. Their haploid genetic state, which allows the phenotypic, observable expression of a genetic trait in the presence of a single mutant gene.
2. Their rapid rate of growth, which permits observation of transmission of a trait through many generations.
3. The availability of large test populations, which allows isolation of spontaneous mutants and their induction by chemical and physical mutagenic agents.
4. Their low cost of maintenance and propagation, which make it possible to perform a large number of experimental procedures.

In the following experiments bacterial test systems are used to demonstrate enzyme induction, screening for chemical carcinogens, and the genetic phenomena of mutation and genetic transfer. The last two mechanisms introduce genetic variability, which is essential for evolutionary survival in asexually reproducing bacterial populations.

Point mutations are permanent, sudden qualitative alterations in genetic material that arise as a result of the addition, deletion, or substitution of one or more bases in the region of a single gene. As a result, one or more amino acid substitutions occur during translation, and a protein that may be inactive, reduced in activity, or entirely different is synthesized. **Spontaneous mutations** are the result of the chemical and physical components in the organism's natural environment. The rate at which they occur is extremely low in all organisms. For example, in *Escherichia coli*, the spontaneous mutation rate at a single locus (specific site on the DNA) is estimated to be in the area of 1×10^{-7}, and the possibility of a mutation at any locus in the genome is approximately 1×10^{-4}. **Induced mutations** are genetic changes resulting from the organism's exposure to an artificial physical or chemical mutagen, that is, an agent capable of inducing a mutation. The resultant mutations are of the same type that occur spontaneously; however, their rate is increased, and in some cases dramatically so.

Transfer of genetic material and its subsequent incorporation into the bacterial genome is also a source of genetic variation in some bacteria. This transfer may occur by means of the following:

1. **Conjugation:** A mating process between "sexually" differentiated bacterial strains that allows unidirectional transfer of genetic material.
2. **Transduction:** A bacteriophage-mediated transfer of genetic material from one cell to another.
3. **Transformation:** A genetic alteration in a cell resulting from the introduction of free DNA from the environment across the cell membrane.

Enzyme Induction

LEARNING OBJECTIVES

Once you have completed this experiment, you should be able to

1. Understand the mechanism of the lactose operon.
2. Understand the factors affecting the expression of the β-galactosidase gene.

PRINCIPLE

Although bacteria possess a single chromosome, each cell is capable of synthesizing hundreds of different enzymes. Studies have shown that these enzymes are not present within the cells in equal concentrations. Some enzymes, called **constitutive enzymes,** are synthesized at a constant rate regardless of conditions in the cell's environment. Synthesis of other enzymes, called **adaptive enzymes,** occurs only when necessary, and is subject to regulatory mechanisms that are dependent on the environment. One such mechanism, **induction,** requires the presence of a substrate, the inducer, in the environment to initiate synthesis of its specific enzyme, called an **inducible enzyme.** An extensively studied inducible enzyme in *E. coli* is **β-galactosidase,** which acts on the disaccharide lactose to yield the monosaccharides glucose and galactose. The gene for β-galactosidase is a member of a cluster of genes, called an **operon,** that is involved in the metabolism of lactose. The member genes of the lactose (Lac) operon function as a unit, all being transcribed only when the inducer, lactose, is present in the surrounding medium. See Figure 57.1.

To illustrate β-galactosidase induction, two test strains of *E. coli* will be used: a prototrophic (wild type) strain (lactose-positive) and an auxotrophic (mutant) strain (lactose-negative), which carries a mutation in the gene for β-galactosidase as well as a mutation in the lactose operon regulatory gene. Both test strains will be grown in the following media:

1. Inorganic synthetic medium lacking an organic carbon and energy source that is required by the heterotrophic *E. coli.*
2. Inorganic synthetic medium plus glucose, which can be utilized by both strains as a carbon and energy source.
3. Inorganic synthetic medium plus lactose, which can be utilized only by the prototrophic strain.

Orthonitrophenyl-β-D-galactoside (ONPG), a colorless analog of lactose, can serve as the substrate for the induction of β-galactosidase synthesis. As the inducer, it is hydrolyzed to galactose and a yellow nitrophenolate ion. Following a short incubation period, growth in all the cultures will be determined by spectrophotometry. Induction of β-galactosidase synthesis and activity will be indicated by the appearance of a yellow color in the medium following addition of ONPG, which occurs only in the presence of the nitrophenolate ion. Absence of this macroscopically visible color change indicates that enzyme induction in the lactose-negative strain did not occur.

MATERIALS

Cultures

25-ml inorganic synthetic broth suspensions of 12-hour nutrient agar cultures of a lactose-positive *E. coli* strain (ATCC™ e 23725) and a lactose-negative *E. coli* strain (ATCC e 23735) adjusted to an O.D. of 0.1 at 600 nm.

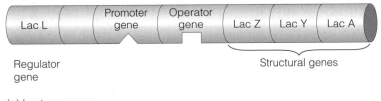

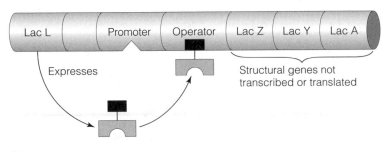

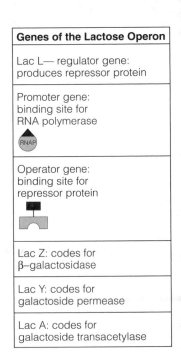

(a) Lactose operon

Regulator gene

Structural genes

(b) No lactose present

Expresses

Structural genes not transcribed or translated

Genes of the Lactose Operon

Lac L— regulator gene: produces repressor protein
Promoter gene: binding site for RNA polymerase
Operator gene: binding site for repressor protein
Lac Z: codes for β–galactosidase
Lac Y: codes for galactoside permease
Lac A: codes for galactoside transacetylase

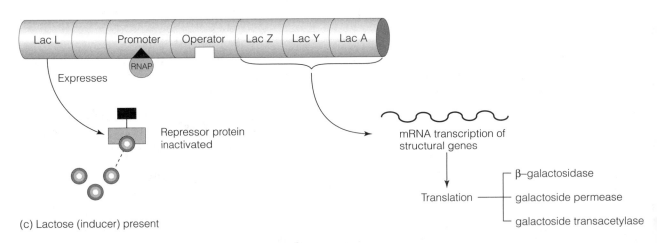

Expresses

Repressor protein inactivated

mRNA transcription of structural genes

Translation ─── β–galactosidase
           galactoside permease
           galactoside transacetylase

(c) Lactose (inducer) present

FIGURE 57.1 Enzyme induction. The mechanism of operation of the lactose operon.

Media

Per designated student group: dropper bottles of sterile 10% glucose, 10% lactose, and water.

Reagents

Dropper bottles of toluene and orthonitrophenyl-β-D-galactoside (ONPG).

Equipment

1-ml and 5-ml sterile pipettes, mechanical pipetting device, six sterile 13- × 100-mm test tubes, test tube racks, six sterile 25-ml Erlenmeyer flasks, Bausch & Lomb Spectronic 20 spectrophotometer, shaking waterbath incubator, and glassware marking pencil.

PROCEDURE

1. Label three sterile test tubes and three sterile 25-ml Erlenmeyer flasks as "Lac⁺" (lactose-positive) and the name of the substrate to be added (glucose, lactose, or water). Similarly label three sterile flasks "Lac⁻" (lactose-negative) for each test organism.

2. Using sterile 5-ml pipettes, aseptically transfer 5 ml of the Lac⁺ and Lac⁻ inorganic synthetic broth cultures to their respectively labeled test tubes.

3. Using a sterile 1-ml pipette, aseptically add 0.5 ml of the glucose and lactose solutions and 0.5 ml of sterile distilled water to the appropriately labeled tubes.

4. Determine the optical density (O.D.) of all cultures at a wavelength of 600 nm.

5. Aseptically transfer each culture to its appropriately labeled flask. *(Note: If side-arm flasks are available, additions and O.D. readings may be made directly.)*

6. Incubate all flasks for 2 hours in a shaking waterbath at 37°C and 100 strokes per minute.

7. Following incubation, transfer all cultures back to their appropriately labeled test tubes.

8. Determine the O.D. for each culture at a wavelength of 600 nm.

9. To each culture, add 5 drops of toluene and shake vigorously (toluene ruptures the cells, releasing intact enzymes).

10. To each culture, add 5 drops of ONPG solution.

11. Incubate all cultures for 40 minutes at 37°C.

12. Observe all cultures for development of yellow color.

Name _____   Date _____

1. Record the initial O.D. of each Lac$^+$ and Lac$^-$ *E. coli* broth culture at 600 nm.
2. Following incubation, determine and record the O.D. of each culture. Based on your observations, indicate whether growth has occurred in each of the cultures.
3. Following the addition of ONPG, observe the cultures for the presence of yellow coloration indicative of β-galactosidase synthesis and activity. Record the colors of your cultures and the presence (+) or absence (−) of the β-galactosidase activity.

Cultures	O.D. at 600 nm		Growth (+) or (−)	Color of Culture with ONPG	β-galactosidase (+) or (−)
	Prior to Incubation	Following Incubation			
Lac$^+$ *E. coli* Glucose					
Lactose					
Water					
Lac$^-$ *E. coli* Glucose					
Lactose					
Water					

4. Explain the absence of growth in some of the cultures.

REVIEW QUESTIONS

1. Distinguish between constitutive enzymes and inducible enzymes.

2. Explain what is meant by an operon.

3. Explain the purpose of the ONPG in the procedure.

4. Compare and contrast the methods for DNA transfer in microbial cells.

5. How can you explain why *Staphylococcus aureus,* which was initially sensitive to penicillin, is now resistant to this antibiotic?

Bacterial Conjugation

LEARNING OBJECTIVE

Once you have completed this experiment, you should be able to

1. Demonstrate genetic recombination in bacteria by the process of conjugation.

PRINCIPLE

Genetic variability is essential for the evolutionary success of all organisms. In diploid eukaryotes, the processes of **crossing over,** (exchange of genetic material between homologous chromosomes) and **meiosis** contribute to this variability. In haploid, asexually reproducing prokaryotic organisms, genetic recombination may occur by **conjugation, transduction,** and **transformation.** In this experiment, only the process of conjugation is considered.

Conjugation is a mating process during which a unidirectional transfer of genetic material occurs at physical contact between two "sexually" differentiated cell types. This differentiation, or existence of different mating strains in some bacteria, is determined by the presence of a **fertility factor,** or **F factor,** within the cell. Cells that lack the F factor are recipients (females) of the genetic material during conjugation and are designated as **F⁻**. Cells possessing the F factor have the ability to act as genetic donors (males) during mating. If this F factor is extrachromosomal (a **plasmid** or **episome**), the cells are designated as **F⁺**; most commonly only the F factor is transferred during conjugation. If this factor becomes incorporated into the bacterial chromosome, there is a transfer of chromosomal genes, although generally not involving the entire chromosome or the F factor. The resulting cells are designated **Hfr,** for **high-frequency recombinants.**

In this experiment, you will prepare a mixed culture representing a cross between an Hfr prototrophic (wild type) strain of *E. coli* that is streptomycin-sensitive and an F⁻ auxotrophic (mutant) *E. coli* strain that requires threonine (thr), leucine (leu), and thiamine (thi) and is streptomycin-resistant (Str-r). Following a short incubation period, isolation of only the threonine and leucine recombinants will be performed by plating the mixed culture on a minimal medium containing streptomycin and thiamine. The streptomycin is incorporated into the medium to inhibit the growth of the wild-type, streptomycin-sensitive (Str-s) parental Hfr cells. The thiamine is required as an essential growth factor for the thiamine-negative (thi⁻) recombinant cells. Because of its distant location on the chromosome, this marker will not be transferred during the short mating period. A genetic map showing the site of origin of transfer and the locations of the relevant markers is given in Figure 58.1.

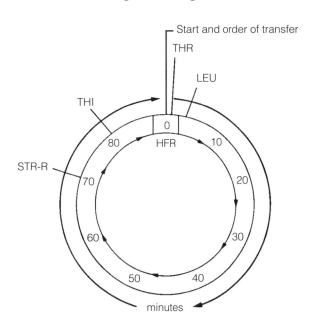

FIGURE 58.1 Genetic map of *Escherichia coli*

MATERIALS

Cultures

12-hour nutrient broth cultures of F⁻ E. coli strain thr⁻, leu⁻, thi⁻, and Str-r (ATCC™ e 23724); and Hfr E. coli strain Str-s (ATCC e 23740).

Media

Per designated student group: three plates of minimal medium plus streptomycin and thiamine.

Equipment

Bunsen burner, beaker with 95% ethyl alcohol, L-shaped bent glass rod, 1-ml sterile pipettes, mechanical pipetting device, sterile 13- × 100-mm test tube, waterbath shaker, and glassware marking pencil.

PROCEDURE

1. With separate sterile 1-ml pipettes, aseptically transfer 1 ml of the F⁻ E. coli culture and 0.3 ml of the Hfr E. coli culture into the sterile 13- × 100-mm test tube.

2. Mix by gently rotating the culture between the palms of your hands.

3. Incubate the culture for 30 minutes at 37°C in a waterbath shaker at the lowest speed setting.

4. Appropriately label two minimal plus streptomycin and thiamine agar plates. Using the spread-plate technique illustrated in Figure 58.2, prepare control plates of the parental Hfr and F⁻ E. coli strains.

   a. Aseptically add 0.1 ml of each E. coli strain to its appropriately labeled agar plate.

   b. With an alcohol-dipped and flamed glass rod, spread the inoculum over the entire surface of the agar plate.

5. Following incubation, *vigorously* agitate the mixed culture to terminate the genetic transfer.

6. Appropriately label a minimal plus streptomycin and thiamine plate. Aseptically add 0.1 ml of the mixed culture. Spread the inoculum over the entire surface with a sterile glass rod.

7. Incubate all plates in an inverted position for 48 hours at 37°C.

(a) Dip the bent glass rod into the beaker of 95% ethyl alcohol.

(b) Sterilize the glass rod by flaming with a Bunsen burner.

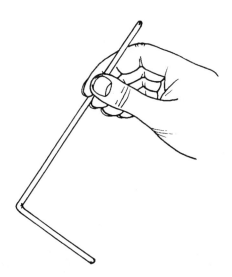

(c) Remove from Bunsen burner, allow flame to extinguish, and cool the glass rod.

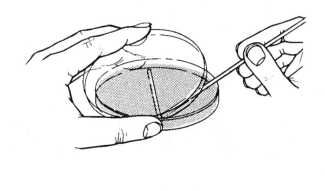

(d) Spread the inoculum over the agar surface by rotating the plate.

FIGURE 58.2 Spread-plate technique

Name _____   Date _____

1. Observe all plates for the presence (+) or absence (−) of colonies. Record your results in the chart.

	Hfr *E. coli* Plate	F⁻ *E. coli* Plate	Mixed-Culture Plate
Growth (+) or (−)			

2. Do you expect any growth to be present on the two parental *E. coli* minimal agar plates? Explain.

3. Did genetic recombination occur? Explain how your observations support your answer.

REVIEW QUESTIONS

1. Explain how genetic variations may be introduced in eukaryotic and prokaryotic cells.

2. Explain the significance of the F factor.

3. Distinguish between F^+ and Hfr bacterial strains.

4. Explain the importance of the streptomycin marker in the parental *E. coli* strains.

Isolation of a Streptomycin-Resistant Mutant

LEARNING OBJECTIVE

Once you have completed this experiment, you should be able to

1. Isolate a streptomycin-resistant mutant in a prototrophic bacterial population by means of the gradient-plate technique.

PRINCIPLE

Mutation, a change in the base sequence of a single gene, although infrequent, is one of the sources of genetic variability in cells. In some instances, these changes enable the cell to survive in an otherwise deleterious environment. An example of such a genetic adaptation is the development of **antibiotic resistance** in a small population of microorganisms prior to the advent and large-scale use of these agents. This microbial characteristic of antibiotic resistance is of major clinical importance because the number of drug-resistant microbial strains continues to increase due to their extensive use and frequent misuse, which over the years have selected for the drug-resistant strains by their microbicidal effects on the sensitive cell forms. These agents select for the resistant mutant and do not act as inducers of the mutation.

In a drug-resistant organism, the mutated gene enables the cell to circumvent the antimicrobial effect of the drug by a variety of mechanisms, including the following:

1. The production of an enzyme that alters the chemical structure of the antibiotic, as in penicillin resistance.

2. A change in the selective permeability of the cell membrane, as in streptomycin resistance.

3. A decrease in the sensitivity of the organism's enzymes to inhibiting mechanisms, as in the resistance to streptomycin, which interferes with the translation process at the ribosomes.

4. An overproduction of a natural substrate (metabolite) to compete effectively with the drug (antimetabolite), as in the resistance to sulfonamides, which produce their antimicrobial effect by competitive inhibition.

The following procedure is designed to allow you to isolate a streptomycin-resistant mutant from a prototrophic (wild type, streptomycin-sensitive) *Escherichia coli* culture by means of the **gradient-plate technique.** This requires preparation of a double-layered agar plate as illustrated in Figure 59.1. The lower,

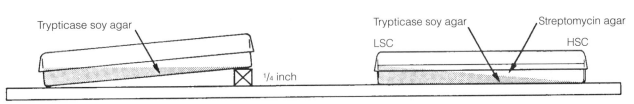

FIGURE 59.1 Preparation of a streptomycin gradient plate

slanted agar-medium layer lacks streptomycin. When poured over the lower slanted layer, the molten agar medium containing the antibiotic will produce a streptomycin concentration gradient in the surface layer. Following a spread-plate inoculation of the prototrophic test culture and incubation, the appearance of colonies in a region of high streptomycin concentration is indicative of streptomycin-resistant mutants.

MATERIALS

Cultures

24-hour nutrient broth culture of *E. coli.*

Media

Per designated student group: two 10-ml trypticase soy agar deep tubes.

Reagent

Stock streptomycin solution (10 mg per 100 ml of sterile distilled water).

Equipment

Sterile Petri dish (100 × 15 mm), sterile 1-ml pipettes, mechanical pipetting device, bent glass rod, beaker with 70% ethanol, waterbath, and glassware marking pencil.

PROCEDURE

1. In a hot waterbath, melt two trypticase soy agar tubes. Cool and maintain at 45°C.

2. Place a pencil under one end of a sterile Petri dish, pour in a sufficient amount of the molten agar medium to cover the entire bottom surface, and allow to solidify in the slanted position.

3. Using a sterile 1-ml pipette, add 0.1 ml of the streptomycin solution to a second tube of molten trypticase soy agar. Mix by rotating the tube between the palms of your hands.

4. Place the dish in a horizontal position, pour in a sufficient amount of the molten agar medium containing streptomycin to cover the gradient agar layer, and allow to solidify.

5. With a sterile 1-ml pipette, add 0.2 ml of the *E. coli* test culture. With an alcohol-dipped and flamed bent glass rod, spread the culture over the entire agar surface as illustrated in Figure 58.2.

6. Incubate the appropriately labeled culture in an inverted position for 48 hours at 37°C.

7. Following incubation, select one or two isolated colonies present in the middle of the streptomycin concentration gradient. With a sterile inoculating loop, streak the selected colonies toward the high-concentration end of the plate.

8. Incubate the plate in an inverted position for 48 hours at 37°C.

Name _____   Date _____

1. Following the initial incubation, observe the plate for the appearance of discrete colonies and indicate their positions in the diagram (LSC = low streptomycin concentration; HSC = high streptomycin concentration).

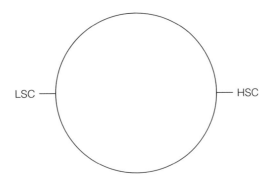

2. Following the second incubation period, observe for a line of growth from the streaked colonies into the area of high streptomycin concentration. Growth in this area is indicative of streptomycin-resistant mutants.

3. Indicate the observed line(s) of growth in the diagram.

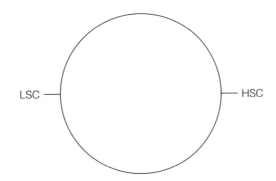

REVIEW QUESTIONS

1. What mechanisms are responsible for antibiotic resistance?

2. Why is it necessary to use an antibiotic gradient-plate preparation for isolation of mutants?

3. 🔍 Why has there been an increase in drug-resistant bacterial strains in recent years?

4. 🔍 Does the streptomycin in the medium cause the mutations? Explain.

The Ames Test: A Bacterial Test System for Chemical Carcinogenicity

LEARNING OBJECTIVE

Once you have completed this experiment, you should be able to

1. Screen for potential chemical carcinogens using a bacterial test system.

PRINCIPLE

Our exposure to a wide variety of chemical compounds has increased markedly over the past decades. Oncological epidemiologists strongly suspect that the intrusion of these chemicals in the form of industrial pollutants, pesticides, food additives, hair dyes, cigarette smoke, and the like may play a significant role in the induction of malignant transformations in humans. From a genetic aspect there is strong evidence linking **carcinogenicity** to **mutagenicity.** Research indicates that approximately 90% of the chemicals proved to be carcinogens are mutagens; they cause cancer by inducing mutations in somatic cells. These mutations are most frequently a result of base substitutions, the substitution of one base for another in the DNA molecule, and frameshift mutations, a shift in the reading frame of a gene because of the addition or deletion of a base.

In view of the rapid advent of new products and new industrial processes with their resultant pollutants, it is essential to determine their potential genetic hazards. Despite the fact that mammalian cell structure and human enzymatic pathways differ from those in bacteria, the chemical nature of DNA is common to all organisms; this permits the use of bacterial test systems for the rapid detection of possible mutagens and therefore possible carcinogens.

The **Ames test** is a simple and inexpensive procedure that uses a bacterial test organism to screen for mutagens. The test organism is a histidine-negative (his⁻) and biotin-negative (bio⁻) auxotrophic strain of *Salmonella typhimurium* that will not grow on a medium deficient in histidine unless a back mutation to his⁺ (histidine-positive) has occurred. It is recognized that the mutagenic effect of a chemical is frequently influenced by the enzymatic pathways of an organism, whereby nonmutagens are transformed into mutagens and vice versa when introduced into human systems. In mammals, this toxification or detoxification frequently occurs in the liver. The Ames test generally requires the addition of a liver homogenate, S-9, which serves as a source of activating enzymes, to make this bacterial system more comparable to a mammalian test system.

In the Ames test by means of the spot method (see Figure 60.1), molten agar containing the test organism, S-9 mix, and a trace of histidine to allow the bacteria to undergo the several cell divisions necessary for mutation to occur is poured on a minimal agar plate. A disc impregnated with the test chemical is then placed in the center of the test plate. Following diffusion of the test compound from the disc, a concentration gradient of the chemical is established. Following incubation, a qualitative indication of the mutagenicity of the test chemical can be determined by noting the number of colonies present on the plate. Each colony represents a his⁻ → his⁺ revertant. A positive result, indicating mutagenicity, is obtained when an obvious increase in the number of colonies is evident when compared to the number of spontaneous revertants on the negative control plate.

In the following procedure, you will perform a modified Ames test; you will not use the S-9 mix to test for the mutagenicity of nitro compounds, which, as in humans, are activated by the bacterial nitro reductases. Four minimal agar plates are inoculated with the *S. typhimurium* test organism. One plate, the negative control, is not exposed to a test chemical. Any colonies developing on this plate are representative of spontaneous $his^-\to his^+$ mutations. The second plate, the positive control, is exposed to a known nitro-carcinogen, 2-nitrofluorene. The remaining two plates are used to determine the mutagenicity of two commercial hair dyes.

MATERIALS
Cultures

24-hour nutrient broth cultures of *S. typhimurium,* Strain TA 1538 (ATCC™ e 29631).

Media

Per designated student group: four minimal agar plates and four 2-ml top agar tubes.

Reagents

Sterile biotin–histidine solution, 2-nitrofluorene dissolved in alcohol, and two commercial hair dyes.

Equipment

1-ml sterile pipettes, mechanical pipetting device, sterile discs, forceps, waterbath, Bunsen burner, and glassware marking pencil.

PROCEDURE

⚠ **Wear disposable gloves and a laboratory coat when handling 2-nitrofluorene. For disposal of this chemical, place excess into a sealable container and put it inside a fume hood for subsequent removal according to your institution's policy for disposal of hazardous materials.**

1. Label three minimal agar plates with the name of the test chemical to be used. Label the fourth plate as a negative control.

2. Melt four tubes of top agar in a hot water-bath and maintain the molten agar at 45°C.

3. To each molten top agar tube, aseptically add 0.2 ml of the sterile biotin–histidine solution and 0.1 ml of the *S. typhimurium* test culture. Mix by rotating the test tube between the palms of your hands.

4. Aseptically pour the top agar cultures onto the minimal agar plates and allow to solidify.

5. Using sterile forceps, dip each disc into its respective test chemical solution and drain by touching the side of the container.

6. Place the chemical-impregnated discs in the center of the respectively labeled minimal agar plates. Place a sterile disc on the plate labeled negative control. With the sterile forceps, *gently* press down on the discs so that they adhere to the surface of the agar.

7. Incubate all plates in an inverted position for 24 hours at 37°C.

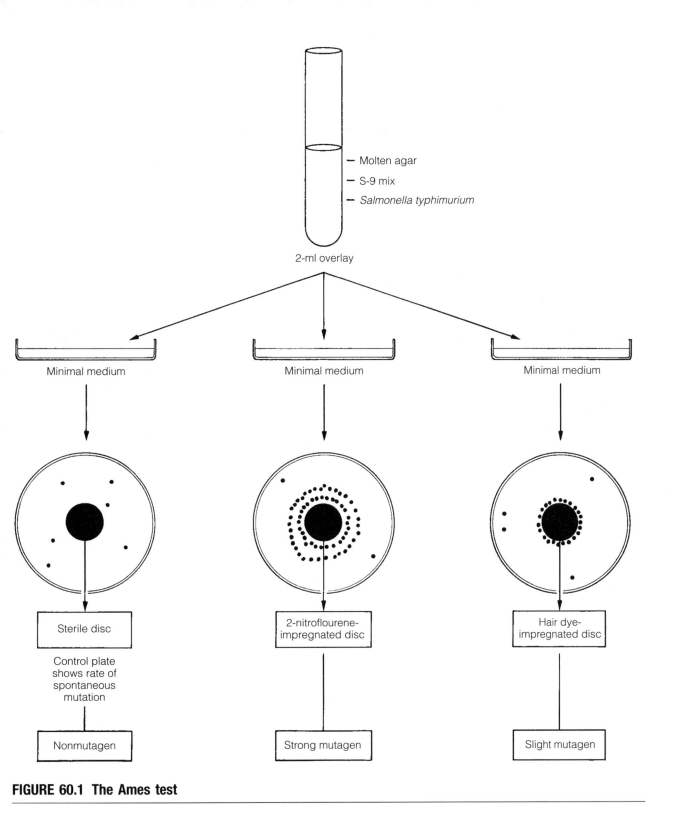

FIGURE 60.1 The Ames test

Name _____   Date _____

1. Count the number of large colonies present on each plate and record on the chart.
2. Determine and record the number of chemically induced mutations by subtracting the number of colonies on the negative control plate, representative of spontaneous mutations, from the number of colonies on each test plate.
3. Determine and record the relative mutagenicity of the test compounds on the basis of the number of induced mutations: If below 10, (−); if greater than 10, (1+); if greater than 100, (2+); and if greater than 500, (3+).

Test Chemical	Number of Colonies	Number of Induced Mutations	Degree of Mutagenicity (−), (1+), (2+), or (3+)
Negative control			
2-nitrofluorene			
Hair dye 1			
Hair dye 2			

REVIEW QUESTIONS

1. What is the purpose of the S-9 in the Ames test?

2. What is the purpose of the biotin–histidine solution in the Ames test?

3. ⚲ What is the relationship between chemical carcinogenicity and mutagenicity?

4. ⚲ What are the advantages of using bacterial systems instead of mammalian systems to test for chemical carcinogenicity?

5. ⚲ What are the disadvantages of using bacterial systems instead of mammalian systems to test for chemical carcinogenicity?

Bacterial Transformation

LEARNING OBJECTIVES

Once you have completed this experiment, you should be able to

1. Transform a competent ampicillin-susceptible strain of *Escherichia coli* into one that is ampicillin resistant by means of a DNA plasmid.
2. Visualize transformed cells using a color marker gene carried in the plasmid.
3. Calculate the efficiency of transformation.

PRINCIPLE

Transformation is a process whereby small pieces of host cell genomic DNA are able to enter a recipient cell and become incorporated into a homologous area on the recipient cell's genome. Historically, transformation had its origin in the pioneering experiments of Fred Griffith in London in the late 1920s. Working with *Streptococcus pneumoniae,* he noted that when an encapsulated smooth (S) strain that was lethal for mice was heat-killed and mixed with a living culture of an avirulent nonencapsulated rough (R) strain and then injected into mice, it proved to be fatal (Figure 61.1). Subsequent isolation of the organisms from the tissues of the dead mice revealed that the rough avirulent strain had been converted to a smooth, encapsulated, and lethal strain of *S. pneumoniae.*

This unusual experiment by Griffith, done long before DNA was determined to be the genetic basis for life, was simply termed by him as a "transformation." Today, in retrospect, we realize that this experiment proved to be the first indication of gene activity and the first demonstration of genetic recombination in bacteria. Later, Avery, McLeod, and McCarty, research scientists at the then Rockefeller Institute, were able to show that the transforming factor in Griffith's experiment was not a protein, as had been previously suspected, but a little-studied organic chemical called deoxyribonucleic acid (DNA).

During the **transformation** process, the donor cells forcibly lyse, releasing small segments of DNA containing 10 to 20 genes. These small segments have the ability to pass through the cell wall and cell membrane of a **competent cell** (a cell that is able to take up DNA from its environment). During naturally occurring transformations, a double-stranded DNA segment passes through the cell wall and into the cell's cytoplasm, and if there is sufficient sequence similarity, the foreign DNA undergoes homologous recombination with the recipient chromosome. The genome of the recipient cell has now been modified to contain DNA with genetic characteristics of the donor cell. Naturally occurring transformations are of great interest medically because they may serve as a vehicle for genetic exchange among pathogenic organisms. Interestingly, it appears that a larger percentage of pathogenic bacteria, such as *Streptococcus pneumoniae, Neisseria gonorrhoeae,* and *Haemophilus influenzae,* is capable of natural transformation than the nonpathogenic bacteria. This raises the intriguing possibility that the exchange of genetic material allows pathogenic cells to acquire the ability to evade a host's bodily defenses.

Not all bacteria are naturally transformable, however, and methods have been developed to produce competency in various types of cells and transform those cells artificially. This process was initiated in the 1970s, when it was shown that treating a recipient cell with a cold calcium chloride ($CaCl_2$) solution allows the passage of donor DNA into the cell.

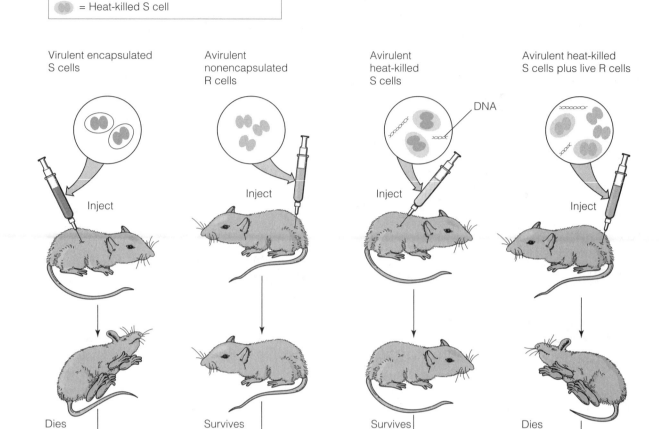

= Encapsulated smooth strain (S cell)
= Nonencapsulated rough strain (R cell)
= Heat-killed S cell

Virulent encapsulated S cells

Avirulent nonencapsulated R cells

Avirulent heat-killed S cells

Avirulent heat-killed S cells plus live R cells

Inject

Inject

Inject

DNA

Inject

Dies

Survives

Survives

Dies

Transformed R → S

Bacteria isolated from mice

FIGURE 61.1 The lethal effect of a transformed avirulent rough strain of _Streptococcus pneumoniae_ transformed with the DNA of a heat-killed virulent smooth strain of _S. pneumoniae_ when mixed together and injected into mice

The porosity of the cell wall is already almost sufficient to allow the passage of intact DNA; it is the cell membrane that is the true barrier, and its permeability is altered by this drastic treatment with $CaCl_2$, allowing DNA to pass through the membrane and into the cell. With our rapidly advancing knowledge in the field of molecular genetics, it is now possible to artificially induce transformations by the use of **plasmids.** Plasmids are small, circular

pieces of extrachromosomal DNA with a length of 5,000 to 100,000 base pairs (bp), capable of autonomous replication in the bacterial cytoplasm. Another membrane-altering method is **electroporation.** In this method, cells are suspended in a DNA solution and subjected to high-voltage electric impulses that destabilize the cell membrane, resulting in increased permeability and enabling DNA to pass into the cells. **Transduction** is a method

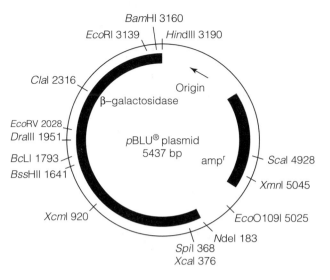

FIGURE 61.2 Plasmid map of pBLU (© 1999 Carolina Biological Supply Company. Used by permission.)

of horizontal passage of genetic material from one bacterial cell to another, by means of a bacteriophage. **Conjugation** occurs when bacterial DNA is transferred from one cell to another via the formation of a protoplasmic bridge, called a conjugative, or sex, pilus.

In the following experiment, the transformation of *E. coli* is artificial because the cells must be treated with a salt concentration and temperature, an environment that is not natural for these cells. You will use a competent strain of *E. coli* in which the Lac operon has been deleted, leaving it devoid of a β-galactosidase gene. The procedure directs you to suspend the *E. coli* in cold $CaCl_2$ and then introduce an ampicillin-resistant plasmid (pBLU®). The plasmid confers ampicillin resistance because it carries the β-lactamase gene amp^r, as shown in Figure 61.2. This suspension is incubated in ice and then heated. The cold shock and heat shock in the presence of $CaCl_2$ alters the permeability of the outer surfaces of the cell and facilitates the passage of the DNA into the cellular cytoplasm (Figure 61.3).

MATERIALS

Note: The cultures, media, reagent, plasmid, and some of the equipment presented here are available in the pBLU Colony Transformation Kits (WW-21-1145X, WW-21-1145Y, WW-21-1146, and WW-21-1146A) available from Carolina Biological Supply Company, Burlington, North Carolina.

Cultures

18- to 24-hour Luria-Bertani (LB) agar base streak plate cultures of *Escherichia coli*.

Media

Per designated student group: four LB agar base plates, three LB agar base plates plus ampicillin (Amp), three LB agar base plates plus ampicillin/X-Gal, and one tube of LB broth. *Note: X-Gal is a colorimetric analog of lactose that is cleaved by β-galactosidase to yield blue-colored colonies.*

Reagent

50 mM $CaCl_2$ solution.

Plasmid pBLU 5437 bp long and has the gene for β-lactamase (penicillinase) and the gene for β-galactosidase.

Equipment

Sterile plastic 13- × 100-mm test tubes or plastic 1.5-ml centrifuge tubes, adjustable micropipette (0.5 to 100 µl) with sterile plastic micropipette tips (10 to 100 µl) (or 1.0-ml graduated, individually wrapped, disposable plastic transfer pipettes), glass beads (6-mm diameter), glassware marking pencil, disposable plastic inoculating loops (standard wire loops may be used), Bunsen burner, waterbath, 500-ml beaker of crushed ice, 500-ml beaker labeled "waste," 500-ml beaker containing disinfectant solution, and a Quebec colony counter (or permanent marker); if the spread-plate method is used, bent glass rod, beaker of alcohol, and turntable.

PROCEDURE

Note: If using a plastic transfer pipette (Figure 61.4), it is essential that you calibrate it to deliver a volume of 100 µl (0.1 ml), required for plasmid transfer. Once calibrated, it should be marked with a permanent glassware marker and retained to be used as a guide in the transformation experiment. Ask your instructor for help if needed.

1. With a glassware marking pencil, label two 13- × 100-mm test tubes, one as "DNA+" and the other as "DNA−." The DNA+ tube will receive the plasmid.

2. Using a sterile pipette, transfer 250 µl (0.25 ml) of ice-cold $CaCl_2$ solution into each tube.

3. Place both tubes in a 500-ml beaker of crushed ice.

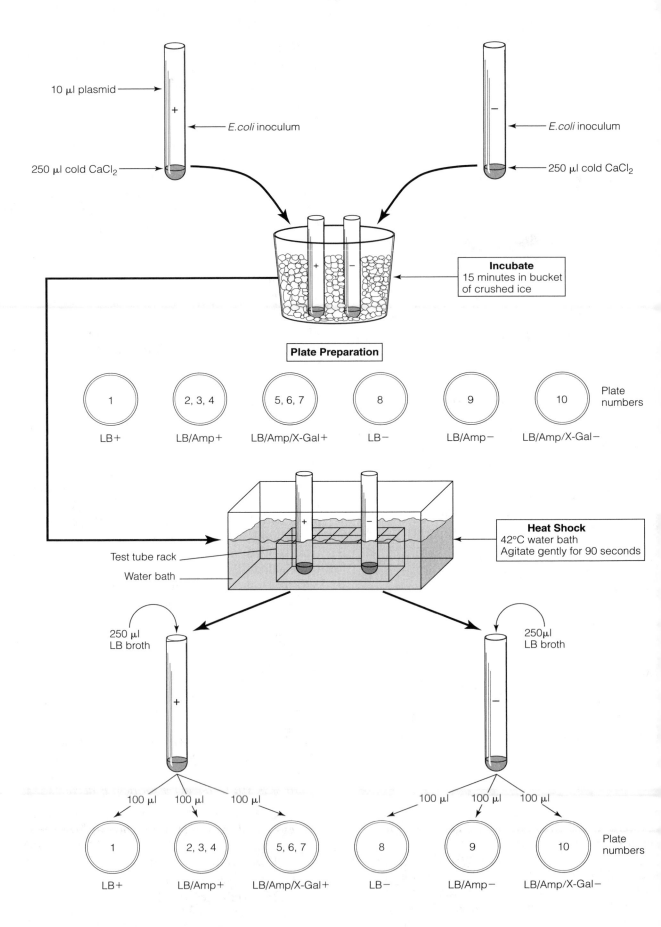

FIGURE 61.3 Transformation procedure

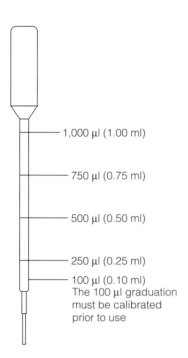

1,000 μl (1.00 ml)

750 μl (0.75 ml)

500 μl (0.50 ml)

250 μl (0.25 ml)

100 μl (0.10 ml)
The 100 μl graduation
must be calibrated
prior to use

FIGURE 61.4 Plastic transfer pipette

4. Using a sterile inoculation loop, obtain a large mass of cells approximately 5 mm in size (about the size of a pencil eraser) from the *E. coli* starter plate, and inoculate the tube labeled "DNA+." *Note: Be sure to immerse the loop directly into the CaCl₂ and shake the loop vigorously to dislodge the inoculum. Discard the plastic loops in the beaker labeled "waste" or sterilize the wire loop by flaming it.*

5. Disperse the cells by gently tapping the tubes with your finger until a uniform milky-white translucent cell suspension is obtained.

6. Repeat Steps 4 and 5 to inoculate the tube marked "DNA−," using an equal amount of inoculum and a sterile inoculating loop.

7. Incubate both tubes in the crushed-ice beaker for 15 minutes.

8. Using a sterile pipette, deliver 10 μl (0.1 ml) of pBLU plasmid into the DNA+ tube. Tap the tube several times with your finger to ensure complete mixing of the plasmid and cell suspension. *Note: Discard the plastic tip or disposable pipette into a beaker containing disinfectant solution.*

9. Return the DNA+ tube to the crushed-ice beaker and incubate for 15 minutes. During this time, label your agar plates as described in Step 10.

10. Label the 10 LB agar plates as follows:
Plate 1: LB+
Plates 2–4: LB/Amp+
Plates 5–7: LB/Amp/X-Gal+
Plate 8: LB−
Plate 9: LB/Amp−
Plate 10: LB/Amp/X-Gal−

11. Remove both tubes from ice after 15 minutes; place them in a test tube rack and immediately into a 42°C waterbath with *gentle agitation* for 90 seconds (heat shocking).

12. Return both tubes to the crushed-ice beaker for 1 minute.

13. With a sterile pipette, add 250 μl (0.25 ml) of LB broth to both the DNA+ and DNA− tubes. Tap the tubes with your finger to achieve uniform cell suspension. (These are the transformation tubes.)

14. Incubate both tubes in a test tube rack for 15 minutes. *Note: This is the **recovery period**, when the cells convert their newly modified genotype into a functionally ampicillin-resistant phenotype.*

15. Using a new plastic micropipetter tip or disposable plastic transfer pipette for each inoculation, inoculate 100 μl (0.1 ml) of cells from the DNA+ transformation tube onto the surface of LB plates 1–7, and inoculate 100 μl (0.1 ml) of cells from the DNA− transformation tube onto plates 8–10.

16. Place six sterile 6-mm glass beads on the surface of each inoculated plate. Replace the cover and spread the cell suspension by gently moving the plate up and down and then side to side a few times. *Note: Do not swirl or rotate the plate.* If the spread-plate technique is used, refer to Experiment 2 (page 14) for the correct procedure.

17. Repeat Steps 15 and 16 for the remaining plates.

18. Allow the plates to set for a few minutes so that the inoculum may be absorbed by the agar.

19. Remove the glass beads from the plate by lifting the cover slightly while holding the plate vertically over the beaker of disinfectant, allowing the beads to leave the plate. *Note: This step may be eliminated if the spread-plate procedure is used.*

20. Incubate all plates at 37°C for 24 to 36 hours or at room temperature for 48 to 72 hours.

Name _____   Date _____

1. In the following chart, predict whether each plate will experience growth or no growth. Use a plus (+) sign for growth and a minus (−) for no growth.
2. Without removing the cover of the Petri plates, observe the colonies through the bottom of each plate.
3. Perform a colony count on each plate using an electronic colony counter if available, or use a permanent marker to mark each colony on the bottom of the plate as it is counted. Plates with more than 300 colonies should be designated as **TNTC (too numerous to count);** plates with fewer than 30 colonies are designated as **TFTC (too few to count).** Record your results in the following chart.
4. For each plate, did transformation occur? Record your results in the following chart.

Plate Number	Designation	Growth + or −	Transformation Yes or No	Number of Colonies
1	LB+			
2	LB/Amp+			
3	LB/Amp+			
4	LB/Amp+			
5	LB/Amp/X-Gal+			
6	LB/Amp/X-Gal+			
7	LB/Amp/X-Gal+			
8	LB−			
9	LB/Amp−			
10	LB/Amp/X-Gal−			

5. Determine the sum of the colonies on the three experimental plates for LB/AMP+ and the three LB/AMP/X-Gal+ plates. Determine the averages and use these figures to calculate the transformation efficiency, using the following protocol.

   Avg. LB/Amp+ colonies _____   Avg. LB/Amp/X-Gal+ colonies _____

   a. The concentration of the plasmid used in this experiment was 0.005 µg/µl. Keep in mind that you used 10 µl.

      Total mass of plasmid = volume X concentration
      Total mass of plasmid = _____

   b. Calculate the total volume of the cell suspension.

      Total volume = volume of $CaCl_2$ solution + volume of plasmid + volume of LB broth
      Total volume = _____

   c. Calculate the fraction of cell suspension spread on each plate.

      Fraction spread = volume of suspension spread on plate / total volume
      Fraction spread = _____

d. Calculate the mass of DNA plasmid in the cell suspension spread on each plate.

   Mass of DNA plasmid spread = Total mass of plasmid × fraction spread

   Mass of DNA plasmid spread = _____

e. Calculate the transformation efficiency (the number of colonies per µg of plasmid DNA).

   Transformation efficiency = Average of colonies counted/mass of plasmid spread

Express your results using scientific notation, as found in Appendix 1.

Transformation Efficiency	
LB/Amp+	LB/Amp/X-Gal+

REVIEW QUESTIONS

1. Why do you think that it is necessary to autoclave transformed cells immediately after the termination of the experiment?

2. What was the purpose of treating *E. coli* cells with ice followed by heat?

3. What does cell competency mean?

4. How would you explain to an untrained neighbor the process scientists use to change the genes (genotype) of an organism to give it useful new traits?

5. At a neighboring lab bench, students find that there are a few white colonies scattered among a few hundred blue colonies on their LB/Amp/X-Gal+ plates.

   a. How would you explain this result?

   b. To confirm your hypothesis, you ask to see their LB/Amp/X-Gal− plate. What do you see that supports your hypothesis?

Medical Microbiology

LEARNING OBJECTIVES

Once you have completed the experiments in this section, you should be familiar with the

1. Characteristics and methodology for isolating and identifying selected pathogenic microorganisms.
2. Indigenous microbial flora of selected human anatomical sites.

INTRODUCTION

Although microorganisms are ubiquitous and their benefits to humans have been recorded, a diminutive population of organisms remains an enigma: They are the pathogens, whose existence makes medical or clinical microbiology an especially important science.

That living agents are capable of inducing infections *(contagium viva)* was first put forward by the monk Fracastoro in Verona about 500 years ago. In 1659 Kircher reported the presence of minute motile organisms in the blood of plague victims. Two hundred years after Fracastoro developed his initial concept, the germ theory of disease was formulated by Plenciz based on Leeuwenhoek's revolutionary microscopic observation of microorganisms. Perhaps the most important contributions to microbiology were made by Pasteur, Koch, and Lister during the **Golden Era of Microbiology** from 1870 to 1920. During this time these investigators and their students recorded the observations and discoveries that cemented the cornerstone of medical microbiology. The body of knowledge that has accrued since these early years has been instrumental in making clinical microbiology a major component of laboratory or diagnostic medicine. The major responsibility of this science is isolating and identifying infectious pathogens to enable physicians to treat patients with infectious disease prudently, intelligently, and rapidly.

Many of the experiments so far described have application in the field of clinical microbiology. Among these are isolation and identification of unknown cultures, the use of selective and differential media, and biochemical tests used to separate and identify various microorganisms. Although studying all of the bacterial pathogens responsible for human illness is not possible here, routine experiments for isolating and identifying some of the most frequently encountered infectious organisms and microorganisms that constitute the indigenous flora of the human body are included. The pathogens chosen are pyogenic cocci, members of the genera *Staphylococcus* and *Streptococcus*, the Enterobacteriaceae, and the organisms suspected in formation of dental caries. Experimental procedures designed for the detection and presumptive identification of microorganisms in blood and urine, which are normally sterile body fluids, have also been incorporated into this section. Organisms that naturally reside in or on body surfaces and constitute the body's **normal flora** are also examined.

The need for the expeditious detection and identification of pathogens has led to the development of rapid testing methods. These are microbiologically and immunologically based and can be performed quickly and without the need for sophisticated and expensive equipment. Some prototypic experiments using these rapid methods are included along with the traditional procedures.

⚠ Many of the organisms that are used, although attenuated by having been subcultured on artificial complex media for many generations, must be viewed as potential pathogens and therefore handled with respect. At this point in your training, your manipulative skills should be sufficiently developed, allowing you to perform aseptically in any medical, hospital, or clinical laboratory setting to prevent infection of yourself and others.

Microbial Flora of the Mouth: Determination of Susceptibility to Dental Caries

LEARNING OBJECTIVES

Once you have completed this experiment, you should be

1. Familiar with the organisms responsible for dental caries.

2. Able to perform experiments that demonstrate the host's susceptibility to formation of caries.

PRINCIPLE

A variety of microorganisms are known to be involved in the formation of dental caries, including *Lactobacillus acidophilus*, *Streptococcus mutans*, and *Actinomyces odontolyticus*. These organisms in the oral flora produce organic acids, particularly lactic acid, by fermenting carbohydrates that adhere to the surface of the teeth. In the continued presence of lactic acid, dental enamel undergoes decalcification and softening, which result in the formation of tiny perforations called dental caries.

The actual mechanism of action of these organisms is still unclear. However, it has been noted that *S. mutans* excretes an enzyme called **dextransucrase** (glycosyl transferase), which is capable of polymerizing sucroses into a large polymer, dextran, plus the monosaccharide fructose. This polysaccharide clings tenaciously to the teeth and forms dental plaque, in which streptococci reside and ferment fructose with the formation of lactic acid (Figure 62.1).

Similarly, *L. acidophilus* produces lactic acid as an end product of carbohydrate fermentation. Oral lactobacilli are capable of metabolizing glucose found in the mouth, producing organic acids that reduce the oral acid concentration to a pH of less than 5. At this

pH, decalcification occurs and dental decay begins.

One of the best microbiological methods for determining susceptibility to dental caries is the **Snyder test.** This test measures the amount of acid produced by the action of the lactobacilli on glucose. The test employs a differential medium, Snyder agar (pH 4.7), which contains glucose and the pH indicator bromcresol green that gives the medium a green color.

Following incubation, Snyder agar cultures containing lactobacilli from the saliva will show glucose fermentation with the production of acid, which tends to lower the pH to 4.4, the level of acidity at which dental caries form. At this pH the green medium turns yellow. A culture demonstrating a yellow color within 24 to 48 hours is suggestive of the host's susceptibility to the formation of dental caries. A culture that does not change color is indicative of lower susceptibility.

MATERIALS

Cultures

Organisms of the normal oral flora present in saliva.

Media

Per designated student group: two Snyder test agar deep tubes.

Equipment

Bunsen burner, 1-in square blocks of paraffin, sterile 1-ml pipettes, mechanical pipetting device, sterile test tubes, and glassware marking pencil.

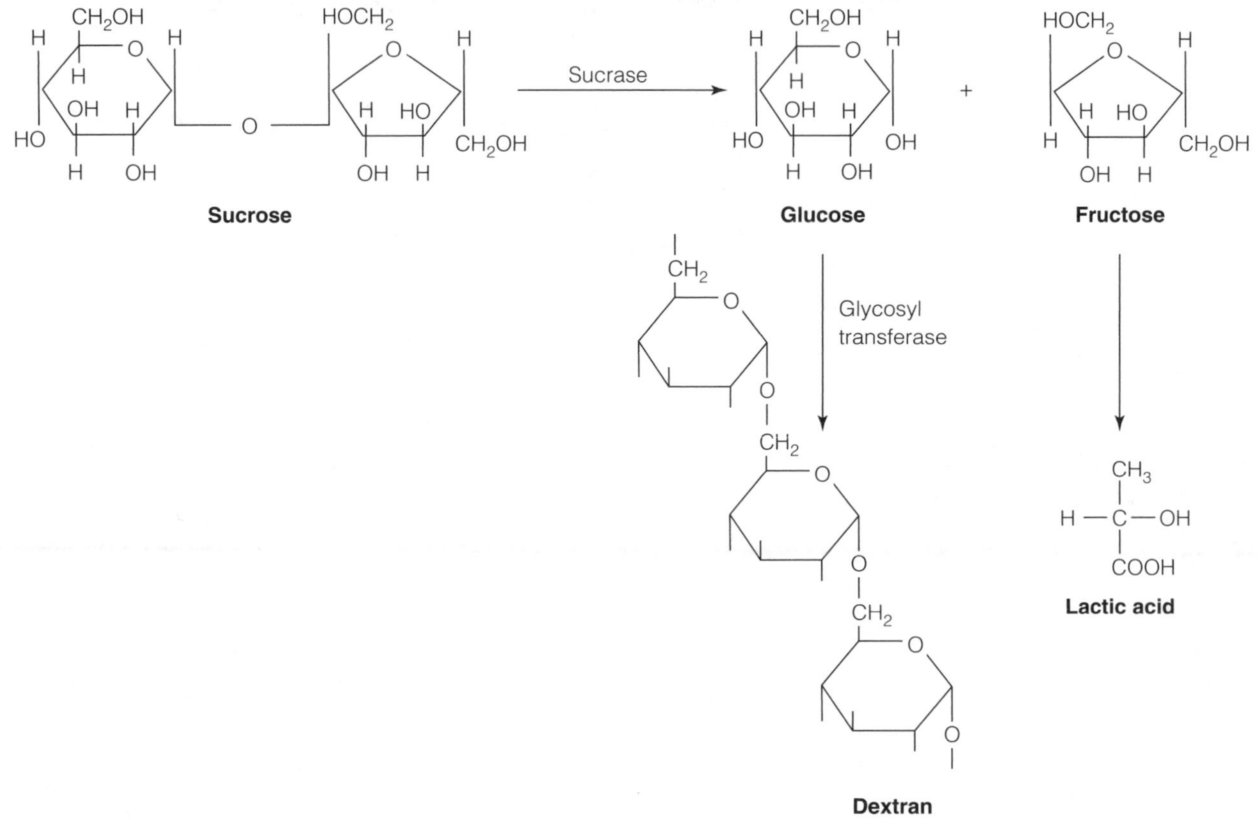

FIGURE 62.1 Degradation of sucrose and subsequent conversion of glucose into dextran by *Streptococcus mutans*

PROCEDURE

1. Melt two appropriately labeled Snyder agar deep tubes and cool to 45°C.

2. Chew one square of paraffin for a period of 3 minutes *without swallowing the saliva.* As saliva develops, collect it in a sterile test tube.

3. Vigorously shake the collected saliva sample and transfer 0.2 ml of saliva with a sterile pipette into one of the Snyder test medium tubes that have been cooled to 45°C. *Note: Don't let the pipette touch the sides of the tubes or the agar.*

4. Mix the contents of the tube thoroughly by rolling the tube between the palms of your hands or by tapping it with your finger.

5. Rapidly cool the inoculated tube of Snyder agar in an ice-water bath.

6. Repeat Steps 3 through 5 to inoculate the second tube.

7. Incubate both tubes for 72 hours at 37°C. Observe cultures at 24, 48, and 72 hours.

Name   Date

1. Examine the Snyder test cultures daily during the 72-hour incubation period for a change in the color of the culture medium. Use an uninoculated tube of the medium as a control. Record the color of the cultures in the chart.

2. Using Table 62.1 to interpret your observations, record your findings about susceptibility to caries in the chart.

 Refer to photo number 68 in the color-plate insert for illustration of this reaction.

TABLE 62.1 Assessment of Susceptibility to Dental Caries

Caries Activity	Hours of Incubation		
	24	48	72
Marked	Positive	. . .	. . .
Moderate	Negative	Positive	. . .
Slight	Negative	Negative	Positive
Negative	Negative	Negative	Negative

Source: Courtesy of Difco Laboratories, Inc., Detroit, Michigan.
Positive: Complete color change; green is no longer dominant.
Negative: No color change or a slight color change; medium retains green color throughout.

	Color of Snyder Test Cultures			
Tube Number	24 hr.	48 hr.	72 hr.	Caries Susceptibility (Yes) or (No)

REVIEW QUESTIONS

1. How would you explain the differential nature of the Snyder agar medium as used for the detection of dental caries?

2. How would you explain the mechanism responsible for the formation of dental caries by resident microorganisms?

3. What is the function of the paraffin in this procedure?

4. Based on your results, what is your tendency to form dental caries?

   Is this result consistent with your dental history?

5. Are all members of the resident flora of the mouth capable of initiating dental caries? Explain.

6. What is the ideal time of day to perform this procedure? Why?

Normal Microbial Flora
of the Throat and Skin

LEARNING OBJECTIVE

Once you have completed this experiment, you should be able to

1. Identify microorganisms that normally reside in the throat and skin.

PRINCIPLE

The normal flora are regularly found in specific areas of the body. This specificity is far from arbitrary and depends on environmental factors such as pH, oxygen concentration, amount of moisture present, and types of secretions associated with each anatomical site. Native microbial flora are broadly located as follows:

1. **Skin:** Staphylococci (predominantly *S. epidermidis*), streptococci (α-hemolytic, nonhemolytic, and enterococci), diphtheroid bacilli, yeasts, and fungi.

2. **Eye conjunctiva:** Staphylococci, streptococci, diphtheroids, and neisseriae.

3. **Upper respiratory tract:** Staphylococci, streptococci (α-hemolytic, nonhemolytic, enterococci, and *S. pneumoniae*), diphtheroids, spirochetes, and members of the genera *Branhamella, Neisseria,* and *Haemophilus.*

4. **Mouth and teeth:** Anaerobic spirochetes and vibrios, fusiform bacteria, staphylococci, and anaerobic levan-producing and dextran-producing streptococci responsible for dental caries.

5. **Intestinal tract:** In the upper intestine, predominantly lactobacilli and enterococci. In the lower intestine and colon, 96% to 99% is composed of anaerobes such as members of the genera *Bacteroides, Lactobacillus, Clostridium,* and *Streptococcus,* and 1% to 4% is composed of aerobes, including coliforms, enterococci, and a small number of *Proteus, Pseudomonas,* and *Candida* species.

6. **Genitourinary tract:** Staphylococci, streptococci, lactobacilli, gram-negative enteric bacilli, clostridia, spirochetes, yeasts, and protozoa such as *Trichomonas* species.

In this exercise, you will study the resident flora of the throat and skin. Since these sites represent sources of mixed microbial populations, you will perform streak-plate inoculations, as outlined in Experiment 2, to effect their separations. The discrete colonies, thus formed, can then be studied morphologically, biochemically, and microscopically to identify the individual genera of these mixed flora.

The procedure used to identify the native flora of the throat involves the following steps:

1. A *blood agar plate* is inoculated to demonstrate the α-hemolytic and β-hemolytic reactions of some streptococci and staphylococci. A distinction between these two genera can be made based on their colonial and microscopic appearances. The streptococci typically form pinpoint colonies on blood agar, whereas the staphylococci form larger pinhead colonies that might show a golden coloration. When viewed under a microscope, the streptococcal cells form chains of varying lengths, whereas the staphylococci are arranged in clusters.

2. A *chocolate agar plate* is inoculated to detect *Neisseria* spp. by means of the oxidase test. Members of this genus are recognized when the colonies develop coloration that is pink to dark purple upon addition of *p*-aminodimethylaniline oxalate following incubation.

3. A *Mueller-Hinton tellurite* or *Tinsdale agar plate* is inoculated to demonstrate the presence of diphtheroids, which appear as black, pinpoint colonies on this medium. This coloration is due to the diffusion of

the tellurite ions into the bacterial cells and their subsequent reduction to tellurium metal, which precipitates inside the cells.

The procedure used to identify the native flora of the skin involves the following steps:

1. A *blood agar plate* inoculated to determine the presence of hemolytic microorganisms, specifically the staphylococci and streptococci: Differentiation between these two genera may be made as previously described.

2. A *mannitol salt agar plate* inoculated for the isolation of the staphylococci: The generally avirulent staphylococcal species can be differentiated from the pathogenic *S. aureus* because the latter is able to ferment mannitol, causing yellow coloration of this medium surrounding the growth.

3. A *Sabouraud agar plate* inoculated to detect yeasts and molds: Yeast cells will develop pigmented or nonpigmented colonies that are elevated, moist, and glistening. Mold colonies will appear as fuzzy, powdery growths arising from a mycelial mat in the agar medium.

MATERIALS

Media

Per designated student group: two blood agar plates, two mannitol salt agar plates, one chocolate agar plate, one Mueller-Hinton tellurite or Tinsdale agar plate, one Sabouraud agar plate, and two 5-ml sterile saline tubes.

Reagents

Crystal violet, Gram's iodine, safranin, 1% *p*-aminodimethylaniline oxalate, and lactophenol–cotton-blue.

Equipment

Sterile cotton swabs, tongue depressors, desiccator jar with candle, microscope, glass slides, Bunsen burner, glassware marking pencil, and disposable gloves.

PROCEDURE

⚠ **You must wear disposable gloves in Steps 1–3.**

1. Place a tongue depressor on the extended tongue and with a sterile cotton swab, obtain a specimen from the pharyngeal tonsil by rotating the swab vigorously over its surface without touching the tongue, as illustrated.

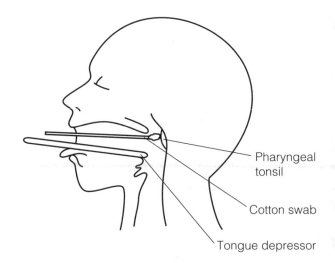

2. Inoculate a tube of sterile saline with the swab and mix until you have a uniform suspension.

3. Using a sterile inoculating loop, inoculate one plate each of blood agar, chocolate agar, mannitol salt agar, and Mueller-Hinton tellurite or Tinsdale agar, all previously labeled with the source of the specimen, by means of a four-way streak inoculation as described in Experiment 2.

4. Using a sterile cotton swab moistened in sterile saline, obtain a specimen from the skin by rubbing the swab vigorously against the palm of your hand.

5. Inoculate a tube of sterile saline with the swab and mix the solution.

6. Inoculate one plate each of blood agar, mannitol salt agar, and Sabouraud agar, as described in Step 3.

7. Incubate the inverted chocolate agar plate in a CO_2 incubator, in a CO_2 incubation bag, or in a candle jar. If you use the candle jar, place a lighted candle in a desiccator jar and cover the jar tightly to effect a 5% to 10% CO_2 environment required for the growth of the *Neisseria*. Incubate the jar for 48 hours at 37°C.

8. Incubate the inverted Sabouraud agar plate for 48 hours at 25°C and the remaining plates for 48 hours at 37°C.

Name _____   Date _____

1. Examine the blood agar plate cultures for zones of hemolysis (refer to Experiment 15 for an explanation of the hemolytic reactions).

2. Add the *p*-aminodimethylaniline oxalate to the surface of the growth on the chocolate agar plate. Observe for the appearance of a pink-to-purple color on the surface of any of the colonies.

3. Examine the Mueller-Hinton tellurite or Tinsdale agar plate for the presence of black colonies.

4. Examine the Sabouraud agar plate for the appearance of moldlike growth.

5. Examine the mannitol salt agar plate for the presence of growth that is indicative of staphylococci. Then examine the color of the medium surrounding the growth. A yellow color is indicative of *S. aureus* (refer to Experiment 15 on the differentiation of staphylococcal species on this medium).

6. Record your observations in the chart and indicate the types of organisms that may be present in each specimen.

📷 *Refer to photo numbers 19–24, 57, 59, 69, and 70 in the color-plate insert for illustration of these reactions.*

Culture	Throat Specimen	Skin Specimen
Blood agar: Type of hemolysis		
Chocolate agar: (+) or (−) pink-to-purple colonies		
Mueller-Hinton tellurite or Tinsdale: (+) or (−) black colonies		
Sabouraud agar: (+) or (−) fungal colonies		
Mannitol salt agar: (+) or (−) growth		
Color of medium		
Types of organisms present		

7. Prepare two Gram-stained smears from each of the blood agar cultures, choosing well-isolated colonies that differ in their cultural appearances and demonstrate hemolytic activity. Observe microscopically for the Gram reaction and the size, shape, and arrangement of the cells. In the chart, draw a representative field, describe the morphology of the cells, and attempt to identify each isolate.

8. Prepare two lactophenol–cotton-blue–stained smears of organisms obtained from discrete colonies that differ in appearance on the Sabouraud agar culture (refer to Experiment 36). Observe microscopically, draw a representative field, and attempt to identify the fungi by referring to Experiment 38.

Skin Specimen	Isolate 1	Isolate 2
	○	○
Gram reaction		
Morphology		
Organism		
Throat Specimen	Isolate 1	Isolate 2
	○	○
Gram reaction		
Morphology		
Organism		
Sabouraud Agar Colonies Specimen	Isolate 1	Isolate 2
	○	○
Morphology		
Organism		

REVIEW QUESTION

1. Why are some microorganisms termed normal flora, and of what value are they to the well-being of the host?

2. A 6-year-old female is taken to her pediatrician for a checkup. As the doctor takes the child's history, her mother reports that the child had a severe sore throat several weeks earlier that regressed without treatment. Upon examination the pediatrician notes that the child has a systolic heart murmur consistent with mitral insufficiency and suspects that she has rheumatic fever.

   a. How was the earlier pharyngitis related to the subsequent development of rheumatic fever?

   b. Rheumatic fever is diagnosed on clinical and serological findings. What test should be done to diagnose rheumatic fever?

   c. How are rheumatic fever patients treated?

3. A 35-year-old female underwent serious abdominal surgery involving extensive bowel resection. She was maintained postoperatively on a regimen of intravenous broad-spectrum antibiotics. Three days postoperative she spike a fever without a clear source. She complains of vaginal discomfort. Blood cultures reveal the presence of an ovoid cell that reproduced by budding.

   a. Based on this observation, what do you think this organism is?

   b. Is it part of the normal flora in humans?

   c. How did the treatment with broad-spectrum antibiotics predispose the patient to infection with this organism?

Identification of Human Staphylococcal Pathogens

LEARNING OBJECTIVES

Once you have completed this experiment, you should understand

1. The medical significance of the staphylococci.
2. Selected laboratory procedures designed to differentiate among the major staphylococcal species.

PRINCIPLE

The genus *Staphylococcus* is composed of both pathogenic and nonpathogenic organisms. They are gram-positive cocci and occur most commonly as irregular clusters of spherical cells. They are mesophilic nonspore-formers; however, they are generally highly resistant to drying, especially when sequestered in organic matter such as blood, pus, and tissue fluids. They are capable of surviving outside of the body for extended periods of time, even up to several months. Many staphylococci are indigenous to skin surfaces and mucous membranes of the upper respiratory tract. Breaks in the skin and mucous linings may serve as portals of entry into underlying tissues, with the possibility of infection by virulent strains.

The three major species are *S. aureus*, *S. saprophyticus*, and *S. epidermidis*. Strains of the last two species are generally avirulent; however, under special circumstances in which a suitable portal of entry is provided, *S. epidermidis* may be the etiological agent for **skin lesions** and **endocarditis,** and *S. saprophyticus* has been implicated in some **urinary tract infections.**

Infections are primarily associated with *S. aureus* pathogenic strains that are often responsible for the formation of **abscesses,** localized pus-producing lesions. These lesions most commonly occur in the skin and its associated structures, resulting in **boils, carbuncles, acne,** and **impetigo.** Infections of internal organs and tissues are not uncommon, however, and include **pneumonia, osteomyelitis** (abscesses in bone and bone marrow), **endocarditis** (inflammation of the endocardium), **cystitis** (inflammation of the urinary bladder), **pyelonephritis** (inflammation of the kidneys), **staphylococcal enteritis** due to enterotoxin contamination of foods, and, on occasion, **septicemia.**

Strains of *S. aureus* produce a variety of metabolic end products, some of which may play roles in the organisms' pathogenicity. Included among these are **coagulase,** which causes clot formation; **leukocidin,** which lyses white blood cells; **hemolysins,** which are active against red blood cells; and **enterotoxin,** which is responsible for a type of gastroenteritis. Additional metabolites of a nontoxic nature are **DNase, lipase, gelatinase,** and the fibrinolysin **staphylokinase.**

When there is a possibility of staphylococcal infection, isolation of *S. aureus* is of clinical importance. These virulent strains can be differentiated from other staphylococci and identified by a variety of laboratory tests, some of which are illustrated in Table 64.1.

In this exercise you will distinguish among the staphylococcal species by performing a computer-assisted multitest procedure, traditional test procedures, or both. The traditional and computer-based procedures are described below.

The traditional procedures involve the following steps:

1. **Mannitol salt agar:** This medium is selective for salt-tolerant organisms such as staphylococci. Differentiation among the staphylococci is predicated on their ability

TABLE 64.1 Laboratory Tests for Differentiation of Staphylococcal Species

Test	S. aureus	S. epidermidis	S. saprophyticus
Mannitol salt agar			
Growth	+	+	+
Fermentation	+	−	−
Colonial pigmentation	Generally golden yellow	White	White
Coagulase	+	−	−
DNase	+	−	−
Hemolysis	Generally beta	−	−
Novobiocin sensitivity	Sensitive	Sensitive	Resistant

to ferment mannitol. Following incubation, mannitol-fermenting organisms, typically *S. aureus* strains, exhibit a yellow halo surrounding their growth, and non-fermenting strains do not. (Refer to Experiment 15 for a more detailed discussion on the use of this medium.)

2. **Coagulase test:** Production of coagulase is indicative of an *S. aureus* strain. The enzyme acts within host tissues to convert fibrinogen to fibrin. It is theorized that the fibrin meshwork that is formed by this conversion surrounds the bacterial cells or infected tissues, protecting the organism from nonspecific host resistance mechanisms such as phagocytosis and the anti-staphylococcal activity of normal serum. In the coagulase tube test for bound and free coagulase, a suspension of the test organism in citrated plasma is prepared and the inoculated plasma is then periodically examined for fibrin formation, or coagulation. Clot formation within 4 hours is interpreted as a positive result and indicative of a virulent *S. aureus* strain. The absence of coagulation after 24 hours of incubation is a negative result, indicative of an avirulent strain.

3. **Deoxyribonuclease (DNase) test:** Generally, coagulase-positive staphylococci also produce the hydrolytic enzyme DNase; thus this test can be used to reconfirm the identification of *S. aureus*. The test organism is grown on an agar medium containing DNA. Following incubation, DNase activity is determined by the addition of

0.1% toluidine blue to the surface of the agar. DNase-positive cultures capable of DNA hydrolysis will show a rose-pink halo around the area of growth. The absence of this halo is indicative of a negative result and the inability of the organism to produce DNase.

4. **Novobiocin sensitivity:** This test is used to distinguish *S. epidermidis* from *S. saprophyticus*. It requires inoculation of a Mueller-Hinton agar plate with the test organism and application of a 30-μg novobiocin antibiotic disc to the surface of the agar. Following incubation, the sensitivity of an organism to the antibiotic is determined by the Kirby-Bauer method as described in Experiment 44.

A computer-assisted procedure is the API® (Analytical Profile Index) STAPH-IDENT® system (developed by Analytab Products, Division of Sherwood Medical, Plainview, New York). STAPH-IDENT is a rapid, computer-based micromethod for the separation and identification of the newly proposed 13 species of staphylococci. The system consists of 10 microcupules containing dehydrated substrates for the performance of conventional and chromogenic tests. The addition of a suspension of the test organism serves to hydrate the media and to initiate the biochemical reactions. The identification of the staphylococcal species is made with the aid of the differential charts or the STAPH-IDENT Profile Register that is part of the system (Table 64.2), or both.

TABLE 64.2 API STAPH-IDENT Profile Register

Profile	Identification		Profile	Identification	
0 040	STAPH CAPITIS		3 000	STAPH EPIDERMIDIS	
0 060	STAPH HAEMOLYTICUS		3 040	STAPH EPIDERMIDIS	
0 100	STAPH CAPITIS		3 140	STAPH EPIDERMIDIS	
0 140	STAPH CAPITIS		3 540	STAPH HYICUS (An)	
0 200	STAPH COHNII		3 541	STAPH INTERMEDIUS (An)	
0 240	STAPH CAPITIS		3 560	STAPH HYICUS (An)	
0 300	STAPH CAPITIS		3 601	STAPH SIMULANS	NOVO
0 340	STAPH CAPITIS			STAPH SAPROPHYTICUS	NOVO R
0 440	STAPH HAEMOLYTICUS				
0 460	STAPH HAEMOLYTICUS		4 060	STAPH HAEMOLYTICUS	
0 600	STAPH COHNII		4 210	STAPH SCIURI	
0 620	STAPH HAEMOLYTICUS		4 310	STAPH SCIURI	
0 640	STAPH HAEMOLYTICUS		4 420	STAPH HAEMOLYTICUS	
0 660	STAPH HAEMOLYTICUS		4 440	STAPH HAEMOLYTICUS	
			4 460	STAPH HAEMOLYTICUS	
1 000	STAPH EPIDERMIDIS		4 610	STAPH SCIURI	
1 040	STAPH EPIDERMIDIS		4 620	STAPH HAEMOLYTICUS	
1 300	STAPH AUREUS		4 660	STAPH HAEMOLYTICUS	
1 540	STAPH HYICUS (An)		4 700	STAPH AUREUS	COAG+
1 560	STAPH HYICUS (An)			STAPH SCIURI	COAG−
			4 710	STAPH SCIURI	
2 000	STAPH SAPROPHYTICUS	NOVO R			
	STAPH HOMINIS	NOVO S	5 040	STAPH EPIDERMIDIS	
2 001	STAPH SAPROPHYTICUS		5 200	STAPH SCIURI	
2 040	STAPH SAPROPHYTICUS	NOVO R	5 210	STAPH SCIURI	COAG+
	STAPH HOMINIS	NOVO S	5 300	STAPH AUREUS	COAG−
2 041	STAPH SIMULANS			STAPH SCIURI	
2 061	STAPH SIMULANS		5 310	STAPH SCIURI	
2 141	STAPH SIMULANS		5 600	STAPH SCIURI	
2 161	STAPH SIMULANS		5 610	STAPH SCIURI	COAG+
2 201	STAPH SAPROPHYTICUS		5 700	STAPH AUREUS	COAG−
2 241	STAPH SIMULANS			STAPH SCIURI	
2 261	STAPH SIMULANS		5 710	STAPH SCIURI	
2 341	STAPH SIMULANS		5 740	STAPH AUREUS	
2 361	STAPH SIMULANS				
2 400	STAPH HOMINIS	NOVO S	6 001	STAPH XYLOSUS	XYL+ ARA+
	STAPH SAPROPHYTICUS	NOVO R		STAPH SAPROPHYTICUS	XYL− ARA−
2 401	STAPH SAPROPHYTICUS		6 011	STAPH XYLOSUS	
2 421	STAPH SIMULANS		6 021	STAPH XYLOSUS	
2 441	STAPH SIMULANS		6 101	STAPH XYLOSUS	
2 461	STAPH SIMULANS		6 121	STAPH XYLOSUS	
2 541	STAPH SIMULANS		6 221	STAPH XYLOSUS	
2 561	STAPH SIMULANS		6 300	STAPH AUREUS	
2 601	STAPH SAPROPHYTICUS		6 301	STAPH XYLOSUS	
2 611	STAPH SAPROPHYTICUS		6 311	STAPH XYLOSUS	
2 661	STAPH SIMULANS		6 321	STAPH XYLOSUS	
2 721	STAPH COHNII (SSP1)		6 340	STAPH AUREUS	COAG+
2 741	STAPH SIMULANS			STAPH WARNERI	COAG−
2 761	STAPH SIMULANS		6 400	STAPH WARNERI	

➡

TABLE 64.2 API STAPH-IDENT Profile Register (continued)

Profile	Identification		Profile	Identification	
6 401	STAPH XYLOSUS	XYL+ ARA+	7 141	STAPH INTERMEDIUS (An)	
	STAPH SAPROPHYTICUS	XYL− ARA−	7 300	STAPH AUREUS	
6 421	STAPH XYLOSUS		7 321	STAPH XYLOSUS	
6 460	STAPH WARNERI		7 340	STAPH AUREUS	COAG−
6 501	STAPH XYLOSUS		7 401	STAPH XYLOSUS	
6 521	STAPH XYLOSUS		7 421	STAPH XYLOSUS	
6 600	STAPH WARNERI		7 501	STAPH INTERMEDIUS (An)	COAG+
6 601	STAPH SAPROPHYTICUS	XYL− ARA−		STAPH XYLOSUS	COAG−
	STAPH XYLOSUS	XYL+ ARA+	7 521	STAPH XYLOSUS	
6 611	STAPH XYLOSUS		7 541	STAPH INTERMEDIUS (An)	
6 621	STAPH XYLOSUS		7 560	STAPH HYICUS (An)	
6 700	STAPH AUREUS		7 601	STAPH XYLOSUS	
6 701	STAPH XYLOSUS		7 621	STAPH XYLOSUS	
6 721	STAPH XYLOSUS		7 631	STAPH XYLOSUS	
6 731	STAPH XYLOSUS		7 700	STAPH AUREUS	
			7 701	STAPH XYLOSUS	
7 000	STAPH EPIDERMIDIS		7 721	STAPH XYLOSUS	
7 021	STAPH XYLOSUS		7 740	STAPH AUREUS	
7 040	STAPH EPIDERMIDIS				

Source: STAPH-IDENT® is a registered trademark of Analytab Products, Division of Sherwood Medical, Plainview, New York.

MATERIALS

Cultures

24-hour trypticase soy agar slant cultures of *Staphylococcus epidermidis*, *Staphylococcus saprophyticus* (ATCC™ 15305), and *Staphylococcus aureus* (ATCC™ e 27660). Number-coded, 24-hour blood agar cultures of the above organisms for the STAPH-IDENT system.

Media

Per designated student group: three mannitol salt agar plates, one DNA agar plate, three Mueller-Hinton agar plates, and the STAPH-IDENT system.

Reagents

Citrated human or rabbit plasma, 0.1% toluidine blue, and McFarland barium sulfate standards.

Equipment

Bunsen burner, inoculating loop, 13 × 100-mm test tubes, 15 × 150-mm test tubes, sterile Pasteur pipettes, 1-ml sterile pipettes, mechanical pipetting device, sterile cotton swabs, 30-μg novobiocin antibiotic discs, glassware marking pencil, metric ruler, forceps, and beaker with 95% alcohol.

PROCEDURE

1. Preparation of DNA agar plate culture:
   a. With a glassware marking pencil, divide the bottom of the plate into three sections. Label each section with the name of the organism to be inoculated.
   b. Aseptically make a single line of inoculation of each test organism in its respective sector on the agar plate.
2. Preparation of agar plate cultures for novobiocin sensitivity determination:
   a. Label the three Mueller-Hinton agar plates with the name of the test organism to be inoculated. Inoculate each plate with its respective organism according to the Kirby-Bauer procedure as outlined in Experiment 44.
   b. Using alcohol-dipped and flamed forceps, aseptically apply a novobiocin antibiotic disc to the surface of each inoculated plate. *Gently* press the discs down with sterile forceps to ensure that they adhere to the agar surface.

3. Preparation of mannitol salt agar plate cultures: Aseptically make a single line of inoculation of each test organism in the center of the appropriately labeled agar plates.

4. Incubation of all plate cultures: Incubate them in an inverted position for 24 to 48 hours at 37°C.

5. Coagulase test procedure:

   a. Label three 13 × 100-mm test tubes with the name of the organism to be inoculated.

   b. Aseptically add *0.5 ml of a 1:4 dilution of citrated rabbit or human plasma* and 0.1 ml of each test culture to its appropriately labeled test tube.

   c. At the end of the laboratory session, incubate all tubes that are coagulase-negative for 20 hours at 37°C.

STAPH-IDENT system procedure:

1. Prepare strip:

   a. Dispense 5 ml of tap water into incubation tray.

   b. Place API strip in incubation tray.

2. Prepare inoculum:

   a. Add 2 ml of 0.85% saline (pH 5.5–7.0) to a sterile 15 × 150-mm test tube.

   b. Using a sterile swab, pick up a sufficient amount of inoculum to prepare a saline suspension with a final turbidity that is equivalent to a No. 3 McFarland ($BaSO_4$) turbidity standard. *Note: Be sure to use suspension within 15 minutes of preparation.*

3. With a sterile Pasteur pipette, add 2 or 3 drops of the inoculum to each microcupule.

4. Place plastic lid on tray and incubate for 5 hours at 37°C.

Experiment 64 / Observations and Results

Name _____   Date _____

1. Examine the bacterial plasma suspensions for clot formation 5 minutes, 20 minutes, 1 hour, and 4 hours after inoculation by holding the test tubes in a slanted position. Place all coagulase-negative suspensions in an incubator at 37°C for observation 24 hours after inoculation. Record your observations and results in the chart.

Staphylococcal Species	Appearance of Plasma Clotted (+) or Unclotted (−)					Coagulase (+) or (−)
	5 min.	20 min.	1 hr.	4 hr.	24 hr.	
S. aureus						
S. epidermidis						
S. saprophyticus						

2. Examine the mannitol salt agar plate. Note and record the following in the chart:
   a. Presence (+) or absence (−) of growth of each test organism.
   b. Color of the medium surrounding the growth of each test organism.
   c. Whether each test organism is a mannitol fermenter (+) or nonfermenter (−).
3. Flood the DNA agar plate with 0.1% toluidine blue. Observe for the delayed development of a rose-pink coloration surrounding the growth of each test organism. Record your color observation and indicate the presence (+) or absence (−) of DNase activity in the chart.
4. With a metric ruler, measure the size of the zone of inhibition, if present, surrounding each of the novobiocin discs on the agar plates. A zone of inhibition of 17 mm or less is indicative of novobiocin resistance, whereas a zone greater than 17 mm indicates that the organism is sensitive to this antibiotic. Record in the chart the susceptibility of each test organism to novobiocin as sensitive (S) or resistant (R).

📷 *Refer to photo numbers 19 and 71–74 in the color-plate insert for illustration of the reactions.*

Procedure	S. aureus	S. epidermidis	S. saprophyticus
Mannitol salt agar:			
Growth	_____	_____	_____
Color of medium	_____	_____	_____
Fermentation	_____	_____	_____
DNA agar:			
Color of medium	_____	_____	_____
DNase activity	_____	_____	_____
Novobiocin sensitivity:			
Growth inhibition in mm	_____	_____	_____
Susceptibility—(R) or (S)	_____	_____	_____

5. Interpret your STAPH-IDENT system reactions on the basis of the observed color changes in each of the microcupules described in the chart below. Record your color observation and result as (+) or (−) for each test in this chart.

Microcupule			Interpretation of Reactions		Reaction Results	
No.	Substrate		Positive	Negative	Color	(+) or (−)
1	PHS	p-nitrophenyl-phosate, disodium salt	Yellow	Clear or straw-colored		
2	URE	Urea	Purple to red-orange	Yellow or yellow-orange		
3	GLS	p-nitrophenyl-β-D-glucopyranoside	Yellow	Clear or straw-colored		
4	MNE	Mannose	Yellow or yellow-orange	Red or orange		
5	MAN	Mannitol				
6	TRE	Trehalose				
7	SAL	Salicin				
8	GLC	p-nitrophenyl-β-D-glucuronide	Yellow	Clear or straw-colored		
9	ARG	Arginine	Purple to red-orange	Yellow or yellow-orange		
10	NGP	2-naphthyl-β-D-galactopyranoside	Add 1–2 drops of STAPH-IDENT reagent			
			Plum-purple (mauve)	Yellow or colorless		

6. Construct a four-digit profile for your unknown organism as follows: A four-digit profile is derived from the results obtained with STAPH-IDENT. The 10 biochemical tests are divided into four groups, as follows:

PHS   MNE   SAL   NGP
URE   MAN   GLC
GLS   TRE   ARG

Only positive reactions are assigned a numerical value. The value depends on the location within the group.

A value of 1 for the first biochemical in each group (e.g., PHS, MNE)
A value of 2 for the second biochemical in each group (e.g., URE, MAN)
A value of 4 for the third biochemical in each group (e.g., GLS, TRE)
A value of 0 for all negative reactions

A four-digit number is obtained by totaling the values of each of the groups.

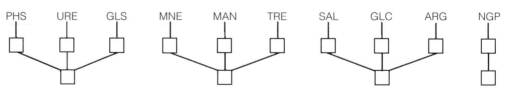

Using Table 64.2 on pages 417–418 and your four-digit profile number, identify your organism.

Unknown organism: _____

Identification of Human Streptococcal Pathogens

LEARNING OBJECTIVES

Once you have completed this experiment, you should understand

1. The medical significance of streptococci.
2. Selected laboratory procedures designed to differentiate streptococci on the basis of their hemolytic activity and biochemical patterns associated with the Lancefield group classifications.

PRINCIPLE

Members of the genus *Streptococcus* are perhaps responsible for a greater number of infectious diseases than any other group of microorganisms. Morphologically, they are cocci that divide in a single plane forming chains. They form circular, translucent to opaque, pinpoint colonies on solid media. All members of this group are gram-positive, and many are nutritionally fastidious, requiring enriched media such as blood for growth.

The streptococci are classified by means of two major methods: (1) their **hemolytic activity,** and (2) the **serologic classification of Lancefield.** The observed hemolytic reactions on blood agar are of the following three types:

1. **(α) Alpha-hemolysis,** an incomplete form of hemolysis, produces a green zone around the colony. α-hemolytic streptococci, the *Streptococcus viridans* species, are usually nonpathogenic opportunists. In some instances, however, they are capable of inducing human infections such as **subacute endocarditis,** which may precipitate valvular damage and heart failure if untreated. *Streptococcus pneumoniae,* the causative agent of **lobar pneumonia,** will be studied in a separate experiment.

2. **(β) Beta-hemolysis,** a complete destruction of red blood cells, exhibits a clear zone of approximately 2 to 4 times the diameter of the colony. The streptococci capable of producing β-hemolysins are most frequently associated with pathogenicity.

3. **(γ) Gamma-hemolysis** is indicative of the absence of any hemolysis around the colony. Most commonly, γ-hemolytic streptococci are avirulent.

Lancefield classified the streptococci into 20 **serogroups,** designated **A** through **V,** omitting **I** and **J,** based on the presence of an antigenic group-specific hapten called the **C-substance.** This method of classification generally implicates the members of Groups A, B, C, and D in human infectious processes.

β-hemolytic streptococci belonging to **Group A,** and collectively referred to as *Streptococcus pyogenes,* are the human pathogens of prime importance. Members of this group are the main etiological agents of human respiratory infections such as **tonsillitis, bronchopneumonia,** and **scarlet fever,** as well as skin disorders such as **erysipelas** and **cellulitis.** In addition, these organisms are responsible for the development of complicating infections, namely **glomerulonephritis** and **rheumatic fever,** which may surface when primary streptococcal infections either go untreated or are not completely eradicated by antibiotics. The β-hemolytic streptococci found in **Group B** are indigenous to the vaginal mucosa and have been shown to be responsible for **puerperal fever** (childbirth fever), a sometimes-fatal **neonatal meningitis,** and **endocarditis.** Members of **Group C** are also β-hemolytic and have been implicated in **erysipelas, puerperal fever,** and **throat infections. Group D** streptococci generally

TABLE 65.1 Laboratory Differentiation of Streptococci

Group	A	B	C	D		K, H, N
Organisms	*S. pyogenes*	*S. agalactiae*	*S. equi*	*E. faecalis* Enterococci	*S. bovis* Nonenterococci	*S. salivarius* *S. sanguis* *S. miltis*
Hemolysis	β	β	β	α → γ	α → γ	α
Bacitracin test	+	−	−	−	−	−
CAMP test	−	+	−	−	−	−
Bile esculin hydrolysis	−	−	−	+	+	−
6.5% NaCl medium	NG	NG	NG	G	NG	NG
Growth at 10°C	NG	NG	NG	G	NG	NG
Growth at 45°C	NG	NG	NG	G	NG or G	NG

NG = no growth; G = growth

exhibit α- or γ-hemolysis on blood agar plates. This group includes the **enterococci** such as *Enterococcus faecalis,* which may be the etiological agent of **urinary tract infections,** and the **nonenterococci** such as *S. bovis,* which is of lesser medical concern in humans. The enterococci tend to be antibiotic-resistant, particularly to penicillin and more recently to vancomycin.

The virulence of the streptococci is associated with their ability to produce a wide variety of extracellular metabolites. Included among these are the **hemolysins** (α and β), **leukocidins** that destroy phagocytes, and the **erythrogenic toxin** responsible for the rash of scarlet fever. Also of medical significance are the metabolites **hyaluronidase** (the spreading factor), which hydrolyzes the tissue cement hyaluronic acid; **streptokinase,** a fibrinolysin; and the **nucleases,** ribonuclease and deoxyribonuclease, which destroy viscous tissue debris. The last three metabolic end products facilitate the spread of the organisms, thereby initiating secondary sites of streptococcal infection.

Although the different groups of streptococci have similar colonial morphology and microscopic appearance, they can be separated and identified by the performance of a variety of laboratory tests. Toward this end, you will perform laboratory procedures to differentiate among the medically significant streptococci on the basis of their Lancefield group classification and their hemolytic patterns. Table 65.1 will aid in this separation.

Identification of Group A streptococci involves the following procedures:

1. **Bacitracin test:** A filter-paper disc impregnated with 0.04 unit of bacitracin is applied to the surface of a blood agar plate previously streaked with the organism to be identified. Following incubation, the appearance of a zone of growth inhibition surrounding the disc is indicative of Group A streptococci. Absence of this zone suggests a non-Group A organism.

2. **Directigen™ test:** A rapid, nongrowth-dependent immunological procedure for the detection of the Group A antigen, developed by Becton Dickinson and Company. In this test, a clinical specimen is subjected to reagents designed to extract the Group A antigen, which is then mixed with a reactive and a negative control latex. Agglutination with the reactive latex is indicative of Group A streptococci.

Group B streptococci are identified with the **CAMP test** (named for Christie, Atkins, and Munch-Peterson). Group B streptococci produce a peptide, the CAMP substance, that acts in concert with the β-hemolysins produced by some strains of *Staphylococcus aureus,* causing an increased hemolytic effect. Following inoculation and incubation, the resultant effect appears as an arrow-shaped zone of hemolysis adjacent to the central streak of *S. aureus* growth. The non-Group B streptococci do not produce this reaction. Figure 65.1 illustrates the CAMP reactions.

Identification of Group D streptococci involves the following:

1. **Bile esculin test:** In the presence of bile, Group D streptococci hydrolyze the glycoside esculin to 6,7-dihydroxy-coumarin that reacts with the iron salts in the medium to produce a brown-to-black coloration of the medium following incubation. Lack of this dark coloration is indicative of a non-Group D organism.

2. **6.5% sodium chloride broth:** The Group D enterococci can be separated from the nonenterococci by the ability of the former to grow in this medium.

Hemolytic activity is identified with a blood agar medium. The pathogenic streptococci, primarily the β-hemolytic, can be separated from the generally avirulent α- and γ-hemolytic streptococci by the type of hemolysis produced on blood agar as previously described.

MATERIALS

Cultures

24-hour blood agar slant cultures of *Streptococcus pyogenes* (ATCC™ 12385), *Enterococcus faecalis*, *Streptococcus bovis*, *Streptococcus agalactiae*, *Streptococcus mitis*, and *Staphylococcus aureus* (ATCC 25923).

Media

Per designated student group: five blood agar plates, three bile esculin agar plates, and three 6.5% sodium chloride broths.

Reagents

Directigen Rapid Group A Strep Test (Becton Dickinson and Company), crystal violet, Gram's iodine, ethyl alcohol, safranin, and Taxo® A discs (0.04 unit of bacitracin).

Equipment

Bunsen burner, inoculating loop, staining tray, lens paper, bibulous paper, microscope, sterile cotton swabs, glassware marking pencil, sterile 12 × 75-mm test tubes, sterile Pasteur pipettes, sterile applicators, 95% alcohol in beaker, forceps, and mechanical rotator.

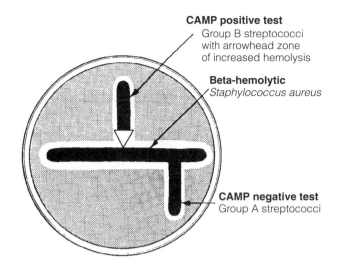

CAMP positive test
Group B streptococci with arrowhead zone of increased hemolysis

Beta-hemolytic
Staphylococcus aureus

CAMP negative test
Group A streptococci

FIGURE 65.1 CAMP reactions

PROCEDURE

1. Prepare a Gram-stain preparation of each streptococcal culture and observe under oil immersion. Record in the chart your observations of cell morphology and Gram reaction.

2. Prepare the blood agar plate cultures to identify the type of hemolysis as follows:
   a. With a glassware marking pencil, divide the bottoms of two blood agar plates to accommodate the five test organisms. Label each section with the name of the culture to be inoculated.
   b. Using sterile inoculating technique, make a single line streak of inoculation of each organism in its respective sector on the blood plates.

3. Prepare the blood agar plate cultures for the bacitracin test as follows:
   a. With a glassware marking pencil, label the covers of two blood agar plates with the names of the organisms to be inoculated, *S. pyogenes* and *S. agalactiae*.
   b. Using a sterile cotton swab, inoculate the agar surface of each plate with its respective test organism by streaking first in a horizontal direction, then vertically to ensure a heavy growth over the entire surface.

c. Using alcohol-dipped and flamed forceps, apply a single 0.04-unit bacitracin disc to the surface of each plate. Gently touch each disc to ensure its adherence to the agar surface.

4. Prepare a blood agar plate culture for the **CAMP test** as follows:

   a. Using a sterile inoculating loop, make a single line of inoculation along the center of the plate using the *S. aureus* culture.

   b. With a sterile loop, inoculate *S. pyogenes* on one side and perpendicular to the central *S. aureus* streak, starting about 5 mm from the central streak and extending toward the periphery of the agar plate.

   c. On the opposite side of the central streak, but not directly opposite the *S. pyogenes* line of inoculation, repeat Step 4b using *S. agalactiae*.

5. Prepare the bile esculin agar plate cultures as follows:

   a. Label the three bile esculin plates with the names of the organisms to be inoculated, *S. bovis*, *S. mitis*, and *E. faecalis*.

   b. Aseptically inoculate each plate with its test organism by making several lines of inoculation on the agar surface.

6. Prepare 6.5% sodium chloride broth cultures as follows:

   a. Label three tubes of 6.5% sodium chloride broth with the names of the organisms to be inoculated, *S. bovis*, *E. faecalis*, and *S. mitis*.

   b. With a sterile loop, inoculate each tube with its organism.

7. Conduct the **Directigen test** procedure as follows:

   a. Label two sterile 12 × 75-mm test tubes as *S. pyogenes* and *S. agalactiae*.

   b. Add 0.3 ml of Reagent 1 to both test tubes.

   c. Using a sterile cotton swab, transfer the test organisms into their respectively labeled test tubes. *Note: These samples will emulate the throat swabs obtained in a clinical solution.*

   d. Add 1 drop of Reagent 2 to each test tube. Mix by rotating the swab against the side of the tube. Allow the swabs to remain in the test tubes for 3 minutes.

   e. Add 1 drop of Reagent 3 to both tubes and mix.

   f. Remove swabs after extracting as much liquid as possible by rolling them against the sides of the tubes.

   g. Place 1 drop of negative antigen control on both circles in Column A of test slide.

   h. Place 1 drop of positive antigen control on both circles in Column B of test slide.

   i. Dispense 1 drop of each streptococcal sample on both circles in Columns C and D, respectively.

   j. Using a new sterile applicator for each specimen, spread each specimen within the confines of both circles in Columns A, B, C, and D.

   k. Add 1 drop of reactive latex to the top row of circles.

   l. Add 1 drop of control latex to the bottom row of circles.

   m. Place the slide on a mechanical rotator for 4 minutes under a moistened humidifying cover.

   n. Compare the agglutination seen in the upper "reactive latex" circles with the consistency of the latex in the bottom "control latex" circles. Any agglutination in the top circles distinct from any background granules seen in the bottom circles indicates Group A streptococci.

8. Incubate all tubes and plates in an inverted position for 24 hours at 37°C.

Name _____   Date _____

1. Microscopically examine the Gram-stained smears of the test organisms. Record your observations of their cellular morphology and Gram reactions.
2. Examine the two blood agar plates for bacitracin activity. Record your observations of the presence (+) or absence (−) of a zone of inhibition of any size surrounding the discs.
3. Examine the blood agar plate for the CAMP reaction. Record your observations of the presence (+) or absence (−) of increased arrow-shaped hemolysis.
4. Examine the bile esculin plates for the presence (+) or absence (−) of a brown-black coloration in the medium and record your observations.
5. Observe the 6.5% sodium chloride broth cultures for the presence (+) or absence (−) of growth and record your observations.
6. Examine the two blood agar plates for the presence and type of hemolysis produced by each of the test organisms. Record your observations of the appearance of the medium surrounding the growth and the type of hemolytic reaction that has occurred—α, β, or γ.
7. Observe the Directigen test slide for the presence (+) or absence (−) of agglutination in the reactive and control latex circles. Based on your observations, indicate the Lancefield group classification of each test organism. Record your results.
8. Based on your observations, classify each test organism according to its Lancefield group.

📷 *Refer to photo numbers 6, 20–22, and 75–77 in the color-plate insert for illustration of these reactions.*

Procedure	S. pyogenes	S. agalactiae	S. bovis	E. faecalis	S. mitis
Gram stain: Morphology	_____	_____	_____	_____	_____
Reaction	_____	_____	_____	_____	_____
Bacitracin test: Zone of inhibition	_____	_____	_____	_____	_____
CAMP test: Increased hemolysis	_____	_____	_____	_____	_____
Bile esculin test: Color of medium	_____	_____	_____	_____	_____
Result: (+) or (−)	_____	_____	_____	_____	_____
6.5% NaCl broth: Growth	_____	_____	_____	_____	_____
Hemolytic activity: Appearance of medium	_____	_____	_____	_____	_____
Type of hemolysis	_____	_____	_____	_____	_____

Directigen test:
  Agglutination (+) or (−) in:
    Reactive circle
	_____	_____	_____	_____	_____

Control circle	_____	_____	_____	_____	_____
Lancefield group	_____	_____	_____	_____	_____
Group classification					

REVIEW QUESTIONS

1. How do the purposes of the bacitracin and CAMP tests differ?

2. What is the mechanism of the bile esculin test?

3. Why is it important medically to distinguish between the enterococci and the nonenterococci?

4. Why can some streptococci produce secondary sites of infection?

5. The streptococci are known to be fastidious organisms that require an enriched medium for growth. How would you account for the fact that a medium enriched with blood (blood agar) is the medium of preference for growth of these organisms?

Identification of *Streptococcus pneumoniae*

LEARNING OBJECTIVE

Once you have completed this experiment, you should understand

1. Laboratory procedures to differentiate between *Streptococcus pneumoniae* and other α-hemolytic streptococci.

PRINCIPLE

The pneumococcus *Streptococcus pneumoniae* is the major α-hemolytic, streptococcal pathogen in humans. It serves as an etiological agent of **lobar pneumonia,** an infection characterized by acute inflammation of the bronchial and alveolar membranes. These organisms are gram-positive cocci, tapered or lancet-shaped at their edges, that occur in pairs or as short, tight chains. The large, thick **capsules** formed in vivo are responsible for antiphagocytic activity, which is believed to enhance the organisms' virulence. In addition, the pneumococci produce **α-hemolysis** on blood agar plates. Because of these properties (short-chain formation, α-hemolysis, and failure of the capsule to stain on Gram staining), the organisms closely resemble *Streptococcus viridans* species. The *S. pneumoniae* can be differentiated from other α-hemolytic streptococci on the basis of the following laboratory tests:

Test	S. pneumoniae	S. mitis
Hemolysis	α	α
Bile solubility	+	−
Optochin	+	−
Inulin fermentation	+	−
Quellung reaction	+	−
Mouse virulence	+	−

Brief descriptions of the tests and their mechanisms follow:

1. **Bile solubility test:** In the presence of surface-active agents such as **bile** and **bile salts** (sodium desoxycholate or sodium dodecyl sulfate), the cell wall of the pneumococcus undergoes lysis. Other members of the α-hemolytic streptococci will not be lysed by these agents and are bile-insoluble. Following incubation, bile-soluble cultures will appear clear, and bile-insoluble cultures will be turbid.

2. **Optochin test:** This is a growth inhibition test in which 6-mm filter-paper discs impregnated with 5 mg of **ethylhydrocupreine hydrochloride** (optochin) and called P-discs are applied to the surface of a blood agar plate streaked with the test organisms. The *S. pneumoniae*, being sensitive to this surface-active agent, are lysed with the resultant formation of a zone of inhibition greater than 15 mm surrounding the P-disc. Nonpneumococcal α-hemolytic streptococci are resistant to optochin and fail to show a zone of inhibition or produce a zone less than 15 mm.

3. **Inulin fermentation:** The pneumococci are capable of fermenting inulin, while most other α-hemolytic streptococci are inulin-nonfermenters. Following incubation, the **acid** resulting from inulin fermentation will change the color of the culture from red to yellow. Cultures that are not capable of fermenting inulin will not exhibit a color change, which is a negative test result.

4. **Quellung** (Neufeld) **reaction:** This **capsular swelling** reaction is a sensitive and accurate method of determining the presence of *S. pneumoniae* in sputum. The reaction of the pneumococcal capsular

polysaccharide, a hapten antigen, with an omnivalent capsular antiserum produces a microscopically visible swollen capsule surrounding the *S. pneumoniae* organisms.

5. **Mouse virulence test:** Laboratory white mice are highly susceptible to infection by *S. pneumoniae* and resistant to other streptococcal infections. Intraperitoneal injection of 0.1 ml of pneumococcus-infected sputum will kill the mouse. Examination of the peritoneal fluid by Gram stain and culture will reveal the presence of *S. pneumoniae*.

In the following experiment, you will use hemolytic patterns, bile solubility, the Quellung reaction, the optochin test, and the inulin fermentation test for laboratory differentiation of *S. pneumoniae* from other α-hemolytic streptococci.

MATERIALS

Cultures

24-hour blood agar slant cultures of *Streptococcus pneumoniae* and *Streptococcus mitis*.

Media

Per designated student group: one blood agar plate, two phenol red inulin broth tubes, and four 13 × 75-mm tubes containing 1 ml of nutrient broth.

Reagents

Crystal violet, Gram's iodine, ethyl alcohol, safranin, methylene blue, 10% sodium desoxycholate, commercially available Taxo® P-discs (5 mg of optochin), and omnivalent pneumococcal antiserum.

Equipment

Bunsen burner, inoculating loop, glass slides, sterile cotton swabs, sterile 1-ml serological pipettes, mechanical pipetting device, 95% alcohol in beaker, forceps, and glassware marking pencil.

PROCEDURE

1. **Bile solubility test**

   a. Label two nutrient broth tubes *S. pneumoniae* and two other tubes *S. mitis*.

   b. Aseptically add 2 loopsful of the test organisms to the appropriately labeled

sterile test tubes to effect a heavy suspension.

   c. Aseptically add 0.5 ml of sodium desoxycholate to one tube of each test culture. The remaining two cultures will serve as controls.

   d. Incubate the tubes in a waterbath at 37°C for 1 hour.

2. **Optochin test**

   a. With a glassware marking pencil, divide the bottom of a blood agar plate into two equal sections and label one section *S. pneumoniae* and the other *S. mitis*.

   b. Using a sterile cotton swab, heavily inoculate the surface of each section with its respective test organism in a horizontal and then vertical direction, being careful to stay within the limits of each section.

   c. Using alcohol-dipped and flamed forceps, apply a single Taxo P-disc (optochin) to the surface of the agar in each section of the inoculated plate. Touch each disc slightly to ensure its adherence to the agar surface.

   d. Incubate the plate in an inverted position for 24 to 48 hours at 37°C.

3. **Inulin fermentation test**

   a. Label two phenol red inulin broth tubes with the name of each test organism to be inoculated.

   b. Using sterile technique and loop inoculation, inoculate each experimental organism in its appropriately labeled tube of medium.

   c. Incubate the tube cultures for 24 to 48 hours at 37°C.

4. **Quellung reaction**

   a. Spread a loopful of each test culture on labeled clean glass slides and allow the slides to air-dry.

   b. Place a loopful of the omnivalent capsular antiserum and a loopful of methylene blue to each of two coverslips.

   c. Place the coverslips over the dried bacterial smears.

5. Prepare a Gram-stain preparation of each test organism and observe under oil immersion. Record your observations of cell morphology and Gram reaction in the chart.

Name _____   Date _____

1. Examine blood agar plates for the presence of hemolysis and optochin activity by measuring the zone of inhibition, if any, surrounding the disc. Indicate in the chart the presence (+) or absence (−) of the 15-mm or greater zone of inhibition surrounding the optochin discs and indicate whether each organism is optochin-sensitive or optochin-resistant.

Refer to photo number 78 in the color-plate insert for illustration of this reaction.

2. For the bile solubility test, examine the tubes after 1 hour of incubation for the presence or absence of turbidity in each culture. Record your observations of the appearance (clear or turbid) and bile solubility of each test organism.

3. Observe the phenol red inulin broth cultures and record the color of each culture and whether it is indicative of a positive (+) or negative (−) result.

4. Examine slides of the Quellung reaction under oil immersion and indicate in the chart the presence (+) or absence (−) of capsular swelling surrounding the blue-stained cells.

Procedure	*S. pneumoniae*	*S. mitis*
Gram stain: Morphology		
Reaction		
Optochin test: Zone of inhibition in mm		
Resistant or sensitive		
Bile solubility test: Appearance of culture		
Bile solubility		
Inulin fermentation: Color of medium		
Fermentation		
Quellung reaction: Capsular swelling		

REVIEW QUESTIONS

1. Why is it clinically important to distinguish *S. pneumoniae* from other α-hemolytic streptococci?

2. How would you separate *S. pneumoniae* from other α-hemolytic streptococci?

3. 🔍 What are secondary pneumonias? Why do they develop most frequently following viral infections?

4. 🔍 Why did it require many years of research to develop an effective, long-term pneumo-coccal vaccine?

Identification of Enteric Microorganisms Using Computer-Assisted Multitest Microsystems

LEARNING OBJECTIVES

Once you have completed this experiment, you should be familiar with

1. The members of the family Enterobacteriaceae.

2. Laboratory procedures designed to identify enteric pathogens using commercial multitest microsystems.

INTRODUCTION

The Enterobacteriaceae are a significant group of bacteria that are endogenous to the intestinal tract or that may gain access to this site via a host's ingestion of contaminated food and water. The family consists of a number of genera whose members vary in their capacity to produce disease. The *Salmonella* and *Shigella* are considered to be pathogenic. Members of other genera, particularly *Escherichia* and *Enterobacter*, and to a lesser extent, *Klebsiella* and *Proteus*, constitute the natural flora of the intestines and are generally considered to be avirulent. Remember, however, that all can produce disease under appropriate conditions.

The Enterobacteriaceae are gram-negative, short rods. They are mesophilic, nonfastidious organisms that multiply in many foods and water sources. They are all nonspore-formers and susceptible to destruction by common physical and chemical agents. They are resistant to destruction by low temperatures and can therefore frequently survive in soil, sewage, water, and many foods for extended periods of time.

From a medical point of view, the pathogenic Enterobacteriaceae are salmonellae and shigellae. Salmonellae are responsible for enteric fevers, **typhoid,** the milder **paratyphoid,** and **gastroenteritis.** In typhoid, *Salmonella*

typhi penetrates the intestinal mucosa and enters the bloodstream, thus infecting organs such as the gallbladder, intestines, liver, kidney, spleen, and heart. Ulceration of the intestinal wall, caused by the release of the lipopolysaccharide endotoxin into the blood over a long febrile period, and enteric symptoms are common. **Gastroenteritis** is caused by a number of *Salmonella* species. Symptoms associated with this type of food poisoning include abdominal pain, nausea, vomiting, and diarrhea, which develop within 24 hours of ingestion of contaminated food and last for several days.

Several shigellae are responsible for **shigellosis,** a bacillary dysentery that varies in severity. Ulceration of the large intestine, explosive diarrhea, fever, and dehydration occur in the more severe cases.

Isolation and identification of enteric bacteria from feces, urine, blood, and fecally contaminated materials are of major importance in the diagnosis of enteric infections. Although the Enterobacteriaceae are morphologically alike and in many ways metabolically similar, laboratory procedures for the identification of these bacteria are based on differences in biochemical activities (Figure 67.1).

In the past, several **multitest systems** have been developed for differentiation and identification of members of the Enterobacteriaceae. They use microtechniques that incorporate a number of media in a single unit. At least six multitest systems are commercially available. The obvious advantages of these units are the need for minimal storage space, the use of less media, the rapidity with which results may be obtained, and the applicability of the results to a computerized system for identification of organisms. There are also certain disadvantages with these systems, including difficulty in obtaining the proper inoculum

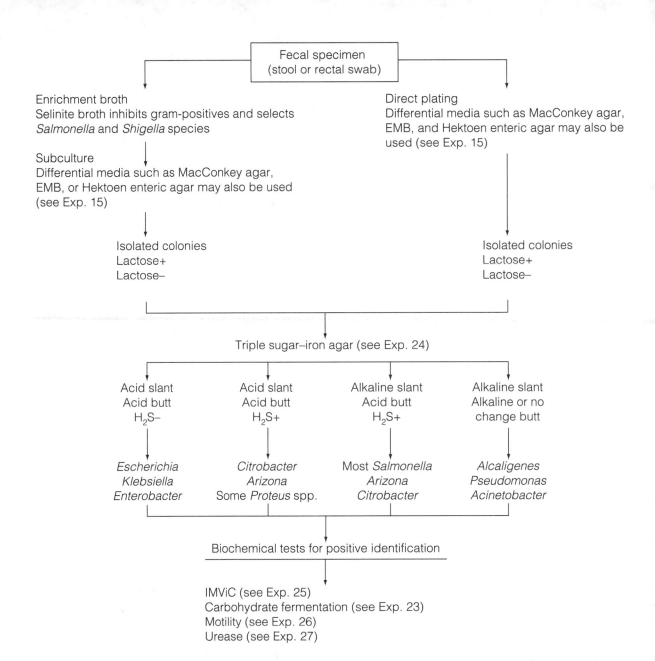

FIGURE 67.1 Conventional laboratory procedures for isolation and identification of enteric microorganisms

size since some media require heavy inoculation while others need to be lightly inoculated, the possibility of media carryover from one compartment to another, and the possibility of using inoculum of improper age. Despite these difficulties, when properly correlated with other properties such as Gram stain and colonial morphology on specialized solid media, these systems are acceptable for the identification of Enterobacteriaceae. The most frequently used systems are discussed.

PART A: Enterotube Multitest System and ENCISE II

The **Enterotube® Multitest System** (Roche Diagnostics, Division of Hoffmann-La Roche, Inc.) consists of a single tube containing 12 compartments (Figure 67.2) and a self-enclosed inoculating needle. This needle can touch a single isolated colony and then in one operation be drawn through all

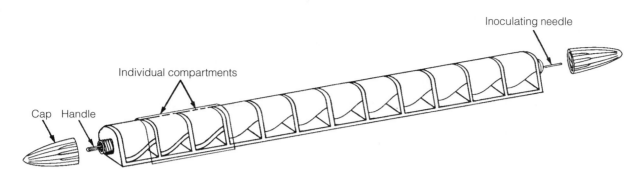

FIGURE 67.2 Enterotube multitest (courtesy of Roche Diagnostics, Division of Hoffmann-La Roche, Inc.)

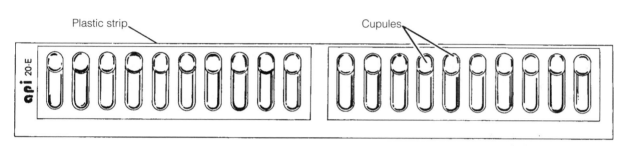

Microtubes containing dehydrated media

FIGURE 67.3 The API 20-E system (courtesy of Analytab Products, Division of Sherwood Medical, Plainview, New York)

12 compartments, thereby inoculating all of the test media. In this manner, 15 standard biochemical tests can be performed in one inoculating procedure. Following incubation, the color changes that occur in each of the compartments are interpreted according to the manufacturer's instructions to identify the organisms. This method has been further refined to permit identification of the enteric bacteria by means of a computer-assisted system called **ENCISE** (Enterobacteriaceae numerical coding and identification system for Enterotube).

PART B: API (Analytical Profile Index) System

The API® 20-E (Analytab Products, Plainview, New York) employs a plastic strip composed of 20 individual microtubes, each containing a dehydrated medium in the bottom and an up-

per cupule as shown in Figure 67.3. The media become hydrated during inoculation of a suspension of the test organism, and the strip is then incubated in a plastic-covered tray to prevent evaporation. In this manner, 22 biochemical tests are performed. Following incubation, identification of the organism is made by using differential charts supplied by the manufacturer or by means of a computer-assisted system called **PRS** (profile recognition system). PRS includes an API coder, profile register, and selector.

In the following experiment, you will inoculate an Enterotube and an API strip with an unknown enteric organism. Following incubation, you will make your identification by two methods: (1) the traditional method of noting the characteristic color changes and interpreting them according to manufacturers' instructions, and (2) the computer-assisted method illustrated in Figure 67.4.

(a) The Enterotube®

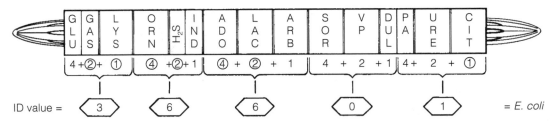

ID value = ⟨3⟩ ⟨6⟩ ⟨6⟩ ⟨0⟩ ⟨1⟩ = *E. coli*

1. Each positive reaction is indicated by circling the number directly below its compartment.
2. The circled numbers in each bracket are added together, and the sum is placed in the box below.
3. The resultant 5-digit number (ID value) is then located in the computer coding manual to identify the organism.

(b) The API® strip

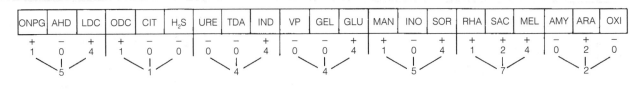

7-digit profile number = | 5 | 1 | 4 | 4 | 5 | 7 | 2 | = *E. coli*

1. The 21 tests are divided into seven groups of three each.
2. A value of 1 is assigned to the first positive test in each group.
3. A value of 2 is assigned to the second positive test in each group.
4. A value of 4 is assigned to the third positive test in each group.
5. A 7-digit number is obtained by totaling the positive values of each of the seven groups of three. This number is located in the analytical profile index to identify the organism.

FIGURE 67.4 Computer assisted techniques for the identification of Enterobacteriacea

MATERIALS

Cultures

Number-coded, 24-hour trypticase soy agar streak plates of *Escherichia coli*, *Salmonella typhimurium*, *Klebsiella pneumoniae*, *Enterobacter aerogenes*, *Shigella dysenteriae*, and *Proteus vulgaris*.

Media

Per designated student group: one Enterotube II, one API 20-E strip, and one 5-ml tube of 0.85% sterile saline.

Reagents

Sterile mineral oil, 10% ferric chloride, Kovac's reagent, VP reagent for API system, nitrate reduction reagents, Barritt's reagent (VP test reagent for Enterotube system), 1.5% hydrogen peroxide, and 1% *p*-aminodimethylaniline oxalate (oxidase reagent).

Equipment

Bunsen burner, inoculating loop, 5-ml pipette, mechanical pipetting device, sterile Pasteur pipettes, glassware marking pencil, API profile recognition system and differential identification charts, and Enterotube ENCISE pads and color reaction charts.

PROCEDURE

Enterotube II System

1. Familiarize yourself with the components of the system: screw caps at both ends, medium-containing compartments, self-enclosed inoculating needle, plastic side bar, and blue-taped section.

2. Label the Enterotube with your name and the number of the unknown culture supplied by the instructor.

3. Remove the screw caps from both ends of the Enterotube. Using the inoculating needle contained in the Enterotube, aseptically pick some inoculum from an isolated colony on the provided streak-plate culture.

4. Inoculate the Enterotube as follows:

   a. Twist the needle in a rotary motion and withdraw it slowly through all 12 compartments.

   b. Replace the needle in the tube and with a rotary motion push the needle into the first three compartments (GLU/GAS, LYS, and ORN). The point of the needle should be visible in the H_2S/IND compartment.

   c. Break the needle at the exposed notch by bending, discard the needle remnant, and replace the caps at both ends. The presence of the needle in the three compartments maintains anaerobiosis, which is necessary for dextrose fermentation, CO_2 production, and the decarboxylation of lysine and ornithine.

5. Remove the blue tape covering the ADO, LAC, ARB, SOR, VP, DUL/PA, URE, and CIT compartments. Beneath this tape are tiny air vents that provide aerobic conditions in these compartments.

6. Place the clear plastic slide band over the GLU/GAS compartment to contain the wax, which may be spilled by the excessive gas production of some organisms.

7. Incubate the tube on a flat surface for 24 hours at 37°C.

API 20-E System

1. Familiarize yourself with the components of the system: incubation tray, lid, and the strip with its 20 microtubes.

2. Label the elongated flap on the incubation tray with your name and the number of the unknown culture supplied by the instructor.

3. With a pipette, add approximately 5 ml of tap water to the incubation tray.

4. Using a sterilized loop, touch an isolated colony on the provided streak-plate culture, transfer the inoculum to a 5-ml tube of sterile saline, and mix well to effect a uniform suspension.

5. Remove the API strip from its sterile envelope and place it in the incubation tray.

6. Tilt the incubation tray. Using a sterile Pasteur pipette containing the bacterial saline suspension, fill the tube section of each compartment by placing the tip of the pipette against the side of the cupule. Fill the cupules in the CIT, VP, and GEL microtubes with the bacterial suspension.

7. Using a sterile Pasteur pipette, fill the cupules of the AHD, LDC, ODC, and URE microtubes with sterile mineral oil to provide an anaerobic environment.

8. Cover the inoculated strip with the tray lid and incubate for 18 to 24 hours at 37°C.

Name   Date

PART A: Enterotube II System

1. Observe all reactions in the Enterotube except IND and VP, and interpret your observations using the manufacturer's instructions. Record your observations and results in the chart.

2. Perform the IND and VP tests as follows:

    a. Place the Enterotube in a rack with the GLU and VP compartments facing downward.

    b. With a needle and a syringe, gently pierce the plastic film of the H_2S/IND compartment and add 2 or 3 drops of Kovac's reagent. Read the results after 1 minute.

    c. As in Step 2b, add 2 drops of Barritt's reagent to the VP compartment and read the results after 20 minutes.

    d. Record your IND and VP observations and results.

3. Based on your results, identify your unknown organism using the manufacturer's color identification charts.

4. Determine and record in the chart the five-digit ID value as described in Figure 67.4 on page 436. Identify your unknown organism by referring to the computer coding manual.

 Refer to photo number 79 in the color-plate insert.

PART B: API 20-E System

1. Observe all reactions in the API strip that do not require addition of a test reagent, and interpret your observations using the manufacturer's instructions. Record your observations and results in the chart.

2. Add the required test reagents in the following order: Kovac's reagent to IND, VP reagent to VP (read the result after 15 minutes), ferric chloride to TDA, nitrate reagents to GLU, and oxidase reagent to OXI. Note color changes and interpret your observations according to the manufacturer's instructions. Record your observations and results in the chart.

3. Based on your results, identify your unknown organism using the differential identification chart.

4. Determine and record in the chart the seven-digit profile number as described in Figure 67.4 on page 436. Identify your unknown organism by referring to the profile recognition system.

 Refer to photo number 80 in the color-plate insert.

		API 20-E		Enterotube	
Code	**Name**	**Appearance** (color)	**Result** (+) or (−)	**Appearance** (color)	**Result** (+) or (−)
ONPG	β-galactosidase				
AHD	Arginine dihydrolase				
LDC/LYS	Lysine decarboxylase				
ODC/ORN	Ornithine decarboxylase				
CIT	Citrate				
H₂S	Hydrogen sulfide				
URE	Urease				
TDA	Tryptophan deaminase				
IND	Indole				
VP	Acetonin				
GEL	Gelatin				
GLU	Glucose				
MAN	Mannitol				
INO	Inositol				
SOR	Sorbitol				
RHA	Rhamnose				
SAC	Sucrose				
MEL	Melibiose				
AMY	Amygdalin				
ARA/ARB	Arabinose				
OXI	Oxidase				
ADO	Adonitol fermentation				
GAS	Gas production				
PHE/PA	Phenylalanine				
LAC	Lactose				
DUL	Dulcitol				
Organism					

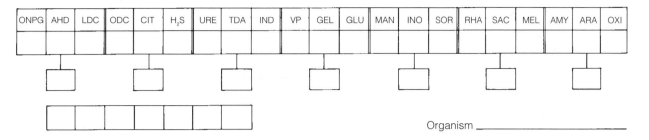

ONPG	AHD	LDC	ODC	CIT	H₂S	URE	TDA	IND	VP	GEL	GLU	MAN	INO	SOR	RHA	SAC	MEL	AMY	ARA	OXI

Organism _____

Determination of API 20-E seven-digit profile number

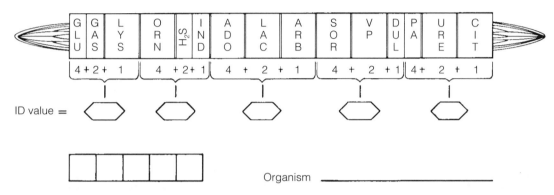

ID value = ⬡ ⬡ ⬡ ⬡ ⬡

Organism _____

Determination of Enterotube five-digit identification number

REVIEW QUESTIONS

1. What are the advantages of multitest systems?

Disadvantages?

2. What Enterobacteriaceae are of medical significance?

   List and describe the infections caused by these organisms.

3. Would similar results be obtained by use of the computer-assisted method and the traditional color-change method?

4. What is the clinical justification for the use of a rapid test procedure such as the Enterotube II System for the identification of enteric microorganisms?

Isolation and Presumptive Identification of *Campylobacter*

LEARNING OBJECTIVE

Once you have completed this experiment, you should understand

1. The laboratory procedures required for the isolation, cultivation, and presumptive identification of the genus *Campylobacter*.

PRINCIPLE

Clinicians are aware of the medical significance of *Campylobacter* strains as the etiological agents of enteric infections. The incidence of enteritis caused by *Campylobacter jejuni* equals or exceeds that of salmonellosis or shigellosis. The clinical syndrome, although varying in severity, is generally characterized by acute gastroenteritis accompanied by the rapid onset of fever, headache, muscular pain, malaise, nausea, and vomiting. Twenty-four hours following this acute phase, diarrhea develops that may be mucoid, bloody, bile-stained, and watery. The precise epidemiology of the infection is not clear; however, contact with animals, water-borne organisms, and oral–fecal transmission remain suspect.

The organisms (*campylo,* curved; *bacter,* rod) were formerly called vibrios because of their curved and spiral morphology. In the early 1980s they were reclassified in the genus *Campylobacter.* They are gram-negative and curved or spiral, with a single flagellum located at one or both poles of the cell. In pure culture, two types of colonies have been recognized and designated as Types I and II. The more commonly observed Type I colonies are large, flat, and spread with uneven margins. They are nonhemolytic, watery, and grayish. Type II colonies are also nonhemolytic, but they are smaller (1 to 2 mm), with unbroken edges. They are convex and glistening.

Initially, the isolation of *Campylobacter* organisms from fecal specimens was difficult be-cause of their microaerophilic nature and their 42°C optimal growth temperature. Furthermore, in the absence of selective media, their growth was masked by the overgrowth of other enteric organisms, and they were often overlooked on primary isolation. This situation has been rectified with the development of selective media that are designed specifically for isolating *Campylobacter* species and that inhibit the growth of other enteric organisms. These media are nutritionally enriched and supplemented with 5% to 10% sheep or horse blood. In addition they contain three to five antimicrobial agents depending on the medium. For example, cephalosporins, one of the antimicrobial agents present in the Campy BAP™ medium, is selective for *C. jejuni* and inhibits the species *C. intestinalis,* which is rarely responsible for enteric infections.

The most essential requirement for cultivating campylobacteria is a microaerophilic incubation atmosphere. High concentrations of oxygen are toxic to these organisms, and an atmosphere of 3% to 10% carbon dioxide and 5% to 10% oxygen is optimal for their growth. The incubation temperature for *C. jejuni* is 42°C. At this temperature the organism grows optimally, while growth of *C. intestinalis* is inhibited.

In the experiment to follow, a simulated fecal specimen (a culture containing an attenuated strain of *C. jejuni* and other enteric organisms) is used. You will attempt to isolate the *Campylobacter* organisms by using the following two procedures:

1. A conventional method uses MacConkey agar directly, circumventing enrichment procedures, using a mixed simulated fecal population as the test culture.

2. A special method employs Campy BAP agar and the Campy Pak™ and GasPak™

jar (BBL® Microbiology Systems, Division of Beckton Dickinson and Company), which are illustrated in Figure 68.1.

Presumptive identification is made on the basis of colonial morphology and the microscopic appearance of the organisms obtained from a typical isolated colony. You may perform the catalase and oxidase tests as described in Experiments 30 and 31 for further presumptive identification. In the case of *C. jejuni*, both tests should be positive.

MATERIALS

Cultures

Mixed saline suspensions of *Campylobacter jejuni* cultured on a sheep blood-enriched medium, *Salmonella typhimurium*, and *Escherichia coli*.

Media

Per designated student group: one Campy BAP agar plate and one MacConkey agar plate.

Reagents

Crystal violet, Gram's iodine, 95% ethyl alcohol, and 0.8% carbol fuchsin.

Equipment

Bunsen burner, inoculating loop, glassware marking pencil, Campy Pak and GasPak jars, and 10-ml pipettes.

PROCEDURE

1. Aseptically perform a four-way streak inoculation as described in Experiment 2 for the isolation of discrete colonies on both appropriately labeled agar plates.

2. Place the inoculated Campy BAP™ agar plate in the GasPak™ jar in an inverted

FIGURE 68.1 Campy Pak and GasPak jar (courtesy of BBL Microbiology Systems, Division of Becton Dickinson and Company)

position. Following the manufacturer's instructions, open the Campy Pak envelope and place it in the jar. With a pipette, add 10 ml of water to each envelope and immediately seal the jar to establish a microaerophilic environment.

3. Incubate the jar for 48 hours at 42°C.

4. Incubate the MacConkey agar plate culture in an inverted position for 48 hours at 37°C.

Name   Date

1. Observe both plate cultures for the presence of discrete colonies. In the chart, diagram the appearance of representative colonies on both plates and describe their colonial characteristics. Also, note and record the color of the medium surrounding the representative colonies on the MacConkey plate. (Refer to Experiment 15 for an explanation of the selective and differential nature of MacConkey agar.)

Plate Culture	Diagram of Colonies	Colonial Characteristics	Color of Medium
Campy BAP agar			
MacConkey agar			

2. Prepare a Gram stain, using 0.8% carbol fuchsin as the counterstain, of a representative colony agar plate culture. Observe microscopically and record the microscopic morphology and Gram reaction of each preparation.

Gram Stain Preparation	Campy BAP Plate Isolate	MacConkey Agar Plate	
		Isolate 1	Isolate 2
Draw a representative field.			
Microscopic morphology			
Gram reaction			

3. Based on your observations, identify your isolates:

Campy BAP agar culture isolate: _____

MacConkey agar culture Isolate 1: _____

MacConkey agar culture Isolate 2: _____

4. Optional: Perform the catalase and oxidase tests on the representative isolates.

REVIEW QUESTIONS

1. How would you describe the clinical syndrome induced by *C. jejuni*?

2. What are the purposes of the antimicrobial agents present in the selective media used for the isolation of *Campylobacter*?

3. How may *C. jejuni* be separated from *C. intestinalis*?

4. Why may members of *Campylobacter* not be isolated from a stool specimen in a diagnostic laboratory?

Microbiological Analysis of Urine Specimens

LEARNING OBJECTIVES

Once you have completed this experiment, you should be familiar with

1. The organisms responsible for infections of the genitourinary tract.
2. Laboratory methods for detection of bacteriuria and identification of microorganisms associated with the urinary tract.

PRINCIPLE

The anatomical structure of the mammalian urinary system is such that the external genitalia and the lower aspects of the urethra are normally contaminated with a diverse population of microorganisms. The tissues and organs that compose the remainder of the urinary system, the bladder, ureters, and kidneys, are sterile, and therefore urine that passes through these structures is also sterile. When pathogens gain access to this system, they can establish infection. Some etiological agents of urinary tract diseases are illustrated on this page.

Urinary tract infections may be limited to a single tissue or organ, or they may spread upward and involve the entire system. Infections such as **cystitis** involve the bladder but may spread through the ureters to the kidneys. Infections limited to the ureters and kidneys are called **pyelitis. Glomerulonephritis** is an inflammation that results in the destruction of renal corpuscles; **pyelonephritis** results in the destruction of renal tubules. Organisms other than bacteria may also act as etiological agents of urogenital infections. *Trichomonas vaginalis,* a pathogenic flagellated protozoan, is commonly found in the vagina, and under appropriate conditions, it is responsible for a severe inflammatory **vaginitis.** *Candida albicans,* a pathogenic yeast, is normally found in low numbers in the intestines. Under suitable conditions, it can enter the urogenital systems, where it gives rise to vaginal infections. *Schistosoma haematobium* is

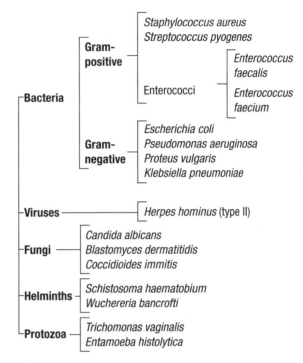

a pathogenic fluke, a helminth, responsible for severe bladder infections.

The initial step in diagnosis of a possible urinary tract infection is laboratory examination of a urine specimen. The sample must be collected midstream in a sterile container following adequate cleansing of the external genitalia. It is imperative to culture the freshly voided, unrefrigerated urine sample immediately to avoid growth of normal indigenous organisms, which may overtake the growth of the more slowly growing pathogens. In this event the infectious organism might be overlooked, resulting in an erroneous diagnosis.

Clinical evaluation of the specimen requires a quantitative determination of the microorganisms per ml of urine. Urine in which the bacterial count per ml exceeds 100,000 (10^5) denotes significant **bacteriuria** and is indicative of a urinary tract infection. Urine in which counts range from 0 to 1,000 per ml are generally normal.

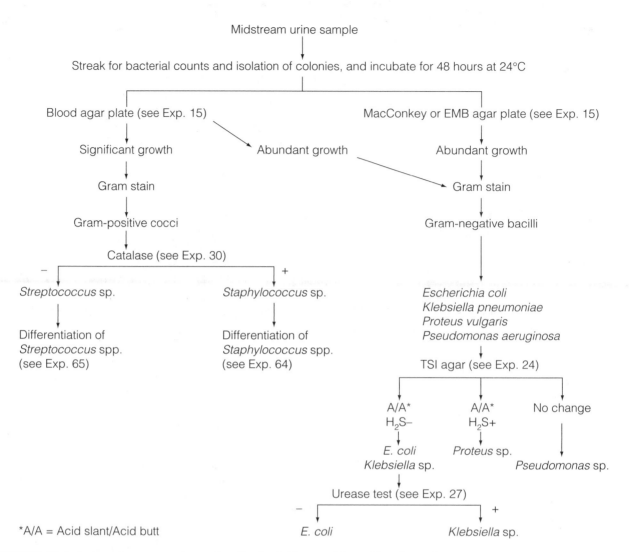

Midstream urine sample

Streak for bacterial counts and isolation of colonies, and incubate for 48 hours at 24°C

Blood agar plate (see Exp. 15)

MacConkey or EMB agar plate (see Exp. 15)

Significant growth

Abundant growth

Abundant growth

Gram stain

Gram stain

Gram-positive cocci

Gram-negative bacilli

Catalase (see Exp. 30)

−
Streptococcus sp.

+
Staphylococcus sp.

Escherichia coli
Klebsiella pneumoniae
Proteus vulgaris
Pseudomonas aeruginosa

Differentiation of
Streptococcus spp.
(see Exp. 65)

Differentiation of
Staphylococcus spp.
(see Exp. 64)

TSI agar (see Exp. 24)

A/A*
H₂S−

A/A*
H₂S+

No change

E. coli
Klebsiella sp.

Proteus sp.

Pseudomonas sp.

Urease test (see Exp. 27)

−
E. coli

+
Klebsiella sp.

*A/A = Acid slant/Acid butt

FIGURE 69.1 Laboratory procedures for the isolation and identification of urinary tract pathogens

In the conventional method, a urine sample is streaked over the surface of an agar medium with a special loop calibrated to deliver a known volume. Following incubation, the number of isolated colonies present on the plate is determined and multiplied by a factor that converts the volume of urine to 1 ml. The final calculation is then equal to the number of organisms per ml of sample.

Example: Twenty-five colonies were present on a plate inoculated with a loop calibrated to deliver 0.01 ml of a urine specimen.

Number of colonies	×	Factor that converts 0.01 ml to 1 ml	=	Organisms per ml
25	×	100	=	2,500 organisms per ml

If the specimen is turbid, dilution is necessary prior to culturing. In this case, conventional 10-fold dilutions are prepared in physiological saline to effect a final dilution of 1:1,000 (see Experiment 20). Each of the dilutions (10^{-1}, 10^{-2}, and 10^{-3}) is then streaked on the surface of a suitable agar plate medium for isolation of colonies. Following incubation, the number of microorganisms per ml of sample is determined by the following formula:

Organisms per ml = Number of colonies
× Factor that converts the volume of urine to 1 ml
× Dilution factor

Example: Twenty-five colonies were counted on 10^{-2} dilution plate inoculated with a loop calibrated to deliver 0.01 ml of urine. Therefore $25 \times 100 \times 100 = 250{,}000$ organisms per ml.

On determination of bacteriuria, identification of the infectious organism can be accomplished by the laboratory procedures outlined in Figure 69.1.

A newer, less conventional, and less time-consuming method uses a diagnostic urine-culture tube, Bacturcult®, devised by Wampole Laboratories (Figure 69.2). Bacturcult is a sterile, disposable plastic tube coated on the interior with a special medium that allows detection of the bacteriuria and a presumptive class identification of urinary bacteria.

Following incubation of the Bacturcult urine culture, bacteriuria can be detected with a bacterial count. This is performed by placing the counting strip around the Bacturcult tube over an area of even colony distribution and counting the number of colonies within the circle. The average number of colonies counted is interpreted in Table 69.1.

For the presumptive identification of bacteria, the medium contains two substrates, lactose and urea, and the pH indicator phenol red. Depending on the organism's enzymatic action on these substrates, differentiation of urinary bacteria into three groups following incubation is possible based on observable color changes that occur in the culture:

Group I: *E. coli* and *Enterococcus*—yellow.

Group II: *Klebsiella, Staphylococcus,* and *Streptococcus*—rose to orange.

Group III: *Proteus* and *Pseudomonas*—purplish-red.

Mixed cultures do not always produce clear-cut color changes, however. Therefore, if additional testing is required, the discrete colonies that develop on the medium can be used as the source for subculturing into other media.

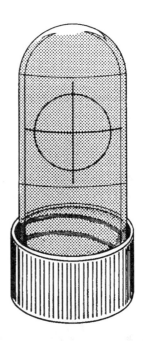

FIGURE 69.2 Bacturcult culture tube (Wampole Laboratories Division, Carter-Wallace, Inc., Cranbury, NJ 08512. Reprinted with permission.)

In this experiment, seeded saline cultures will be used to simulate urine specimens. This is done to minimize the risk of using a potentially infectious body fluid, urine, as the test sample. The conventional procedure performed with the calibrated loop will be used to determine the number of cells in the specimens. The Bacturcult tube will be used for enumeration and presumptive group identification. If your instructor desires to emulate more closely a clinical evaluation of urine, then a mixed seeded culture must be used. Representative colonies isolated from the

TABLE 69.1 Bacturcult: Interpretation of Colony Counts

Average Number of Colonies Within Circle	Approximate Number of Bacteria per ml	Diagnostic Significance
< 25	< 25,000	Negative bacteriuria
25 to 50	25,000 to 100,000	Suspicious*
> 50	> 100,000	Positive bacteriuria

Source: Wampole Laboratories Division, Carter-Wallace, Inc., Cranbury, NJ 08512. Reprinted with permission.
*Additional testing recommended.

blood agar streak-plate culture for detection of bacteriuria can then be identified following the schema in Figure 69.1.

MATERIALS

Cultures

Six saline cultures, each seeded with one of the following 24-hour cultures: *Enterococcus faecalis, Staphylococcus aureus, Proteus vulgaris, Escherichia coli, Pseudomonas aeruginosa,* and *Klebsiella pneumoniae.* Optional: Saline culture seeded with a gram-positive and a gram-negative organism.

Media

Per designated student group: six blood agar plates, three sterile 9-ml tubes of saline, and six Bacturcult culture tubes.

Equipment

Bunsen burner, calibrated 0.01-ml platinum loop, glassware marking pencil, sterile 1-ml pipettes, and mechanical pipetting device.

PROCEDURE

Bacturcult Procedure

1. Label each Bacturcult tube with the name of the bacterial organism present in the urine sample.
2. Fill each tube almost to the top with urine.
3. Immediately pour the urine out of each tube, allowing all the fluid to drain for several seconds. Replace the screw cap securely.
4. Immediately prior to incubation, loosen the cap on each tube by turning the screw cap counterclockwise for one-half turn.

5. Incubate the tubes with the caps down for 24 hours at 37°C.

Calibrated Loop Procedure for Bacterial Counts

1. Label the three 9-ml sterile saline tubes and the three blood agar plates 10^{-1}, 10^{-2}, and 10^{-3}, respectively.
2. Using the three 9-ml saline blanks, aseptically prepare a 10-fold dilution of the urine sample to effect 10^{-1}, 10^{-2}, and 10^{-3} dilutions.
3. With a calibrated loop, aseptically add 0.01 ml of the 10^{-1} urine dilution to the appropriately labeled blood agar plate and streak for isolation of colonies as illustrated.

4. Repeat Step 3 to inoculate the remaining urine sample dilutions.
5. Incubate all plates in an inverted position for 24 hours at 37°C.

Name   Date

1. Determine the number of colonies in each of the Bacturcult urine cultures as follows:

   a. Place the counting strip around the tube over an area of even colony distribution and count the number of colonies within the circle.

   b. Repeat the count in another area of the tube.

   c. Average the two counts.

   d. Record in the chart the average number of colonies counted within the circle.

2. Based on your colony count, determine and record in the chart the approximate number of bacteria per ml of each sample and its diagnostic significance as negative bacteriuria, suspicious, or positive bacteriuria.

3. Observe and record in the chart the color of the medium in each of the urine cultures and the presumptive bacterial group.

Urine Culture	Number of Colonies	Number of Bacteria per ml	Diagnostic Significance	Color of Medium	Presumptive Group
E. faecalis					
S. aureus					
K. pneumoniae					
P. vulgaris					
P. aeruginosa					
E. coli					

4. Determine the number of colonies on each blood agar culture plate and calculate the number of organisms per ml of the urine. Record your results in the chart.

Urine Sample Dilution	Number of Colonies	Organisms per ml of Sample	Bacteriuria (+) or (−)
10^{-1}			
10^{-2}			
10^{-3}			

REVIEW QUESTIONS

1. What types of urinary infections may be caused by different microorganisms?

2. How is a clinical diagnosis of a bacteriuria established?

3. If five colonies were counted on a 10^{-3} dilution plate streaked with 0.01 ml of urine, what was the number of organisms per ml of the original specimen, and is this count indicative of bacteriuria? Explain.

4. How accurate is a laboratory analysis of a 24-hour, unrefrigerated, nonmidstream urine sample? Explain.

5. A male patient is diagnosed as having a urinary tract infection. A urine culture is ordered by his physician. She requests that a voided specimen be used rather than a catheterized sample. Why does she make this request?

Microbiological Analysis of Blood Specimens

LEARNING OBJECTIVES

Once you have completed this experiment, you should be familiar with

1. The microorganisms most frequently associated with septicemia.
2. Laboratory methods for the isolation and presumptive identification of the etiological agents of septicemia.

PRINCIPLE

Blood is normally a sterile body fluid. This sterility may be breached, however, when microorganisms gain access into the bloodstream during the course of an infectious process. The transient occurrence of bacteria in the blood is designated as **bacteremia** and implies the presence of nonmultiplying organisms in this body fluid.

Bacteremias may be encountered in the course of some bacterial infections such as pneumonia, meningitis, typhoid fever, and urinary tract infections. A bacteremia of this nature does not present a life-threatening situation because the bacteria are present in low numbers and the activity of the host's nonspecific immune system is generally capable of preventing further systemic invasion of tissues. A more dangerous and clinically alarming syndrome is **septicemia,** a condition characterized by the rapid multiplication of microorganisms, with the possible elaboration of their toxins into the bloodstream. The clinical picture frequently present in septicemia is that of septic shock, which is recognized by a severe febrile episode with chills, prostration, and a drop in blood pressure.

A large and diverse microbial population has been implicated in septicemia. The major offenders include the following:

1. Gram-negative bacteria, because of their endotoxic properties, are the most frequently encountered etiological agents that present the more serious complications of septicemia. Included among these agents are *Hemophilus influenzae, Neisseria meningitidis, Serratia marcescens, Escherichia coli, Pseudomonas aeruginosa,* and *Salmonella* spp. Less frequently implicated are *Francisella tularensis* and members of the genera *Campylobacter* and *Brucella.*
2. Gram-positive bacteria that generally do not produce the presenting signs of septic shock include primarily members of the genera *Streptococcus* and *Staphylococcus.*
3. *Candida albicans* is the major fungal invader of the bloodstream.

In the clinical setting, to facilitate the rapid initiation of effective chemotherapy, a culture of the suspect blood sample is required for the isolation and identification of the offending organisms. A blood sample is drawn and cultured in an appropriate medium under both aerobic and anaerobic conditions. Over a period of 3 to 7 days, the cultures are observed for turbidity and Gram-stained smears are prepared to ascertain the presence of microorganisms in the blood. Upon detection of microbial growth in the cultures, transfers onto a variety of specialized agar media are made for the identification of the infectious agent. The schema for this protocol is shown in Figure 70.1.

Two methods are outlined in this exercise. Either method or both methods may be used for the isolation and presumptive identifica-

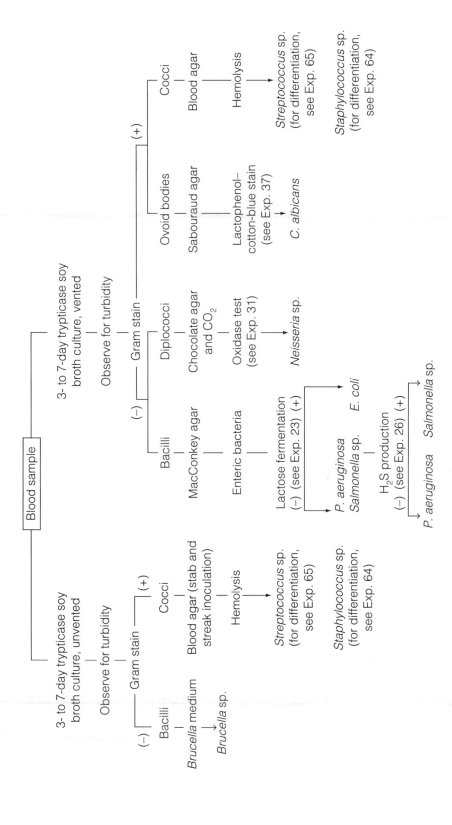

FIGURE 70.1 Schema for the isolation and identification of the etiological agents of septicemia

tion of the microorganisms in the experimental culture. Both procedures use a simulated blood specimen, a prepared culture containing blood previously seeded with selected microorganisms. The traditional method is a modification of the schema shown in Figure 70.1. This procedure requires the preparation of Gram-stained smears for the morphological study of the organisms and the inoculation of selected agar media for their isolation and preliminary identification. The alternative method uses the commercially available **BBL™ Septi-Chek System** (Roche Diagnostic Systems, Division of Hoffmann-La Roche, Inc.), a single unit composed of the Septi-Chek culture bottle and the Septi-Chek slide as illustrated in Figure 70.2. The culture-bottle component permits the qualitative determination of the presence of microorganisms in the blood sample, and the slide component is designed for the simultaneous subculturing of the organisms onto a plastic slide containing three differential media (chocolate, MacConkey, and malt agar). Differential growth on these media provides preliminary information as to the nature of the infectious agent and isolated colonies for further study.

MATERIALS

Culture

48- to 72-hour simulated blood culture prepared as follows: 10 ml of citrated blood, obtained from a blood bank, or 10 ml of saline seeded with 2 drops each of *Escherichia coli*, *Neisseria perflava*, and *Saccharomyces cerevisiae*, each adjusted to an O.D. of 0.1 at 600 nm, in 90 ml of trypticase soy broth containing 0.05% of sodium polyanetholesulfonate (SPS) used to prevent clotting of the blood sample.

Media

Per designated student group: one each of blood agar plate, MacConkey agar plate, chocolate agar plate, Sabouraud agar plate, and Septi-Chek system.

Reagents

Crystal violet, Gram's iodine, 95% ethyl alcohol, safranin, lactophenol–cotton-blue stain, 70% isopropyl alcohol, and 1% *p*-aminodimethylaniline oxalate.

FIGURE 70.2 Septi-Chek System (Roche Diagnostic Systems, Division of Hoffmann-La Roche Inc.)

Equipment

Sterile 20-gauge, 1½-inch needles; sterile 1-ml and 10-ml syringes; Bunsen burner; staining tray; inoculating loop; lens paper; bibulous paper; microscope; glassware marking pencil; and disposable gloves.

PROCEDURE

⚠ **Use gloves throughout the procedure.**

1. Swab the rubber stopper of the blood-culture bottle with 70% isopropyl alcohol and allow to air-dry.
2. Using a sterile needle and 1-ml syringe, aseptically remove 0.5 ml of the blood culture by penetrating the rubber stopper.

⚠ **Dispose of the needle and syringe, as a single unit, into the provided puncture-proof receptacle.**

3. To prepare a smear, place a small drop of the culture on a clean glass slide and spread evenly with an inoculating loop.
4. Place 1 drop of culture in one corner of the appropriately labeled blood agar plate and prepare a four-way streak inoculation as described in Experiment 2.
5. Repeat Step 4 to inoculate the MacConkey, chocolate, and Sabouraud agar plates.
6. Incubate the agar plate cultures in an inverted position for 24 to 48 hours as follows: Sabouraud agar culture at 25°C,

chocolate agar culture in a 10% CO_2 atmosphere at 37°C, and the remaining cultures at 37°C.

7. Follow the Septi-Chek system procedure as follows:

   a. Remove the protective top of the screw cap of the culture bottle, disinfect the rubber stopper with 70% isopropyl alcohol, and allow to air-dry.

   b. Using the 10-ml syringe, aseptically transfer 10 ml of the experimental culture to the appropriately labeled Septi-Chek culture bottle.

   c. Aseptically vent the bottle for aerobic incubation.

   d. Replace the protective top of the screw cap on the bottle.

   e. Gently invert the bottle two or three times to disperse the blood evenly throughout the medium.

   f. Incubate the culture for 4 to 6 hours at 37°C.

   g. Attach the Septi-Chek slide according to the manufacturer's instructions.

   h. Tilt the combined system to a horizontal position and hold until the liquid medium enters the slide chamber and floods the agar surfaces. While maintaining this position, rotate the entire system one complete turn to ensure that all agar surfaces have come in contact with the liquid medium. Return the system to an upright position.

   i. Incubate the system in an upright position at 37°C.

   j. Check the culture bottle daily for turbidity and the slide for visible colony formation.

Name _____ Date _____

1. Examine the blood agar plate culture for the presence (+) or absence (−) of hemolytic activity. If hemolysis is present, determine the type observed. Record your observations in the chart.

2. For the performance of the oxidase test, add *p*-aminodimethylaniline oxalate to the surface of the growth on the chocolate agar plate. The presence of pink-to-purple colonies is indicative of *Neisseria* spp. Record your observations and the oxidase test results in the chart.

3. Examine the MacConkey agar plate culture for determination of lactose fermentation. Lactose fermenters exhibit a pink-to-red halo in the medium, a red coloration on the surface of their growth, or both a halo and red coloration. Record your observations and indicate the presence or absence of lactose fermenters in the chart.

4. Examine the Sabouraud agar plate culture for the presence of growth. Prepare a lactophenol–cotton-blue–stained smear from an isolated colony (see Experiment 37). Examine the smear microscopically for the presence of large ovoid bodies indicative of the yeast cells. Record your morphological observations in the chart.

📷 *Refer to photo numbers 16, 20–22, 47, 57, and 59 in the color-plate insert for illustration of the above reactions.*

5. Observe the Septi-Chek slide system for the presence of growth on the three agar surfaces. If growth is present on:

a. Medium 1 (MacConkey agar), examine for fermentative patterns as described in Step 3 and record your observations in the chart.

b. Medium 2 (chocolate agar), perform the oxidase test as described in Step 2 and record your observations in the chart.

c. Medium 3 (malt agar), prepare and examine microscopically a lactophenol–cotton-blue–stained smear as described in Step 4. Record your observations in the chart.

Culture	Traditional Procedure	Septi-Chek System
Blood agar Hemolysis: (+) or (−)	_____	_____
Type of hemolysis	_____	_____
Chocolate agar Color of colonies	_____	_____
Oxidase test: (+) or (−)	_____	_____
MacConkey agar Color of colonies	_____	_____
Color of medium	_____	_____
Lactose fermentation: (+) or (−)	_____	_____
Sabouraud or malt agar Cell morphology	_____	_____
Presumptive identification **of organisms present**	_____	_____

REVIEW QUESTIONS

1. Differentiate between septicemia and bacteremia, and explain the medical significance of each.

2. Why are blood samples cultured in both vented and unvented systems?

3. A 15-year-old boy is admitted to the hospital and presents the following symptoms: chills, fever, increased pulse rate, and a drop in blood pressure. The patient indicates that these symptoms have occurred intermittently. The physician suspects a bacteremia and orders a series of three blood cultures over a 24-hour period. Explain the following:

   a. Why did the physician order more than one blood culture?

   b. Why does blood culture medium contain an anticoagulant?

4. Prior to the introduction of antibiotic therapy, what was the prognosis for patients with septicemia? What significant factors played roles in recovery in the absence of antibiotics?

Species Identification of Unknown Bacterial Cultures

LEARNING OBJECTIVE

Once you have completed this experiment, you should be able to

1. Identify an unknown bacterial species by the use of dichotomous keys and *Bergey's Manual of Systematic Bacteriology*.

PRINCIPLE

At this point in the course, you have developed the manipulative skills and the cognitive microbiological knowledge to identify microorganisms beyond their genus classification to the level of their species identification. Therefore, in this experiment, you will use dichotomous keys, *Bergey's Manual of Systematic Bacteriology*, and information accrued from previously performed laboratory procedures to help identify the species of an unknown culture.

In Experiment 33, "Genus Identification of Unknown Bacterial Cultures," you were required to use a variety of biochemical tests to successfully accomplish the experimental purpose. Your review of the required procedures and ensuing results should indicate that only a few of these tests were actually necessary, in most instances, for the identification of the unknown culture. Similarly, species identification can be accomplished by using a limited number of carefully selected laboratory procedures. Notice that what appears to be a spurious result in some cases, one that departs from the expected norm for a particular species, may be attributable to strain differences within the given species. These nonconforming results may be verified by the use of *Bergey's Manual* to ascertain the existence of variable biochemical test results for the particular species being studied.

In this experimental procedure, you will receive a mixed culture containing a gram-positive and a gram-negative organism. The proto-col will require (1) Gram staining, (2) streak plating for observation of colonial characteristics, (3) use of selective media for the preparation of pure cultures, (4) the performance of appropriate biochemical tests as indicated in the dichotomous keys outlined in Figures 71.1 on pages 460–461 and 71.2 on pages 462–463, and (5) information in *Bergey's Manual*.

MATERIALS

Cultures

Per student: number-coded, 24- to 48-hour mixed trypticase soy broth cultures each containing a gram-positive and a gram-negative organism selected from the species listed in Figures 71.1 and 71.2.

Media

Per student: one trypticase soy agar plate, two trypticase soy agar slants, one trypticase soy broth, one phenylethyl alcohol agar plate, and one MacConkey agar plate.

Required media for the biochemical tests listed in Figures 71.1 and 71.2 should be available on your request.

Reagents

Crystal violet, Gram's iodine, 95% ethyl alcohol, safranin, and required reagents for the interpretation of the biochemical reactions listed in Figures 71.1 and 71.2.

Equipment

Bunsen burner, inoculating loop and needle, staining tray, immersion oil, lens paper, bibulous paper, microscope, and glassware marking pencil.

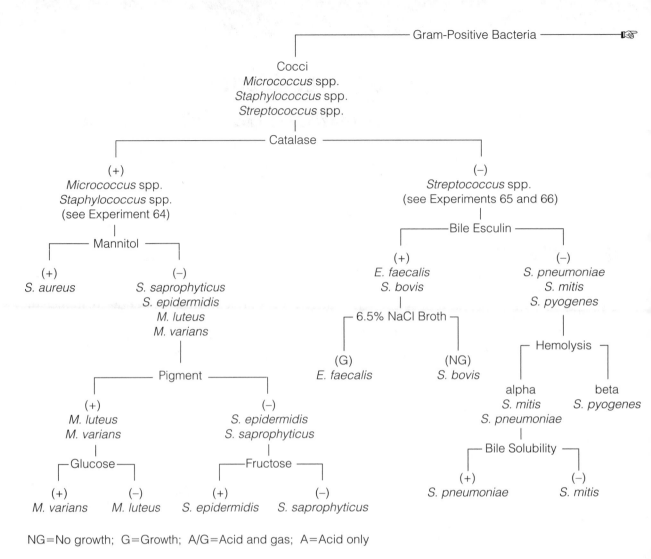

NG=No growth; G=Growth; A/G=Acid and gas; A=Acid only

FIGURE 71.1 Schema for the identification of gram-positive bacteria

PROCEDURE

Session 1: Separation of the Bacteria in Mixed Unknown Culture

1. Prepare a trypticase soy agar broth subculture of the unknown and refrigerate following incubation. You will use this culture if contamination of the test culture is suspected during the identification procedure.

2. Prepare a Gram-stained smear of the original unknown culture.

3. Prepare four-way streak inoculations (see Experiment 2) on the following media for the separation of the microorganisms in the mixed cultures:

   a. Trypticase soy agar for observation of colonial characteristics.

   b. Phenylethyl alcohol agar for isolation of gram-positive bacteria.

   c. MacConkey agar for isolation of gram-negative bacteria.

4. Incubate all the plates in an inverted position and then subculture for 24 to 48 hours at 37°C.

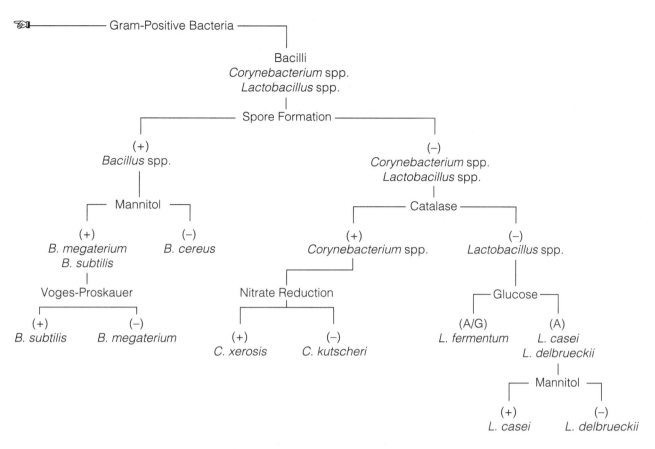

NG= No growth; G= Growth; A/G= Acid and gas; A= Acid only

FIGURE 71.1 (continued) Schema for the identification of gram-positive bacteria

Session 2: Preparation of Pure Cultures

1. Isolate a discrete colony on both the phenylethyl alcohol agar plate and the MacConkey agar plate and aseptically transfer each onto a trypticase soy agar slant (see Experiment 2).

2. Incubate the trypticase soy agar slants for 24 to 48 hours at 37°C.

Session 3: Identification of Unknown Bacterial Species

1. Prepare a Gram-stained smear from each of the trypticase soy agar slant cultures to verify their purity by means of the Gram reaction and cellular morphology.

2. If each Gram-stained preparation is not solely gram-positive or gram-negative, repeat the steps in Sessions 1 and 2 using the refrigerated trypticase soy agar subculture as the test culture.

3. If the isolates are deemed to be pure on the basis of their cultural and cellular morphologies, continue with the identification procedure. During this period and in subsequent sessions, use the dichotomous keys in Figures 71.1 and 71.2 to select and perform the necessary biochemical tests on each of your isolates for identification of their species. Incubate all cultures for 24 to 48 hours at 37°C prior to making your observations.

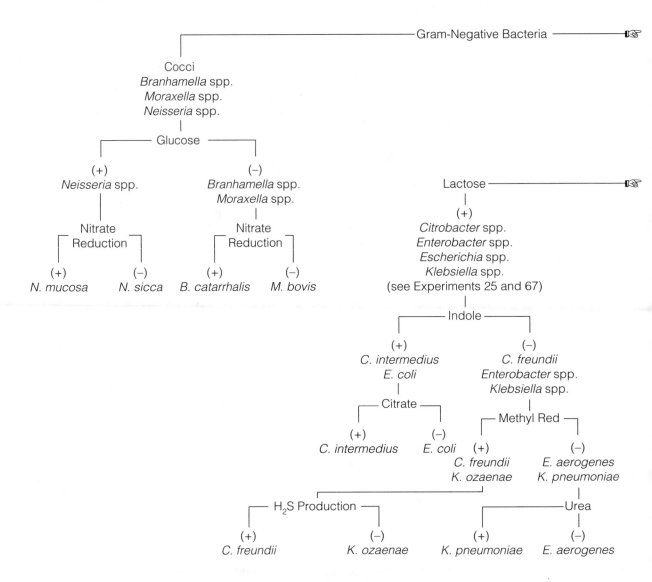

NG=No growth; G=Growth; A/G=Acid and gas; A=Acid only

FIGURE 71.2 Schema for the identification of gram-negative bacteria

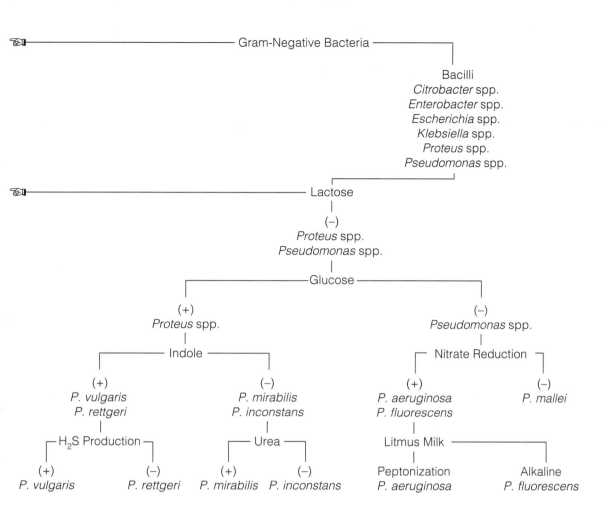

NG= No growth; G= Growth; A/G= Acid and gas; A= Acid only

FIGURE 71.2 (continued) Schema for the identification of gram-negative bacteria

Experiment 71 / Observations and Results

Name Date

Session 1

Examine the Gram-stained smear of your mixed unknown culture. Record your observations in the chart below.

Organism	Cellular Morphology	Gram Reaction
1		
2		

Session 2

1. Examine the trypticase soy agar plate for the appearance of discrete colonies. Select two colonies that differ in appearance and record their colonial morphologies.

   Isolate 1:

   Isolate 2:

2. Examine the phenylethyl alcohol and MacConkey agar plates. Record your observations in the chart below.

Medium	Growth (+) or (+)	Colonial Morphology	Coloration of Medium
Phenylethyl alcohol agar			
MacConkey agar			

Session 3

1. Examine the Gram-stained smears of the trypticase soy agar cultures obtained from the phenylethyl alcohol and MacConkey agar plates. Record your results in the chart below.

Agar Slant From	Cellular Morphology	Gram Reaction
Phenylethyl alcohol agar plate culture		
MacConkey agar plate culture		

2. Examine all of the biochemical test cultures. Record your observations and results in the charts below.

Gram-Positive Isolate		
Biochemical Test	**Observation**	**Result**

Unknown gram-positive organism: _____

Gram-Negative Isolate		
Biochemical Test	**Observation**	**Result**

Unknown gram-negative organism: _____

PART XV

Immunology

LEARNING OBJECTIVES

Once you have completed the experiments in this section, you should be familiar with

1. The basic principles of nonspecific (innate) and specific (acquired) immunity.

2. Serological procedures that demonstrate immunological reactions of agglutination and precipitin formation.

3. Rapid immunodiagnostic screening procedures.

INTRODUCTION

Immunity, or **resistance,** is a state in which a person, either naturally or by some acquired mechanism, is protected from contracting certain diseases or infections. The ability to resist disease may be innate (nonspecific), or it may be acquired (specific) when the disease state is emulated in the host.

Nonspecific immunity is native or natural. It is inborn and provides the basic mechanisms that defend the host against intrusion of foreign substances or agents of disease. This defense is not restricted to a single or specific foreign agent, but it provides the body with the ability to resist many pathological conditions. The mechanisms responsible for this native immunity include the **mechanical barriers,** such as the skin and mucous membranes; **biochemical factors,** such as antimicrobial substances present in the body fluids; and the more sophisticated process of **phagocytosis** and action of the **reticuloendothelial system.**

Specific immunity, **cell-mediated** and **humoral,** is acquired by the host in response to the presence of a single or particular foreign substance, usually protein, called an **antigen** (immunogen). In humoral immunity, antigens that penetrate the mechanical barriers of the host, namely the skin and mucous membranes, stimulate formation of **antibodies.** The function of the antibodies is to bind to the specific antigens that are responsible for their production and to inactivate or destroy them. Antibodies are a group of homologous proteins called **immunoglobulins,** which are found in serum and represent five distinct classes: immunoglobulin G (IgG), immunoglobulin A (IgA), immunoglobulin M (IgM), immunoglobulin D (IgD), and immunoglobulin E (IgE).

The primary immunological complexes (antigen + antibody) are as follows:

1. **Agglutination:** This type of reaction uses specific antibodies, **agglutinins,** that are formed in response to the introduction of particulate antigens into host tissues. When these particulate antigens combine with a homologous antiserum, a three-dimensional mosaic complex occurs. This is called an agglutination reaction and can be visualized microscopically and in some cases macroscopically.

2. **Precipitin formation:** This reaction requires specific antibodies, **precipitins,** that are formed in response to the introduction of soluble, nonparticulate antigens

into host tissues. These antibodies, when present in serum, form a complex with the specific homologous nonparticulate antigen and result in a visible precipitate.

Advances in chemistry, especially immunochemistry, have enabled us to study the interaction of antigens and immunoglobulins outside the body, in a laboratory setting. These advances have provided an immunological discipline known as **serology,** which studies these in vitro reactions that have diagnostic, therapeutic, and epidemiological implications.

In the experiments to follow, you will study several serological procedures based on the principles of agglutination and precipitin formation for the detection of serum antibodies or antigens. The techniques presented in these experiments span a spectrum of methods ranging from basic reactions to more sophisticated forms of antigen and antibody interactions.

⚠ Note that some of the experimental protocols use positive and negative controls provided in the test kits to demonstrate the desired immunological reactions. These controls do not represent the source of potential pathogens capable of inducing infection in students and instructional staff. The rationale for this design is that body fluids, particularly blood of unknown origin, may serve as a major vehicle for the transmission of infectious viral agents. Thus, our concern with the spread of AIDS and hepatitis precludes the use of blood as a test specimen in a college laboratory.

It is further suggested that your instructor present experiments that use positive and negative control test kits as laboratory demonstrations. This will reduce the cost of the required materials, which may otherwise be prohibitive at many academic institutions, but will still allow you and your fellow students to observe the advances in immunological serology.

Precipitin Reaction: The Ring Test

LEARNING OBJECTIVE

Once you have completed this experiment, you should be able to

1. Demonstrate a precipitin reaction by means of the ring test.

PRINCIPLE

The **ring** or **interfacial test** is a simple serological technique that illustrates the precipitin reaction in solution. This antigen–antibody reaction can be demonstrated by the formation of a visible precipitate, a flocculent or granular turbidity, in the test fluid. Antiserum is introduced into a small-diameter test tube, and the antigen is then carefully added to form a distinct upper layer. Following a period of incubation of up to 4 hours, a ring of precipitate forms at the point of contact (interface) in the presence of the antigen–antibody reaction. The rate at which the visible ring forms depends on the concentration of antibodies in the serum and the concentration of the antigen.

To detect the precipitin reaction, a series of dilutions of the antigen is used because both insufficient (zone of antibody excess) and excessive (zone of antigen excess) amounts of antigen will prevent the formation of a visible precipitate (zone of equivalence), as shown in Figure 72.1. In addition, you will be able to determine the optimal antibody:antigen ratio by the presence of a pronounced layer of granulation at the interface of the antiserum and antigen solution. This immunological reaction is illustrated in Figure 72.2.

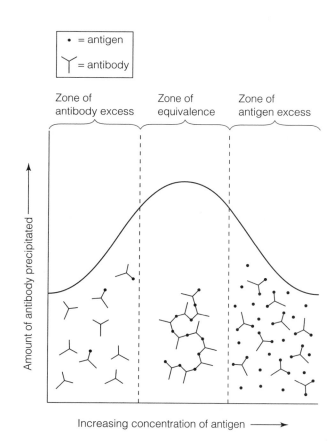

FIGURE 72.1 The precipitin reaction

and 1:75 with physiological saline. The normal bovine serum contains the antigen (bovine globulin), to which antibodies were made commercially in another animal species and provided as antiserum to bovine globulin.

MATERIALS

Reagents

Physiological saline (0.85% NaCl), and commercially available bovine globulin antiserum and normal bovine serum diluted to 1:25, 1:50,

Equipment

Serological test tubes (8 × 75 mm), 0.5-ml pipettes, serological test tube rack, mechanical pipetting device, glassware marking pencil, and 37°C incubator.

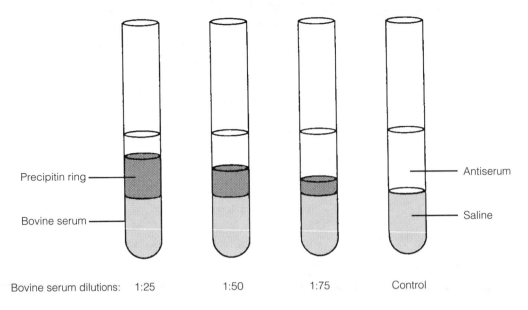

Precipitin ring

Bovine serum

Antiserum

Saline

Bovine serum dilutions: 1:25   1:50   1:75   Control

FIGURE 72.2 Ring test: Precipitin reactions

PROCEDURE

1. Label three serological test tubes according to the antigen dilution to be used (1:25, 1:50, and 1:75) and the fourth test tube as a saline control.

2. Using a different 0.5-ml pipette, transfer 0.3 ml of each of the normal bovine serum dilutions into its appropriately labeled test tube.

3. Using a clean 0.5-ml pipette, transfer 0.3 ml of saline into the test tube labeled as control.

4. Carefully overlay all four test tubes with 0.3 ml of bovine globulin antiserum. To prevent mixing of the sera, tilt the test tube and allow the antiserum to run down the side of the test tube.

5. Incubate all test tubes for 30 minutes at 37°C.

Name _____   Date _____

1. Examine all test tubes for the development of a ring of precipitation at the interface. Indicate the presence or absence of a ring in the chart.
2. Determine and indicate the antigen dilution that produced the greatest degree of precipitation that is indicative of the optimal antibody:antigen ratio.

	Antigen Dilutions			
	1:25	**1:50**	**1:75**	**Saline Control**
Presence of interfacial ring: (+) or (−)				

Dilution showing optimal antibody:antigen ratio . is _____.

REVIEW QUESTIONS

1. How do precipitin and agglutination reactions differ?

2. How would you determine the optimal antigen:antibody ratio by means of the ring test?

3. 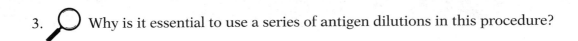 Why is it essential to use a series of antigen dilutions in this procedure?

4. How would you explain the absence of visible precipitate?

Precipitin Reaction: Immunodiffusion

LEARNING OBJECTIVES

Once you have completed this experiment, you should be able to perform a gel diffusion experiment to

1. Demonstrate the characteristic band precipitation patterns on an Ouchterlony plate.
2. Determine the optimal concentration (equivalence) of antibody and antigen necessary for gel diffusion.

PRINCIPLE

Immunodiffusion is an immunological procedure in which optimum concentrations of antigen and antibody join to produce visible bands of precipitation following their diffusion through a nonnutritional, clarified gel medium. The principal applications of this technique include:

1. Identification of the individual antigen and antibody components in the antigen–antibody reaction.
2. Identification of serologically related antigens.
3. Diagnosis of infectious diseases such as hepatitis, some types of typhus, and histoplasmosis.

Gel diffusion analyses may be carried out by means of two separate methods. The **Oudin tube method** is a one-dimensional system that allows diffusion of the antigen when it is layered onto the surface of an antibody-containing agar column in a small-bore precipitin tube. The **Ouchterlony plate method** is a two-dimensional system in which both components, antigen and antibody, are able to diffuse through the gel medium in a Petri dish. In the latter procedure, which you will use in this exercise, a layer of clarified agar ¼-inch thick is poured into a Petri dish and allowed to solidify. When the agar layer is hardened, wells are cut into the agar and filled with the proper dilutions of antibody and antigen. During incubation, these components diffuse toward each other until they reach optimal concentrations, forming precipitin bands in characteristically identifiable patterns as illustrated in Figure 73.1.

In this experiment, you will prepare two Ouchterlony plates. You will use one to determine the optimum antigen and antibody concentrations (equivalence point). You will use the second to demonstrate the precipitin reaction patterns.

MATERIALS

Antigens

Monkey, rabbit, horse, and human sera.

Antiserum

Antihuman serum.

Media

Per designated student group: two diffusion gel plates, four 9-ml 0.85% saline blanks, and one 25-ml tube of 0.85% saline.

Equipment

No. 2 cork borer, suction apparatus, 1-ml and 5-ml pipettes, mechanical pipetting device, Pasteur pipettes with rubber bulbs, glassware marking pencil, and 13 × 100-mm test tubes.

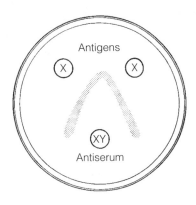

(a) Reaction of identity: demonstrates the homogeneity of both antigens (X) by the development of continuous bands of precipitation forming an angle as both antigen and antiserum diffuse toward each other

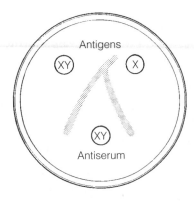

(b) Reaction of partial identity: demonstrates the incomplete homogeneity of antigens (XY and X) by the development of one spur in the band of precipitation extending toward the antigen of lesser prominence

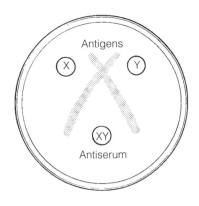

(c) Reaction of nonidentity: demonstrates the lack of homogeneity between the antigens (X and Y) by the formation of two separate bands of precipitation that cross each other, forming a double spur

FIGURE 73.1 Precipitin reaction patterns on Ouchterlony plates

PROCEDURE

> ⚠ **This experiment makes use of human serum antigens. Use disposable gloves throughout the procedure and be careful when handling human or animal sera.**

1. Using the template, prepare two Ouchterlony plates by boring five wells 7 to 10 mm apart with a No. 2 cork borer. With a suction apparatus, remove the agar plugs. Label the wells on the bottoms of the plates as indicated on the following template:

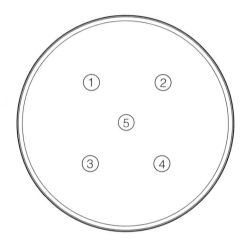

2. Prepare the plate for equivalency determination as follows:

    a. Using separate 1-ml pipettes and the four 9-ml saline blanks, perform a 10-fold serial dilution of the human serum antigen to effect 10^{-1}, 10^{-2}, 10^{-3}, and 10^{-4} dilutions.

    b. Label the top of the plate "equivalency." On the bottom of the plate, label wells 1 through 4 with the antigen dilutions.

    c. Using separate Pasteur pipettes, add the antigen dilutions to their respective wells and the human antiserum to well 5. *Note: Do not overfill the wells.*

3. Prepare the plate for demonstration of precipitin reaction patterns as follows:

    a. Prepare 1:50 dilutions of each of the four antigens using 0.1 ml of serum and 4.9 ml of 0.85% saline.

    b. Label the top of the plate "precipitin reactions." On the bottom of the plate, label wells 1 through 4 with the serum antigen to be used.

    c. Using separate Pasteur pipettes, add the diluted serum antigens to their respectively labeled wells and the human antiserum to well 5, being careful not to overfill the wells.

4. Place the prepared plates, top sides up, in a large beaker containing water-saturated filter paper. Cover the top of the beaker loosely with foil to retain moisture.

5. Place the beaker in a refrigerator for 7 days.

Name _____   Date _____

1. Observe the plate for equivalence determination for development of precipitin bands. Draw the observed bands in the plate diagram and indicate which antigen dilutions produced zones of equivalence.

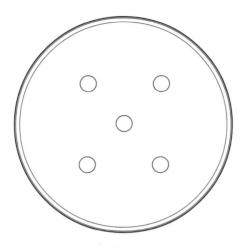

Equivalence determination
Ouchterlony plate

Antigen dilution(s) that produced equivalence zone(s): _____

2. Observe the plate for precipitin band patterns. Draw and identify the observed precipitin bands.

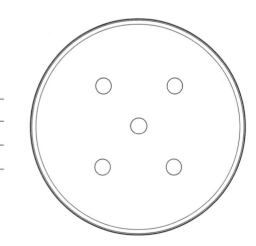

Monkey serum: _____

Rabbit serum: _____

Horse serum: _____

Human serum: _____

REVIEW QUESTIONS

1. What is the equivalence zone?

2. What is the basic difference between gel diffusion occurring in Oudin tubes and on Ouchterlony plates?

3. Why do you think it is necessary to use clarified gels to demonstrate immunodiffusion?

4. Why does the immunodiffusion medium resist contamination with bacteria?

Agglutination Reaction: The Febrile Antibody Test

LEARNING OBJECTIVE

Once you have completed this experiment, you should be able to

1. Demonstrate the agglutination reaction by means of the febrile antibody test and an antibody titer determination.

PRINCIPLE

The **febrile antibody test** is used in the diagnosis of diseases that produce febrile (fever) symptoms. Some of the microorganisms responsible for febrile conditions are salmonellae, brucellae, and rickettsiae. **Febrile antigens,** such as endotoxins, enzymes, and other toxic end products, elaborated by these organisms are used specifically to detect or exclude the homologous antibodies that develop in response to these antigens during infection.

In this procedure, the antigen is mixed on a slide with the serum being observed. Cellular clumping is indicative of the presence of homologous antibodies in the serum; the absence of homologous antibodies is indicated when there is no visible clumping. Only the febrile antigens and antibodies of *Salmonella* spp. will be used.

The second part of this experiment is designed to illustrate that agglutination reactions such as the febrile antibody test can be used to identify an unknown microorganism through serotyping. A specific antiserum prepared in a susceptible, immunologically competent laboratory animal is mixed with a variety of unknown bacterial antigen preparations on slides. The bacterial antigen that is agglutinated by the antiserum is identified and confirmed to be the agent of infection.

These tests are strictly qualitative. A quantitative result can be obtained by performing the **antibody titer test** (Figure 74.1), which

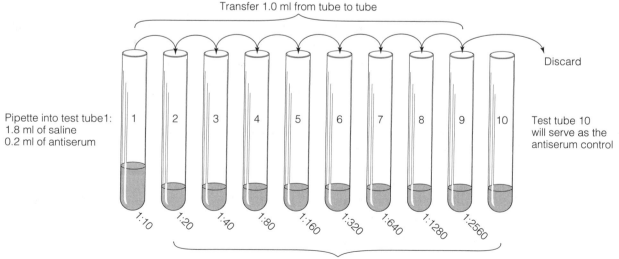

FIGURE 74.1 Antibody titer test: Serial dilution of *Salmonella typhimurium* H antibody

measures the concentration of an antibody in the serum and thus allows the physician to follow the course of an infection. The patient's serum is titrated (diluted), and the decreasing concentrations of the antiserum are mixed with a constant concentration of homologous antigen. The end point of the test will occur in the test tube containing the serum having the highest dilution showing agglutination.

MATERIALS

Cultures

Number-coded, washed saline suspensions of *Escherichia coli*, *Proteus vulgaris*, *Salmonella typhimurium*, and *Shigella dysenteriae*.

Reagents

Physiological saline (0.85% NaCl), commercial preparations of *Salmonella typhimurium* H antigen, and *Salmonella typhimurium* H antiserum.

Equipment

Bunsen burner, inoculating loop, glass microscope slides, 13 × 100-mm test tubes, sterile 1-ml pipettes, mechanical pipetting device, applicator sticks, glassware marking pencil, and microscope.

PROCEDURE

Febrile Antibody Test

1. With a glassware marking pencil, make two circular areas about ½ inch in diameter on a microscope slide. Label the circles A and B.
2. To Area A, add 1 drop of *S. typhimurium* H antigen and 1 drop of 0.85% saline. Mix the two with an applicator stick.
3. To Area B, add 1 drop of *S. typhimurium* H antigen and 1 drop of *S. typhimurium* H antiserum. Mix the two with a clean applicator stick.
4. Pick up the slide, and with two fingers of one hand, rock the slide back and forth.
5. Observe the slide both macroscopically and microscopically, under low power, for cellular clumping (agglutination).

Serological Identification of an Unknown Organism

1. Prepare two microscope slides as in the previous procedure. Label the four areas on the slides with the numbers of your four unknown cultures.
2. Into each area on both slides, place 1 drop of *S. typhimurium* H antiserum.
3. With a sterile inoculating loop, suspend a loopful of each number-coded unknown culture in the drop of antiserum in its appropriately labeled area on the slides.
4. Pick up the slides and slowly rock them back and forth.
5. Observe both slides macroscopically and microscopically, under low power, for agglutination.

Determination of Antibody Titer

1. Place a row of 10 test tubes (13 × 100 mm) in a rack and number the tubes 1 through 10.
2. Pipette 1.8 ml of 0.85% saline into the first tube and 1 ml into each of the remaining nine tubes.
3. Into Tube 1, pipette 0.2 ml of *Salmonella typhimurium* H antiserum. Mix thoroughly by pulling the fluid up and down in the pipette. *Note: Avoid vigorous washing.* The antiserum has now been diluted 10 times (1:10).
4. Using a clean pipette, transfer 1 ml from Tube 1 to Tube 2 and mix thoroughly as described. Using the same pipette, transfer 1 ml from Tube 2 to Tube 3. Continue this procedure through Tube 9.
5. Discard 1 ml from Tube 9. Tube 10 will serve as the antigen control and therefore will not contain antiserum.
6. The antiserum has been diluted during this twofold dilution to give final dilutions of 1:10, 1:20, 1:40, 1:80, 1:160, 1:320, 1:640, 1:1280, and 1:2560.
7. Add 1 ml of the *Salmonella typhimurium* H antigen suspension adjusted to an optical density of 0.5 at 600 nm to all tubes.
8. Mix the contents of the test tubes by shaking the rack vigorously.
9. Incubate the test tubes in a 55°C waterbath for 2 to 3 hours.

Name   Date

Febrile Antibody Test

In the chart:

1. Indicate the presence or absence of macroscopic and microscopic agglutination.
2. Draw a representative field of Areas A and B following microscopic observation for agglutination.

 Refer to photo number 81 in the color-plate insert for illustration of this reaction.

	Area A	Area B
Draw the appearance of antigen–antibody mixture.	◯ Saline *S. typhimurium* H antigen	◯ *S. typhimurium* H. antiserum *S. typhimurium* H antigen
Macroscopic agglutination (+) or (−)		
Microscopic agglutination (+) or (−)		

Serological Identification of an Unknown Organism

In the chart:

1. Indicate the presence or absence of macroscopic and microscopic agglutination in each of the suspensions.
2. Indicate the suspension that is indicative of a homologous antigen–antibody reaction.

		Agglutination		
Cell Antigen	Antiserum	Macroscopic (+) or (−)	Microscopic (+) or (−)	Homologous Antigen–Antibody Reaction
Unknown No: ____	*S. typhimurium* H			
Unknown No: ____	*S. typhimurium* H			
Unknown No: ____	*S. typhimurium* H			
Unknown No: ____	*S. typhimurium* H			

Determination of Antibody Titer

In the chart:

1. Indicate the presence or absence of agglutination in each of the antiserum dilutions.
2. Indicate the end point of the reaction.

Tube	Dilution	Agglutination	Titer
1	1:10		
2	1:20		
3	1:40		
4	1:80		
5	1:160		
6	1:320		
7	1:640		
8	1:1280		
9	1:2560		
10	Antigen control		

REVIEW QUESTIONS

1. What are febrile antibodies?

   What is their clinical significance?

2. What is the purpose of determining an antibody titer?

3. Why does the antibody titer determination use twofold dilutions of the antiserum rather than 10-fold dilutions?

Immunofluorescence

LEARNING OBJECTIVE

Once you have completed this experiment, you should be able to

1. Perform an antigen–antibody reaction using fluorescein-labeled antibodies.

PRINCIPLE

The **fluorescent antibody technique,** introduced by Coons in the mid-1950s, is a rapid and reliable procedure to demonstrate agglutination reactions. It has significant merit for use in the identification of microorganisms or their resultant antibodies. For example, the **fluorescent treponemal antibody-absorption (FTA–ABS) test** is used to diagnose syphilis using *Treponema pallidum* as the antigen for detection of syphilitic antibodies in the patient's serum.

This technique requires the use of a specific antibody that has been tagged at the Fc region with a fluorescent dye, such as fluorescein isocyanate or fluorescein isothiocyanate. When the tagged antibody is applied to the antigen preparation, as in the direct method, or to an antigen–antibody complex, as in the indirect method, a microprecipitate forms at the site of the antigen and exhibits a yellow-green fluorescence when viewed under a fluorescence microscope.

In this experiment you will use the direct immunofluorescence method (Figure 75.1) to demonstrate the antigen–antibody reaction and to identify an unknown antigen. The fluorescein-tagged antibody is added to a heat-fixed bacterial smear. A physiologically buffered saline wash removes any uncombined fluorescent antibody. The resultant agglutination reaction, if present, will be demonstrated by a green fluorescence when viewed under a fluorescence microscope.

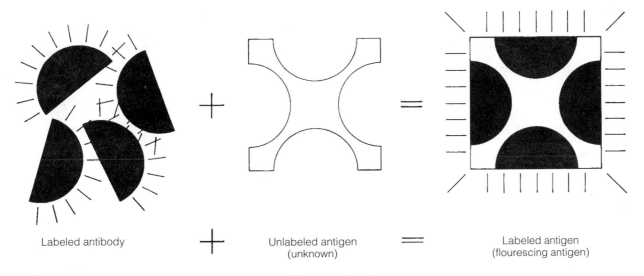

Labeled antibody + Unlabeled antigen (unknown) = Labeled antigen (flourescing antigen)

FIGURE 75.1 Fluorescent antibody technique—direct method

MATERIALS

Cultures

24-hour brain–heart infusion broth cultures of Group A *Streptococcus pyogenes* and Group D *Enterococcus faecalis;* numbered, unknown mixed broth cultures of Group A *S. pyogenes/ Escherichia coli* and Group D *E. faecalis/ Escherichia coli.*

Reagents

Fluorescent antibody *Streptococcus* Group A (Difco 2318-56-6), fluorescent antibody *Enterococcus* Group D (Difco 2319-56-5), phosphate-buffered saline, and buffered glycerol.

Equipment

Microscope slides, Petri dishes, U-shaped glass rods to fit into Petri dishes, filter paper, Coplin jar, glassware marking pencil, and fluorescence microscope.

PROCEDURE

1. With a glassware marking pencil, label two slides *S. pyogenes* and *E. faecalis,* respectively. Divide the third slide in half and label as mixed unknown.

2. Prepare a heat-fixed smear of each known test organism on its appropriately labeled slide. On the slide labeled mixed unknown, make a smear on each half of the slide using the unknown mixed culture.

3. On slides labeled *S. pyogenes* and *E. faecalis,* add 1 drop of each respective fluorescent antibody and spread gently over the surface of the smear. On the slide of the mixed unknown smears, label one side FA-A and the other side FA-D, add 1 drop of each fluorescent antibody to its respectively labeled smear, and allow to spread evenly over the smears.

4. Place moistened filter paper in the Petri dishes, insert the U-shaped glass rod (for slide support), and place the prepared slides on the slide supports. Cover the Petri dishes and incubate for 35 minutes at 25°C.

5. Remove the slides from the Petri dishes and wash away excess antibody with 1% phosphate-buffered saline.

6. Immerse the slide in a Coplin jar containing 1% phosphate-buffered saline for 10 minutes at 25°C.

7. Blot the slides dry with bibulous paper.

8. To each slide add 1 drop of buffered glycerol, cover with a coverslip, and examine under a fluorescence microscope as directed by your instructor.

Name _____   Date _____

1. Based on your microscopic observations, indicate in the chart the color of each smear and the presence (+) or absence (−) of fluorescence.

    📷 *Refer to photo number 82 in the color-plate insert for illustration of this reaction.*

Culture	Color	Fluorescence (+) or (−)
Group A: *S. pyogenes*		
Group D: *E. faecalis*		
Mixed unknown with Group A antibodies		
Mixed unknown with Group D antibodies		

2. Based on your observations, identify your mixed unknown culture.

    Culture Number:

    Organisms:

REVIEW QUESTIONS

1. What does the presence of fluorescence indicate?

2. Why is it necessary to wash away excess labeled antibody before viewing the preparation microscopically?

3. Briefly explain the direct fluorescent antibody procedure.

4. Label all parts of the antibody below, and show where it is tagged with fluorscein isothiocyanate.

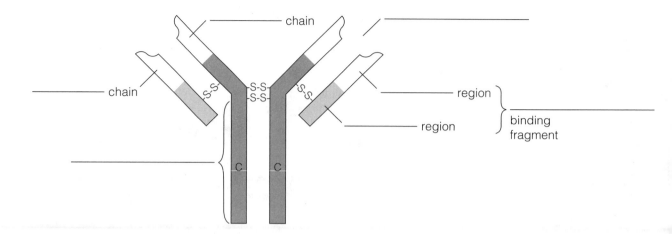

5. Aside from syphilis, what diseases can be diagnosed through immunofluorescent techniques.

6. How would you use immunofluorescense to identify Group A *Streptococcus pyogenes* organisms isolated from a patient and differentiate them from other morphologically similar streptococci and staphylococci?

Latex Agglutination Test

LEARNING OBJECTIVE

Once you have completed this experiment, you should be able to

1. Detect soluble antigens by means of a latex agglutination reaction.

PRINCIPLE

A comparison of the immunodiagnostic precipitation and agglutination procedures indicates that the major difference between the two types of techniques is based on the state of the antigen, that is, whether it is soluble or particulate. Formerly, soluble antigens were detectable only by such techniques as the interfacial and immunodiffusion tests, which are based on a precipitation reaction. Presently, soluble antigens can be detected by procedures that utilize nonspecific, inert particles or red blood cells (erythrocytes) for the adsorption of the soluble antigens onto their surfaces. Red blood cells treated with tannic acid readily adsorb soluble protein antigens, which are then capable of binding with antibodies to produce an agglutination reaction. More recently, synthetic particles such as polystyrene spheres (latex particles) have been used in place of the cells. This procedure allows the adsorption of both protein and nonprotein antigens onto the surfaces of the latex beads, which then can be used to detect antibodies by observing the agglutination of the antigen-carrying latex particles. The value of these newer methods is that the immunocomplexes are visually more discernible than they are by means of the less-sensitive precipitin reaction procedures.

The latex agglutination test can be used for the detection of antibodies produced in response to bacterial, mycotic, and helminthic infections. For example, the procedure is used for the diagnosis of tularemia, typhoid fever, infectious mononucleosis, syphilis, and, more recently, antibiotic-associated pseudomembranous colitis.

Prior to the development of antibiotics, pseudomembranous colitis was a clinically recognized pathological syndrome. However, since the introduction of these "wonder drugs," it has become apparent that this disease could be induced by antibiotic therapy, primarily with clindamycin, ampicillin, and the cephalosporins. The inhibition of the normal residential flora of the intestine allows the expression of *Clostridium difficile* in the colon. Since the colonic environment is anaerobic, rapid multiplication of this gram-positive, spore-forming, toxigenic bacillus will occur. *C. difficile* elaborates two toxins, A and B. Toxin A is suspected of playing a major role in inflammation and damage to the intestinal wall. If the intestinal damage is minimal, diarrhea is the only symptom, and the disease becomes self-limiting. However, if a large amount of Toxin A is produced, the damage to the colon progresses, leading to tissue necrosis, which may be followed by intestinal perforation and sepsis.

Conversely, this test procedure may also be used for the detection of antigens by using antibody-coated latex as in the **Culturette™ Clostridium difficile Test (CDT)** (Becton Dickinson and Company). This is a rapid latex slide agglutination test for the detection of *Clostridium difficile* antigens, which are soluble and therefore present in the supernatant of a stool specimen. Latex beads coated with antibody are then allowed to react with *C. difficile*-associated antigens in the stool supernatant to form a lattice of visibly agglutinated particles.

In this exercise, you will perform a modified procedure. To circumvent the use of a potentially biohazardous stool specimen, you will use a culture of *Salmonella typhimurium* to simulate the clinical sample. The supernatant of this culture will be used in the test and negative circles on the test slide.

MATERIALS

Cultures

Culturette CDT positive test control; 24- to 48-hour trypticase soy broth culture of *S. typhimurium.*

Equipment

Per designated student group or demonstration: Culturette CDT kit, vortex mixer, clinical centrifuge, centrifuge tube, 1-ml pipette, mechanical pipetting device, and wooden applicator sticks.

PROCEDURE

This test is amenable to be performed as a demonstration for economic reasons or conservation of laboratory time.

1. With a 1-ml pipette and a mechanical pipetting device, transfer 0.5 ml of the *S. typhimurium* culture into a centrifuge tube. Add 0.5 ml of the test buffer and mix for 30 seconds on a vortex mixer. Centrifuge for 15 minutes at $\geq 1500 \times$ g.

2. Using the supplied micropipette, place 1 drop of the *S. typhimurium* supernatant in the *test* circle and 1 drop in the *negative* circle on the test slide.

3. Add 1 drop of **detection reagent** (Reagent 1) to the *test* circle. Mix with an applicator stick and spread over the surface of the circle.

4. Add 1 drop of **negative control** (Reagent 2) to the *negative* circle. Mix with an applicator stick and spread over the surface of the circle.

5. Add 1 drop of **positive control** (Reagent 3) to the *positive* circle. Add 1 free-falling drop of detection reagent (Reagent 1). Mix with an applicator stick and spread over the surface of the circle.

6. Rock the agglutination slide for *exactly* 3 minutes. A mechanical slide rotator may be used at 90 to 140 rpm.

7. Immediately read the slide for agglutination. *Note: You must use an incandescent light source.*

Name _____ Date _____

1. Observe the three circles on the agglutination slide for the presence or absence of an agglutination lattice.
2. Record your results below based on the following interpretation of your observations:

Positive test (+): A test that shows more agglutination in the *test* circle than in the *negative* circle.

Negative test (−): A test that shows no agglutination in either the *test* or *negative* circles. The mixture will remain milky in appearance.

Nonspecific agglutination (+/−): A test that shows equal agglutination in both the *test* and *negative* circles.

Slide Circle	Result
Test	
Negative control	
Positive control	

REVIEW QUESTION

1. Why are inert particles preferred over red blood cells in the latex agglutination test?

Enzyme-Linked Immunoabsorbent Assay

LEARNING OBJECTIVE

Once you have completed this experiment, you should be able to

1. Demonstrate a method for the identification of either an antigen or an antibody by use of an enzyme-labeled antibody test procedure.

PRINCIPLE

The **enzyme-linked immunoabsorbent assay (ELISA)** procedure is a widely accepted method that is used for the detection of specific antigens or antibodies. The procedure is predicated on the use of an enzyme-linked (labeled) specific antibody to demonstrate the agglutination reaction for the interpretation of the test result. This test can be performed as a double-antibody technique or as an indirect immunoabsorbent assay. The former method is used for the detection of test antigens; the latter is used for the detection of the test antibodies. In both methods the reactions are carried out in a well of a plastic microtiter plate.

The double-antibody system requires that the unlabeled antibody be allowed to adsorb to the inner surface of the plastic well in the microtiter plate. Any unbound antibody is then washed away, and a specific test antigen is added to the well. If the antigen binds with the antibody adhering to the walls of the well, this immunocomplex will not be removed by the subsequent washing for the removal of any unbound antigen. An enzyme-linked antibody, specific for the antigen, is now added. If the antigen is present in the well, this labeled antibody binds to the antigen, forming an antibody-antigen-antibody complex. Any unbound enzyme-linked antibody is again removed by washing. This is followed by the addition of a substrate that is capable of producing a col-

ored end product upon its reaction with the enzyme. The resultant enzymatically produced color change may be observed by eye or spectrophotometrically.

The indirect immunoabsorbent test procedure is similar to the double-antibody technique in that it requires the use of an enzyme-linked antibody. However, an antigen, rather than an antibody, is adsorbed onto the inner surface of the well.

Enzyme-linked immunoabsorbent assays are used extensively for the diagnosis of human infectious diseases. Included among these are viral infections such as AIDS, influenza, respiratory syncytial viral infection, and rubella. Bacterial infections such as syphilis, brucellosis, salmonellosis, and cholera can also be ascertained by means of this technique. This procedure also can be used for the detection of drugs in blood or tissues.

In this experiment you will use the **Directigen® Flu-A Test** (Becton Dickinson and Company) to demonstrate the application of an in vitro enzyme immunoassay. This rapid, qualitative test employs an enzyme immuno-membrane filter assay to detect influenza A antigen extracted from nasopharyngeal or pharyngeal specimens of symptomatic patients. These specimens are added to a Color-PAC™ test device, and any influenza A antigen present is nonspecifically bound to the membrane surface. Detector enzyme conjugated to monoclonal antibodies specific for the influenza A nucleoprotein antigen are bound to the trapped antigen following their addition to the ColorPAC membrane. Two substrates are then added sequentially and allowed to incubate for 5 to 30 minutes prior to determination of the result.

In the experimental procedure to be followed, the positive control will simulate the nasopharyngeal specimen of a symptomatic patient and will be indicative of a positive result. A pharyngeal swab sample of an asymptomatic individual will be used to illustrate a negative result.

MATERIALS

Cultures

Directigen Flu-A positive control and pharyngeal swab specimen from an asymptomatic individual.

Media

Per designated student group or demonstration: one 2-ml sterile saline tube.

Equipment

Directigen Flu-A Test kit, sterile cotton swabs, sterile ml (0.02-ml) pipette, mechanical pipetting device, and disposable gloves.

PROCEDURE

This test is amenable to be performed as a demonstration for economic reasons or conservation of laboratory time.

⚠ **Wear disposable gloves during the procedure.**

Preparation of Negative Result by Use of a Pharyngeal Specimen

1. Using a sterile cotton swab, obtain a specimen from the pharyngeal tonsil by rotating the swab vigorously over its surface.
2. Immerse the cotton swab into a test tube containing 2 ml of sterile saline. Mix well. Remove as much liquid from the swab as possible by pressing the swab against the inner surface of the tube. Discard the swab into a container of disinfectant.
3. Using a μl (0.02-ml) pipette and a mechanical pipetting device, transfer 124 μl of the pharyngeal specimen into a DispensTube™ provided in the kit.

4. Gently mix and add 8 drops of Reagent A into the DispensTube. Mix well.
5. Insert a tip, provided in the kit, into the DispensTube. Dropwise, dispense all of the extracted specimen into the ColorPAC test well with the sealed flow controller in position. Allow for complete adsorption.
6. Gently mix and rapidly add drops of Reagent 1 until the test well is filled. Allow sufficient time for complete adsorption.
7. Remove the flow controller from the ColorPAC well and discard it into a container of disinfectant.
8. Gently mix and add 4 drops of Reagent 2 onto the ColorPAC membrane. Allow sufficient time for complete adsorption.
9. Gently mix and add 4 drops of Reagent 3 onto the ColorPAC membrane. Allow sufficient time for complete adsorption. Let stand for 2 minutes.
10. Rapidly add enough drops of Reagent 4 to fill the ColorPAC well. Allow sufficient time for complete adsorption.
11. Gently mix and add 4 drops of Reagent 5 onto the ColorPAC membrane. Allow sufficient time for complete adsorption.
12. Gently mix and add 4 drops of Reagent 6 onto the ColorPAC membrane. Allow sufficient time for complete adsorption. *Note: The membrane will turn yellow.*
13. Gently mix and add 4 drops of Reagent 7 onto the ColorPAC membrane. Allow sufficient time for complete adsorption.
14. Wait at least 5 minutes, but no longer than 30 minutes, and then read the results in a well-lighted area.

Preparation of Positive Result by Use of Positive Control

1. Dispense 4 drops of the positive control, provided in the test kit, into a DispensTube.
2. Repeat Steps 4 through 14 as outlined above for the preparation of the negative pharyngeal specimen.

Name   Date

1. Observe the appearance of the inner surface of the ColorPAC test wells.
2. Record your results below based on the following interpretations of your observations:

 Positive test (antigen present): The appearance of a purple triangle (of any intensity) on the ColorPAC membrane indicates the presence of the influenza A antigen in the specimen. A purple dot may be evident in the center of the triangle. The background area should be grayish white.

 Negative test (no antigen detected): The appearance of a purple dot on the ColorPAC membrane indicates the absence of the influenza A antigen in the specimen. The background area should be grayish white.

 Uninterpretable test: The absence of a purple dot, a purple triangle, or an incomplete purple triangle indicates an uninterpretable test.

Pharyngeal specimen:

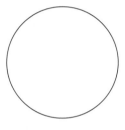

Result

Positive control specimen:

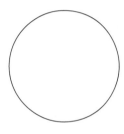

Result

REVIEW QUESTION

1. Why is the ELISA test used to screen human serum for the AIDS virus, while the Western blot procedure is used only as the confirmation test?

Agglutination Reaction: Mono-Test for Infectious Mononucleosis

LEARNING OBJECTIVE

Once you have completed this experiment, you should be able to

1. Detect infectious mononucleosis heterophile antibodies.

PRINCIPLE

Infectious mononucleosis (IM) is an acute, self-limiting infectious disease characterized by mild fever, sore throat, and significant enlargement of lymph nodes and spleen. Hematologically, there is marked leukocytosis in which the number of monocytes is elevated with a concomitant decrease in the number of neutrophils. Clinically, IM may mimic other diseases such as Hodgkin's disease, hepatitis, diphtheria, and lymphatic leukemia. The etiological agent is believed to be the Epstein-Barr virus (EBV).

Diagnosis is made on the basis of the clinical symptoms and the **heterophile antibody test.** Heterophile (heterogenetic) antigens are genetically unrelated but are extremely similar in their chemical compositions. Although not identical per se, antibodies prepared against one antigen will cross-react with others. The most prominent heterophile antigen is the one described by Forssman, who determined that when horse or guinea pig tissues were inoculated into rabbits, the resultant antibodies in the rabbit serum agglutinated sheep red blood cells. It was evident from this result that horse, cat, and guinea pig antigens were all chemically similar to those of sheep. Such antigens have been called **Forssman antigens,** and the resultant antibodies are **Forssman heterophile antibodies.** Although all heterophile antibodies react with sheep red blood cells, the IM antibodies are discernible from other heterophile antibodies because of their inability to be removed from a serum

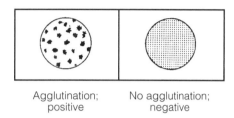

Agglutination;   No agglutination;
positive   negative

FIGURE 78.1 Mono-Test reactions

sample by adsorption with guinea pig antigen. Other heterophile antibodies are easily removed by this method. On this basis, the IM antibodies are **non-Forssman heterophile antibodies.**

Originally, complex presumptive and differential adsorption tests were required to detect IM antibodies. The former used a titration procedure to detect antisheep red blood cell antibodies. A titer greater than 1:56 was indicative of a presumptive diagnosis; a titer of 1:224 with clinical symptoms was considered to be a confirmed diagnosis. A titer higher than 1:224 without clinical symptoms was suggestive of serum sickness resulting from the use of horse antisera. Positive identification of IM antibodies was determined by means of the **differential adsorption test,** devised by Davidsohn, in which boiled guinea pig kidney was used to adsorb all horse and Forssman antibodies, with the exception of the IM antibodies.

In recent years, methods have been devised that do not require this laborious test. In this experiment a 2-minute rapid screening system, called **Mono-Test®** (Wampole Laboratories, Cranbury, New Jersey), will be used to detect IM antibodies. In this one-step procedure, stabilized horse red blood cells will react with IM antibodies, if they are present in the patient's serum, to give a positive agglutination reaction as shown in Figure 78.1.

MATERIALS

Equipment

Mono-Test kit and wooden applicator sticks.

PROCEDURE

This test is amenable to be performed as a demonstration for economic reasons or conservation of laboratory time.

1. Place 1 drop of negative control serum into the circle on the right and 1 drop of positive control serum into the circle on the left.

2. Add 1 drop of Mono-Test reagent to each of the two circles.

3. Using separate wooden applicator sticks, mix and spread each mixture evenly over the entire surface of each circle.

4. Gently rock the card slide in a back-and-forth motion for 1 minute and let stand for another minute.

5. Observe immediately for agglutination.

Name   Date

1. Observe each reaction for the presence or absence of agglutination in each circle. Diagram each reaction and record your results as positive or negative.

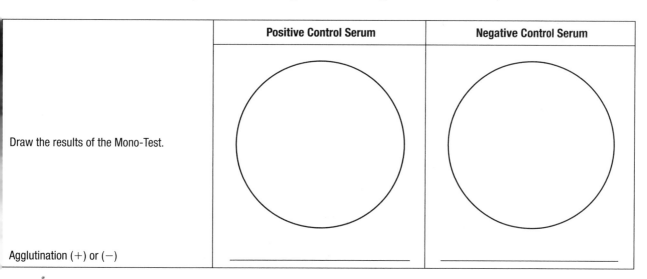

	Positive Control Serum	Negative Control Serum
Draw the results of the Mono-Test.		
Agglutination (+) or (−)	_____	_____

REVIEW QUESTIONS

1. How can IM antibodies be distinguished from other heterophile antibodies?

2. Why are IM antibodies called non-Forssman antibodies?

Sexually Transmitted Diseases: Rapid Immunodiagnostic Procedures

Sexually transmitted diseases (STDs) represent a diverse group of infectious syndromes that share the same mode of transmission, direct sexual contact. Their etiological agents represent a broad spectrum of pathogenic microorganisms that include bacteria, viruses, yeasts, and protozoa. The bacterial STDs include **gonorrhea, syphilis, nongonococcal urethritis,** and **lymphogranuloma venereum.** The representative viral infections are **genital herpes, genital warts, hepatitis B,** and the latest member of this group, **AIDS.** The protozoal and fungal infections, namely **trichomoniasis** and **candidiasis,** are diseases of lesser magnitude in the spectrum of STDs.

The experimental procedures that follow were chosen to demonstrate some of the rapid tests that are currently available for the diagnosis of selected STDs, specifically syphilis, genital herpes, and the chlamydial infections. In the methods that follow, you will perform modified procedures in the absence of clinical specimens. Commercially available positive and negative controls will be used to simulate clinical materials. *It is suggested that any of these tests, if performed, should be done as demonstrations.*

PART A: Rapid Plasma Reagin Test for Syphilis

LEARNING OBJECTIVE

Once you have completed this experiment, you should be able to

1. Perform a rapid screening procedure for diagnosis of syphilis.

PRINCIPLE

Treponema pallidum, the causative agent of **syphilis,** is a tightly coiled, highly motile, delicate spirochete that can be cultivated only in rabbit tissue cultures or rabbit testes. The organisms are resistant to common staining procedures and are best observed under darkfield microscopy.

Syphilis is a systemic infection that, if untreated, progresses through three clinical stages. The first stage, primary syphilis, is characterized by the formation of a painless papule, called a **chancre,** at the site of infection. Secondary syphilis represents the systemic extension of the infection and presents itself in the form of a **maculopapular rash,** malaise, and lymphadenopathy. Following this stage, the disease becomes self-limiting, and the patient appears asymptomatic until the development of tertiary syphilis. In this final stage, life-threatening complications may develop as a result of the extensive cardiovascular and nervous tissue damage that has ensued.

The **rapid plasma reagin (RPR) test,** which has to a large extent replaced the VDRL (Venereal Disease Research Laboratory) agglutination test, determines the presence of **reagin,** the nonspecific antibody present in the plasma of individuals with a syphilitic infection. The reagin appears in the plasma within 2 weeks of infection and will remain at high concentrations until the disease is eradicated. In the RPR test, if the reagin is present in the blood, it will react with a soluble antigen bound to carbon particles to produce a macroscopically visible antigen, or carbon–antibody complex. This procedure has several advantages over the VDRL test:

1. The serum does not require inactivation by heat for 30 minutes.

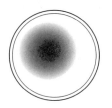

Reactive   Nonreactive

FIGURE 79.1 Test card showing results of the rapid plasma reagin test

2. The serum may be obtained from a finger puncture, unlike the VDRL test, which requires a venous blood sample.

3. The required materials, which include the antigen suspension with a dispensing bottle, diagnostic cards, and capillary pipettes, are all contained in individual kits that do not require additional equipment and are disposable.

In the qualitative form of the RPR test, the patient's blood serum and the carbon-bound antigen suspension are mixed within a circle on the diagnostic card. In the presence of a positive (reactive) serum, the antigen–antibody complex will produce a macroscopically visible black agglutination reaction. The macroscopic appearance of a light-gray suspension, devoid of any form of agglutination, is indicative of a negative (nonreactive) serum (Figure 79.1).

Since this is a nonspecific test, false positive results may be obtained. It is believed that the reagin is an antibody against tissue lipids in general. Therefore it may be present in uninfected individuals due to the release of lipids resulting from normally occurring wear and tear of body tissues. It has also been found that serum levels of reagin are elevated during the course of other infectious diseases such as viral pneumonia, lupus erythematosus, infectious mononucleosis, yaws, and pinta. The serum of patients with a reactive RPR result is subjected to additional serological testing, such as the FTA–ABS (fluorescent treponemal antibody-absorption) test, or the TPI, (*Treponema pallidum* immobilization) test, using the *Treponema pallidum* bacterium as an antigen to detect specific antibodies that are also present in the serum during syphilitic infection.

MATERIALS

Reagents

Commercially prepared syphilitic serum 4+ and nonsyphilitic serum.

Equipment

RPR test kit (Hynson, Westcott and Dunning, Inc., Baltimore, Maryland), disposable gloves, and rotating machine (optional).

PROCEDURE

 Wear disposable gloves.

1. Label circles on the diagnostic plastic card as reactive and nonreactive.

2. Using a capillary pipette with an attached rubber bulb, draw the reactive serum up to the indicated mark (0.05 ml).

3. Expel the serum directly onto the card in the circle labeled reactive serum. With a clean applicator stick, spread the serum to fill the entire circle.

4. Repeat Steps 2 and 3 for the nonreactive serum.

5. Shake the dispensing bottle to mix the suspension. Hold the bottle with attached 20-gauge needle in a vertical position and dispense 1 drop onto each circle containing the test serum.

6. If a mechanical rotator is available, place the card on the rotator set at 100 rpm or rotate the card back and forth manually for 8 minutes.

PART B: Genital Herpes: Isolation and Identification of Herpes Simplex Virus

LEARNING OBJECTIVE

Once you have completed this experiment, you should be able to

1. Perform a tissue culture procedure for the growth and identification of the herpes simplex virus.

PRINCIPLE

The double-stranded herpes simplex virus (HSV) is the etiological agent of a variety of human infections. Included among these are **herpes labialis,** fever blisters around the lips; **keratoconjunctivitis,** infection of the eyes; **herpes genitalis,** eruptions on the genitalia; **herpes encephalitis,** a severe infection of the brain; and **neonatal herpes.** The herpes simplex virus is divided into two antigenically distinct groups, **HSV-1** and **HSV-2.** The former is most frequently implicated with infections above the waist, whereas the latter is predominantly responsible for genital infections.

Primary infection with HSV-2 manifests itself with the appearance of vesicular lesions, characterized by itching, tingling, or burning sensations on or within the male and female genitalia. These vesicles heal spontaneously within 2 weeks. Following this symptomatic phase, however, the virus reverts to a latent state and remains quiescent until exacerbated by some environmental factor. With no chemotherapeutic cure presently available, recurrent genital herpes with subclinical symptoms is common.

Detection of the herpes simplex virus requires the use of tissue culture techniques. The presence of the virus is then determined by the development of cytopathogenic effects in these cultures, such as the detection of intranuclear inclusion bodies. In recent years, these time-consuming, specialized procedures have been greatly facilitated by the availability of immunoenzymatic reagents for the identification of this clinically significant virus.

The **Cellmatics™ HSV Detection System** (Difco Laboratories, Detroit, Michigan) is a self-contained system providing for both the growth and identification of the virus from clinical specimens. In this procedure, the provided tissue culture tubes are inoculated with the clinical sample. Following a 24-hour incubation period and fixation, the presence of HSV antigens is determined by the addition of anti-HSV antibodies, which specifically bind to the HSV antigens. To demonstrate this antigen–antibody complex, a secondary antibody, substrate, and chromogen are added. Following this staining process, HSV-positive cultures viewed microscopically will exhibit brown-black areas of viral infection on a clear background of unstained cells.

In this exercise, you will perform a modified procedure. In the absence of a clinical specimen, the actual culturing and fixation process will not be performed. Instead, the positive and negative commercially available controls will be used to simulate the clinical samples.

MATERIALS

Cultures

Cellmatics HSV Positive and Negative Control Tubes.

Reagents

Cellmatics Immunodiagnostic Reagents Kit, distilled water.

Equipment

5-ml pipettes, mechanical pipetting device, and microscope.

PROCEDURE

1. Warm immunodiagnostic reagents to room temperature.
2. Drain all fluid from the positive and negative control tubes.
3. Using a 5-ml pipette, wash the culture tubes twice with 5 ml of distilled water and drain. *Note: When washing, exercise care to prevent disruption of the monolayer.*
4. Add 10 drops of primary antiserum (Vial 1). *Note: When adding reagents, hold the vial vertically to ensure proper delivery.*
5. Incubate the *tightly capped* tubes in a *horizontal* position for 15 minutes at 37°C. To ensure complete coverage of the monolayer, occasionally rock the tubes gently during incubation.
6. Wash three times with 5 ml of distilled water and drain.
7. Add 10 drops of secondary antibody (Vial 2).
8. Incubate for 15 minutes at 37°C as described in Step 5.
9. Wash three times with 5 ml of distilled water and drain.
10. Add 10 drops of substrate (Vial 3) and 2 drops of chromogen (Vial 4). Mix gently.

11. Incubate for 15 minutes at 37°C as described in Step 5.

12. Wash three times with 5 ml of distilled water and drain.

13. Examine microscopically for the presence of stained cells at 40× and 100× magnifications.

PART C: Detection of Sexually Transmitted Chlamydial Diseases

LEARNING OBJECTIVE

Once you have completed this experiment, you should be able to

1. Perform an immunofluorescent procedure for diagnosis of *Chlamydia* infections.

PRINCIPLE

Members of the genus **Chlamydia** are a group of obligate intracellular parasites. Although once believed to be viruses, their morphologic and physiologic characteristics more closely resemble bacteria, and therefore, they are frequently referred to as small bacteria. Chlamydiae are gram-negative, nonmotile, thick-walled, spherical organisms possessing both DNA and RNA that reproduce by means of binary fission. Their dependence on living tissues for cultivation and their lack of an ATP-generating system emulate the characteristics of viruses, but their bacterial nature is further affirmed by their sensitivity to antibiotic therapy. *Chlamydia trachomatis*, the human pathogen, is now recognized to be responsible for two sexually transmitted diseases, **nongonococcal urethritis** (NGU) and **lymphogranuloma venereum** (LGV). The incidence of both diseases in contemporary society is increasing dramatically.

NGU is a urethritis (inflammation of the urethra) with symptoms similar to, but less severe than, those of gonorrhea. Undiagnosed and untreated infections may lead to **epididymitis** and **proctitis** in men and **cervicitis, salpingitis,** and **pelvic inflammatory disease** in women. LGV, the most severe of the genital chlamydial infections, initially develops with a painless lesion at the portal of entry, the genitalia. Systemic involvement is evidenced by swelling of the regional lymph nodes, which become tender and suppurative before disseminating the organisms to other tissues. In the absence of chemo-therapeutic intervention, scarring of the lymphatic vessels can cause their obstruction, leading to **elephantiasis,** enlargement of the external genitalia in men, and narrowing of the rectum in women.

MicroTrak®, a Direct Specimen Test (Syva Company, Palo Alto, California) is a rapid, immunofluorescent procedure for the detection of *C. trachomatis*. The procedure circumvents the need of culturing these organisms in susceptible tissues prior to their identification. This slide test is designed to detect elementary bodies, the infectious particles produced during the life cycle of this organism, by the use of a staining reagent, a fluorescein-labeled monoclonal antibody specific for the principle *C. trachomatis* outer-membrane protein. In this procedure, a slide smear is prepared from the clinical specimen. Following fixation, when the slide is exposed to the Direct Specimen Reagent, the antibody binds to the organisms. Their presence is then determined by the appearance of apple-green chlamydiae against a red background of counterstained cells when viewed under a fluorescent microscope.

MATERIALS
Cultures

Commercially prepared positive and negative control slides.

Reagents

MicroTrak Direct Specimen Test for *Chlamydia trachomatis*.

Equipment

Fluorescent microscope.

PROCEDURE

1. Stain the positive and negative control slides with the MicroTrak reagent for 15 minutes.

2. Incubate slides for 15 minutes.

3. Rinse the slides with distilled water.

4. Air-dry the slides.

5. Examine the slides under a fluorescent microscope for the presence of apple-green chlamydiae.

Experiment 79 / Observations and Results

Name Date

PART A: Rapid Plasma Reagin Test for Syphilis

1. In the presence of direct light, while tilting the card back and forth, determine the presence or absence of black clumping in each of the serum–antigen mixtures. Record your observations and the reaction as (+) or (−).
2. Complete the following table.

	Reactive Serum	Nonreactive Serum
Appearance of serum-antigen mixture		
Reaction (+) or (−)		
Draw the observed reaction.		

PART B: Genital Herpes: Isolation and Identification of Herpes Simplex Virus

1. Scan the entire stained monolayer of both culture tubes under 40× and 100× magnifications for the presence of brown to blackish-brown stained cells. HSV infection is indicated by the presence of dark-colored cells when viewed against an unstained background of normal cells.
2. Indicate in the chart below the presence (+) or absence (−) of these dark-stained patches.

Control Cultures			
Negative		Positive	
40 ×	100 ×	40 ×	100 ×

PART C: Detection of Sexually Transmitted Chlamydial Diseases

1. Observe the stained control slides for the presence of apple-green particles indicative of chlamydiae. The particles are evident against a reddish background of counterstained cells.

2. Record your results below, indicating the presence (+) or absence (−) of the apple-green chlamydiae on each of the slides.

Positive control slide: _____

Negative control slide: _____

REVIEW QUESTIONS

1. Why is an adult who has a high antibody titer to herpes simplex virus 2 (HSV-2) subject to recurrent genital herpes infections?

2. A 20-year-old college student was informed following a physical examination that her blood test for syphilis was reactive. She indicated that she was a virgin and had never received a blood transfusion. A repeat RPR test was positive, but the TPI test was negative. How would you explain these bizarre results, and what is the clinical status of this patient?

Scientific Notation

Microbiologists are required to perform a variety of laboratory techniques, including preparing and diluting solutions; expressing concentrations of chemicals, antibiotics, and antiseptics in solution; making quantitative determinations of cell populations based on the standard method for plate counting; and making serial dilutions to accommodate the latter procedure. These techniques commonly involve the use of very large or very small numbers (e.g., 9,000,000,000 or 0.0000000009), so it becomes patently clear that the manipulation of such numbers can be cumbersome and at times troublesome. Therefore, it is essential for microbiologists to have a good command of scientific exponential notation known as **scientific notation**.

The basis for this system is predicated on the fact that all numbers can be expressed as the product of two numbers, one of which is the power of the number 10. In scientific notation, the small superscript number next to the 10 is called the **exponent**. Positive exponents tell us how many times the number must be multiplied by 10, while negative exponents indicate how many times a number must be divided by 10 (i.e., multiplied by one-tenth).

For example, a number written using the exponential form designated as scientific notation would appear as 7.5×10^3, meaning that $7.5 \times 10 \times 10 \times 10 = 7500$. Appendix Table 1.1 shows both large and small numbers written in the exponential form.

APPENDIX TABLE 1.1   Scientific (Exponential) Notation

Numbers Greater than One	Numbers Less than One
$1{,}000{,}000{,}000 = 1 \times 10^9$	$0.000\,000\,001 = 1 \times 10^{-9}$
$100{,}000{,}000 = 1 \times 10^8$	$0.000\,000\,01 = 1 \times 10^{-8}$
$10{,}000{,}000 = 1 \times 10^7$	$0.000\,000\,1 = 1 \times 10^{-7}$
$1{,}000{,}000 = 1 \times 10^6$	$0.000\,001 = 1 \times 10^{-6}$
$100{,}000 = 1 \times 10^5$	$0.000\,01 = 1 \times 10^{-5}$
$10{,}000 = 1 \times 10^4$	$0.000\,1 = 1 \times 10^{-4}$
$1000 = 1 \times 10^3$	$0.001 = 1 \times 10^{-3}$
$100 = 1 \times 10^2$	$0.01 = 1 \times 10^{-2}$
$10 = 1 \times 10^1$	$0.1 = 1 \times 10^{-1}$
$1 = 1 \times 10^0$	$1 = 1 \times 10^0$

Note: The exponent to which the power of 10 is raised is equal to the number of zeros to the right of 1.

Note: The exponent to which the power of 10 is raised is equal to the number of zeros to the left of 1 plus 1.

Multiplication

Rule: To multiply two numbers that are written in scientific notation (exponential form) you must **add** the exponents.

Using numbers larger than 1:

$$75 \times 1200 = 90,000$$

Scientific notation: $(7.5 \times 10^1) \times (1.2 \times 10^3) = 9 \times 10^4$

Addition of exponents: $1 + 3 = 4$

Using numbers less than 1:

$$0.75 \times 1200 = 900$$

Scientific notation: $(7.5 \times 10^{-1}) \times (1.200 \times 10^3) = 9 \times 10^2$

Addition of exponents: $(-1 + 3 = 2)$

$$0.75 \times 0.12 = 0.09$$

Scientific notation: $(7.5 \times 10^{-1}) \times (1.2 \times 10^{-1}) = 9 \times 10^{-2}$

Addition of exponents: $(-1) + (-1) = -2$

Division

Rule: To divide two numbers in scientific notation, you must **subtract** the exponents.

$$75,000 \div 1,200,000 = 0.0625$$

Scientific notation:
$$(7.5 \times 10^4) \div (1.2 \times 10^6) = 6.25 \times 10^{-2}$$

Subtraction of exponents: $(4 - 6 = -2)$

$$7,500 \div .012 = 625,000$$

Scientific notation: $(7.5 \times 10^3) \div (1.2 \times 10^{-2}) = 6.25 \times 10^5$

Subtraction of exponents: $(3 - (-2) = 5)$

As the student practices the use of scientific notation with large and small numbers, he/she will become more proficient and more comfortable with this system of scientific calculation.

Methods for the Preparation of Dilutions

In microbiology laboratories as in other science laboratories, solutions must be diluted to achieve a desired final concentration of the active material contained in that solution. A **solution** may be defined as a mixture of two or more substances (**solute**) in which the molecules of the solute are evenly distributed and will not separate on standing or precipitate from the solution. Solutes are dissolved in a solvent or diluent, such as water, alcohol, or some other vehicle in which the solute is soluble. Solutions are usually referred to as stock solutions and may be diluted by a variety of methods, depending upon the experimental requirements. Some of these methods are listed as follows:

1. A **dilution factor** must be determined first in order to dilute a solution. This dilution factor tells us how many times a solution must be diluted and is calculated by dividing the **initial concentration (IC)** of the solution by the **final concentration (FC)** desired.

   Example: You wish to dilute a 10% stock solution to a final concentration of 2%.

   $$10\% \div 2\% = 5 \text{ (dilution factor)}$$

   Take 1.0 ml of the 10% stock solution plus 4.0 ml of diluent (solvent) which equals a total of 5.0 ml. Thus each ml of the final solution will contain 2% of the original 10% initial solution.

2. Another method is used when a specific volume composed of a specific concentration is required.

   Example: You have a 50% concentrated solution and you need 200 ml of a 5% solution.

   a. $\dfrac{IC}{FC} = \dfrac{50\%}{5\%} = 10 \text{ (dilution factor)}$

   b. $\dfrac{\text{volume needed}}{\text{concentration required}} = \dfrac{200 \text{ ml}}{5\%} = 40 \text{ ml}$

   c. 40 ml of 50% IC + 160 ml of diluent = 200 ml of a solution such that each ml will contain 2% rather than the original 50% in the stock solution.

3. The ability to prepare large dilutions is absolutely essential for work in the microbiology laboratory. This method requires that large dilutions be prepared in two steps.

   Example: A solution contains 1.0 g per ml of an active material and needs to be diluted to a final concentration of 1.0 μg per ml. A 1,000,000 (1×10^6)-fold dilution must be made. It is not feasible to make such a dilution in one step since 999,999 ml of diluent would be required and thus such a dilution is not practical. This type of dilution is made as follows:

   a. Dilute 1 ml of the stock solution 1000 times:

      1.0 ml + 999 ml of diluent = 1000 μg/ml

   b. Dilute the solution containing 1000 μg/ml another 1000 times:

      1 ml of 1000 μg/ml + 999 ml diluent = 1.0 μg/ml

4. When working with large molecules such as proteins, there will be times when you will be required to make large dilutions of the sample to be contained in a specific volume.

   Example: You need to make 50 ml of a $\frac{1}{20,000}$ dilution of albumin.

   $$\dfrac{\text{dilution}}{\text{volume needed}} = \dfrac{20,000}{50} = 400 \text{ (dilution factor)}$$

   a. $\dfrac{1.0 \text{ ml of}}{\text{albumin}} + \dfrac{399 \text{ ml of}}{\text{diluent}} = \frac{1}{400}$ dilution

   b. $\dfrac{1.0 \text{ ml of a}}{\frac{1}{400} \text{ dilution}} + \dfrac{49 \text{ ml of}}{\text{diluent}} = \begin{array}{l}50 \text{ ml of a solution}\\ \text{in which each ml}\\ \text{contains } \frac{1}{20,000} \text{ of}\\ \text{albumin.}\end{array}$

   50 (volume) × 400 (dilution factor) = 20,000

5. Perhaps the most useful type of dilution used in microbiology and immunology is the **serial dilution**. This is essential when small volumes of material are needed. This type of dilution procedure has many uses in the microbiology laboratory especially for the determination of the total number of cells in culture (Experiments 20 and 21), the number of viral plaques found in suspensions of viruses (Experiment 39), the antibody titer (Experiment 74), and in other immunological studies. The procedure requires the use of dilution blanks containing a known volume of diluent (distilled water, saline, etc.) to which a specific volume of the sample is added. To facilitate the ease of calculations, dilutions are usually made in multiples of 10. For example: 1.0 ml of a sample is added to a 9.0 ml dilution blank (1.0 ml + 9.0 ml = 10) and is recorded as a 1:10 dilution.

It has been statistically determined that greater accuracy is achieved with very large dilutions made from a series of smaller dilutions. The procedure for the performance of a serial dilution has been explained and illustrated in Experiment 20. For the convenience of the student, it is illustrated in Appendix Figure 2.1:

1. All dilution blanks contain 9.0 ml of diluent.

2. A fresh pipette is used for each dilution, and the used pipettes are placed in a beaker of disinfectant.

3. After delivery of the sample, the tubes are mixed thoroughly before the next dilution is made.

4. Pippetting by mouth is not allowed. Only mechanical pipette aspirators may be used.

The stock solution in Appendix Figure 2.1 has been diluted one million times. In other words, 1.0 ml from Tube 6 will contain $\frac{1}{1,000,000}$ of the sample contained in the stock solution.

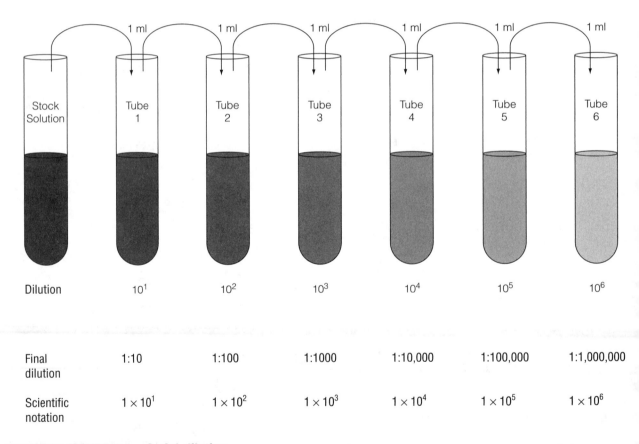

	1 ml	1 ml	1 ml	1 ml	1 ml	1 ml
Stock Solution	Tube 1	Tube 2	Tube 3	Tube 4	Tube 5	Tube 6
Dilution	10^1	10^2	10^3	10^4	10^5	10^6
Final dilution	1:10	1:100	1:1000	1:10,000	1:100,000	1:1,000,000
Scientific notation	1×10^1	1×10^2	1×10^3	1×10^4	1×10^5	1×10^6

APPENDIX FIGURE 2.1   Serial dilution

Microbiological Media

The formulas of the media used in the exercises in this manual are listed alphabetically in grams per liter of distilled water unless otherwise specified. Sterilization of the media is accomplished by autoclaving at 15 lb pressure for 15 minutes unless otherwise specified. Most of the media are available commercially in powdered form, with specific instructions for their preparation and sterilization.

Ammonium sulfate broth (pH 7.3)

Ammonium sulfate	2.0
Magnesium sulfate · 7H$_2$O	0.5
Ferric sulfate · 7H$_2$O	0.03
Sodium chloride	0.3
Magnesium carbonate	10.0
Dipotassium hydrogen phosphate	1.0

Bacteriophage broth 10× (pH 7.6)

Peptone	100.0
Beef extract	30.0
Yeast extract	50.0
Sodium chloride	25.0
Potassium dihydrogen phosphate	80.0

Basal salts agar* and broth (pH 7.0)

0.5 M sodium diphosphate	100.0 ml
1.0 M potassium dihydrogen phosphate	100.0 ml
Distilled water	800.0 ml
0.1 M calcium chloride	1.0 ml
1.0 M magnesium sulfate	1.0 ml
Note: Swirl	
Ammonium sulfate	2.0
*Agar	15.0

Note: Swirl until completely dissolved, autoclave, and cool. Aseptically add 10.0 ml of 1% sterile glucose.

Bile esculin (pH 6.6)

Beef extract	3.0
Peptone	5.0
Esculin	1.0
Oxgall	40.0
Ferric citrate	0.5
Agar	15.0

Blood agar (pH 7.3)

Infusion from beef heart	500.0
Tryptose	10.0
Sodium chloride	5.0
Agar	15.0

Note: Dissolve the above ingredients and autoclave. Cool the sterile blood agar base to 45°C to 50°C. Aseptically add 50 ml of sterile defibrinated blood. Mix thoroughly, avoiding accumulation of air bubbles. Dispense into sterile tubes or plates while liquid.

Brain–heart infusion (pH 7.4)

Infusion from calf brain	200.0
Infusion from beef heart	250.0
Peptone	10.0
Dextrose	2.0
Sodium chloride	5.0
Disodium phosphate	2.5
Agar	1.0

Bromcresol purple dextrose fermentation broth (pH 7.2)

Bacto™ casitone	10
Dextrose	5
Bromcresol purple (0.2%)	0.01

Bromcresol purple (0.2%) is made separately and filter sterilized. 5 ml are aseptically added to the medium.
Note: Autoclave at 12 lb pressure for 15 minutes.

Bromcresol purple lactose fermentation broth (pH 7.2)

Bacto casitone	10
Lactose	5
Bromcresol purple (0.2%)	0.01

Bromcresol purple (0.2%) is made separately and filter sterilized. 5 ml are aseptically added to the above medium.

Note: Autoclave at 12 lbs pressure for 15 minutes.

Bromcresol purple maltose fermentation broth (pH 7.2)

Bacto casitone	10
Maltose	5
Bromcresol purple (0.2%)	0.01

Bromcresol purple (0.2%) is made separately and filter sterilized. 5 ml are aseptically added to the medium.

Note: Autoclave at 12 lb pressure for 15 minutes.

Bromcresol purple sucrose fermentation broth (pH 7.2)

Bacto casitone	10
Sucrose	5
Bromcresol purple (0.2%)	0.01

Bromcresol purple (0.2%) is made separately and filter sterilized. 5 ml are added to the medium aseptically.

Note: Autoclave at 12 lb pressure for 15 minutes.

Campy BAP agar (pH 7.0)

Trypticase peptone	10.0
Thiotone	10.0
Dextrose	1.0
Yeast extract	2.0
Sodium chloride	5.0
Sodium bisulfide	0.1
Agar	15.0
Vancomycin	10.0 mg
Trimethoprim lactate	5.0 mg
Polymyxin B sulfate	2500.0 IU
Amphotericin B	2.0 mg
Cephalothin	15.0 mg
Defibrinated sheep blood	10.0%

Note: Aseptically add the antibiotics and defibrinated sheep blood to the sterile, molten, and cooled agar.

Chocolate agar (pH 7.0)

Proteose peptone	20.0
Dextrose	0.5
Sodium chloride	5.0
Disodium phosphate	5.0
Agar	15.0

Note: Aseptically add 5.0% defibrinated sheep blood to the sterile and molten agar. Heat at 80°C until a chocolate color develops.

Crystal violet agar (pH 7.0)

Bacto beef extract	3
Bacto peptone	5
Bacto crystal violet	0.00014
Bacto agar	15

Note: 1.0 ml of a crystal violet stock solution may be added to the base medium. Stock solution: 14 mg of crystal violet dye dissolved in 100 ml of distilled water.

Decarboxylase broth (Moeller) (pH 6.0)

Peptone	5.0
Beef extract	5.0
Dextrose	0.5
Bromcresol purple	0.01
Cresol red	0.005
Pyridoxal	0.005
Distilled water	1000.0 ml

To make amino acid-specific medium, add one of the amino acids below; dispense in 3- to 4-ml amounts and autoclave at 121°C for 10 minutes.

L-lysine dihydrochloride or L-arginine monohydrochloride or L-ornithine dihydrochloride	10 g/l

Deoxyribonuclease (DNase) agar (pH 7.3)

Deoxyribonucleic acid	2.0
Phytane	5.0
Sodium chloride	5.0
Trypticase	15.0
Agar	15.0

Endo agar (pH 7.5)

Peptone	10.0
Lactose	10.0
Dipotassium phosphate	3.5
Sodium sulfite	2.5
Basic fuchsin	0.4
Agar	15.0

Eosin–methylene blue agar (Levine) (pH 7.2)

Peptone	10.0
Lactose	5.0
Dipotassium phosphate	2.0
Agar	13.5
Eosin Y	0.4
Methylene blue	0.065

Gel diffusion agar

Sodium barbital buffer	100.0 ml
Noble agar	0.8

Glucose acetate yeast sporulation agar (pH 5.5)

Glucose	1
Yeast extract	2
Sodium acetate (with $3H_2O$)	5
Bacto agar	15

Glucose salts broth (pH 7.2)

Dextrose	5.0
Sodium chloride	5.0
Magnesium sulfate	0.2
Ammonium dihydrogen phosphate	1.0
Dipotassium hydrogen phosphate	1.0

Glycerol yeast extract agar supplemented with aureomycin (pH 7.0)

Glycerol	5.0 ml
Yeast extract	2.0
Dipotassium phosphate	1.0
Agar	15.0

Note: Aseptically add aureomycin, 10 μg per ml, to the sterile, molten, and cooled agar.

Grape juice broth

Commercial grape or apple juice	
Ammonium biphosphate	0.25%

Note: Sterilization not required when using a large yeast inoculum.

Hay infusion broth

Hay infusion broth preparations are prepared 1 week ahead of the laboratory session in which they will be used. Into a 2000-ml beaker place about 800 ml of water and two to three handfuls of dry grass or hay (obtained from a farm or storage barn). During the incubation period, the infusion should be aerated by passing air through a rubber tube attached to an air supply. This preparation is sufficient for a class and can be dispensed in 50-ml beakers.

Inorganic synthetic broth (pH 7.2)

Sodium chloride	5.0
Magnesium sulfate	0.2
Ammonium dihydrogen phosphate	1.0
Dipotassium hydrogen phosphate	1.0

KF broth (pH 7.2)

Polypeptone	10.0
Yeast extract	10.0
Sodium chloride	5.0
Sodium glycerophosphate	10.0
Sodium carbonate	0.636
Maltose	20.0
Lactose	1.0
Sodium azide	0.4
Phenol red	0.018

Lactose fermentation broth 1× and 2×* (pH 6.9)

Beef extract	3.0
Peptone	5.0
Lactose	5.0

*For 2× broth use twice the concentration of the ingredients.

Litmus milk (pH 6.8)

Skim milk powder	100.0
Litmus	0.075

Note: Autoclave at 12 lb pressure for 15 minutes.

Luria-Bertani (Miller) agar base (pH 7.0)

Pancreatic digest of casein	10.0
Yeast extract	5.0
Sodium chloride	0.5
Agar	15.0

Luria-Bertani (Miller) broth (pH 7.0)

Tryptone	10.0
Yeast extract	5.0
Sodium chloride	10.0

MacConkey agar (pH 7.1)

Bacto peptone	17.0
Proteose peptone	3.0
Lactose	10.0
Bile salts mixture	1.5
Sodium chloride	5.0
Agar	13.5
Neutral red	0.03
Crystal violet	0.001

Mannitol salt agar (pH 7.4)

Beef extract	1.0
Peptone	10.0
Sodium chloride	75.0
d-Mannitol	10.0
Agar	15.0
Phenol red	0.025

m-Endo broth (pH 7.5)

Yeast extract	6.0
Thiotone peptone	20.0
Lactose	25.0
Dipotassium phoshate	7.0
Sodium sulfite	2.5
Basic fuchsin	1.0

Note: Heat until boiling; do not autoclave.

m-FC broth (pH 7.4)

Biosate peptone	10.0
Polypeptone peptone	5.0
Yeast extract	3.0
Sodium chloride	5.0
Lactose	12.5
Bile salts	1.5
Aniline blue	0.1

Note: Add 10 ml of rosolic acid (1% in 0.2N sodium hydroxide). Heat to boiling with agitation; do not autoclave.

Milk agar (pH 7.2)

Skim-milk powder	100.0
Peptone	5.0
Agar	15.0

Note: Autoclave at 12 lb pressure for 15 minutes.

Minimal agar (pH 7.0)
Minimal agar, supplemented with streptomycin and thiamine*

Solution A (pH 7.0)

Potassium dihydrogen phosphate	3.0
Disodium hydrogen phosphate	6.0
Ammonium chloride	2.0
Sodium chloride	5.0
Distilled water	800.0 ml

Solution B (pH 7.0)

Glucose	8.0
Magnesium sulfate · 7H$_2$O	0.1
Agar	15.0
Distilled water	200.0 ml

Note: Autoclave Solutions A and B separately and combine.

*To Solution B, add 0.001 g of thiamine prior to autoclaving. To the combined sterile and molten medium, add 50 mg (1 ml of 50 mg per ml) sterile streptomycin solution before pouring agar plates.

MR-VP broth (pH 6.9)

Peptone	7.0
Dextrose	5.0
Potassium phosphate	5.0

Mueller-Hinton agar (pH 7.4)

Beef, infusion	300.0
Casamino acids	17.5
Starch	1.5
Agar	17.0

Mueller-Hinton tellurite agar (pH 7.4)

Casamino acids	20.0
Casein	5.0
L-tryptophane	0.05

Potassium dihydrogen phosphate	0.3
Magnesium sulfate	0.1
Agar	20.0

Note: Aseptically add 12.5 ml of tellurite serum to the sterile, 50°C molten agar.

Nitrate broth (pH 7.2)

Peptone	5.0
Beef extract	3.0
Potassium nitrate	5.0

Nitrite broth (pH 7.3)

Sodium nitrite	2.0
Magnesium sulfate · 7H$_2$O	0.5
Ferric sulfate · 7H$_2$O	0.03
Sodium chloride	0.3
Sodium carbonate	1.0
Dipotassium hydrogen sulfate	1.0

Nitrogen-free mannitol agar* and broth (pH 7.3)

Mannitol	15.0
Dipotassium hydrogen phosphate	0.5
Magnesium sulfate	0.2
Calcium sulfate	0.1
Sodium chloride	0.2
Calcium carbonate	5.0
*Agar	15.0

Nutrient agar* and broth (pH 7.0)

Peptone	5.0
Beef extract	3.0
*Agar	15.0

Nutrient gelatin (pH 6.8)

Peptone	5.0
Beef extract	3.0
Gelatin	120.0

Peptone broth (pH 7.2)

Peptone	4.0

Phenol red dextrose broth (pH 7.3)

Trypticase	10.0
Dextrose	5.0
Sodium chloride	5.0
Phenol red	0.018

Note: Autoclave at 12 lb pressure for 15 minutes.

Phenol red inulin broth (pH 7.3)

Trypticase	10.0
Inulin	5.0
Sodium chloride	5.0
Phenol red	0.018

Note: Autoclave at 12 lb pressure for 15 minutes.

Phenol red lactose broth (pH 7.3)

Trypticase	10.0
Lactose	5.0
Sodium chloride	5.0
Phenol red	0.018

Note: Autoclave at 12 lb pressure for 15 minutes.

Phenol red sucrose broth (pH 7.3)

Trypticase	10.0
Sucrose	5.0
Sodium chloride	5.0
Phenol red	0.018

Note: Autoclave at 12 lb pressure for 15 minutes.

Phenylalanine agar (pH 7.3)

Yeast extract	3.0
Dipotassium phosphate	1.0
Sodium chloride	5.0
DL-phenylalanine	2.0
Bacto agar	12.0
Distilled water	1000.0 ml

Note: Completely dissolve ingredients in boiling water. Dispense in tubes, autoclave, and cool in slanted position.

Phenylethyl alcohol agar (pH 7.3)

Trypticase	15.0
Phytane	5.0
Sodium chloride	5.0
β-Phenylethyl alcohol	2.0
Agar	15.0

Potato dextrose agar (pH 5.6)

Infusion from potatoes	200.0
Bacto dextrose	20.0
Bacto agar	15.0

Sabouraud agar (pH 5.6)
Sabouraud agar supplemented with aureomycin*

Peptone	10.0
Dextrose	40.0
Agar	15.0

*Aseptically add aureomycin, 10 μg per ml, to the sterile, molten, and cooled medium.

Salt medium—*Halobacterium*

Sodium chloride	250.0
Magnesium sulfate · 7H$_2$O	10.0
Potassium chloride	5.0
Calcium chloride · 6H$_2$O	0.2
Yeast extract	10.0
Tryptone	2.5
Agar	20.0

Note: The quantities given are for preparation of 1-liter final volume of the medium. In preparation, make up two solutions, one involving the yeast extract and tryptone and the other the salts. Adjust the pH of the nutrient solution to 7. Sterilize separately. Mix and dispense aseptically.

SIM agar (pH 7.3)

Peptone	30.0
Beef extract	3.0
Ferrous ammonium sulfate	0.2
Sodium thiosulfate	0.025
Agar	3.0

Simmons citrate agar (pH 6.9)

Ammonium dihydrogen phosphate	1.0
Dipotassium phosphate	1.0
Sodium chloride	5.0
Sodium citrate	2.0
Magnesium sulfate	0.2
Agar	15.0
Bromthymol blue	0.08

Snyder test agar (pH 4.8)

Tryptone	20.0
Dextrose	20.0
Sodium chloride	5.0
Bromcresol green	0.02
Agar	20.0

Sodium chloride agar, 7.5% (pH 7.0)

Bacto beef extract	3.0
Bacto peptone	5.0
Sodium chloride	7.5
Bacto agar	15.0

Sodium chloride broth, 6.5% (pH 7.0)

Brain–heart infusion broth	100.0 ml
Sodium chloride	6.5

Starch agar (pH 7.0)

Peptone	5.0
Beef extract	3.0
Starch (soluble)	2.0
Agar	15.0

Thioglycollate, fluid (pH 7.1)

Peptone	15.0
Yeast extract	5.0
Dextrose	5.0
L-cystine	0.75
Thioglycollic acid	0.3 ml
Agar	0.75
Sodium chloride	2.5
Resazurin	0.001

Tinsdale agar (pH 7.4)

Proteose peptone, No. 3	20.0
Sodium chloride	5.0
Agar	20.0

Note: Following boiling, distribute in 100-ml flasks. Autoclave, cool to 55°C, add 15 ml of rehydrated Tinsdale enrichment to each 100 ml, and mix thoroughly before dispensing.

Top agar (for Ames test)

Sodium chloride	5.0
Agar	6.0

Tributyrin agar (pH 7.2)

Peptone	5.0
Beef extract	3.0
Agar	15.0
Tributyrin	10.0

Note: Dissolve peptone, beef extract, and agar while heating. Cool to 90°C, add the tributyrin, and emulsify in a blender.

Triple sugar–iron agar (pH 7.4)

Beef extract	3.0
Yeast extract	3.0
Peptone	15.0
Proteose peptone	5.0
Lactose	10.0
Saccharose	10.0
Dextrose	1.0
Ferrous sulfate	0.2
Sodium chloride	5.0
Sodium thiosulfate	0.3
Phenol red	0.024
Agar	12.0

Trypticase nitrate broth (pH 7.2)

Trypticase	20.0
Disodium phosphate	2.0
Dextrose	1.0
Agar	1.0
Potassium nitrate	1.0

Trypticase soy agar (pH 7.3)

Trypticase	15.0
Phytane	5.0
Sodium chloride	5.0
Agar	15.0

Tryptone agar* and broth

Tryptone	10.0
Calcium chloride (reagent)	0.01–0.03 M
Sodium chloride	5.0
*Agar	11.0

Tryptone soft agar

Tryptone	10.0
Potassium chloride (reagent)	5.0 ml
Agar	9.0

Urea broth

Urea broth concentrate (filter-sterilized solution)	10.0 ml
Sterile distilled water	90.0 ml

Note: Aseptically add the urea broth concentrate to the sterilized and cooled distilled water. Under aseptic conditions, dispense 3-ml amounts into sterile tubes.

Yeast extract broth (pH 7.0)

Peptone	5.0
Beef extract	3.0
Sodium chloride	5.0
Yeast extract	5.0

Biochemical Test Reagents

Barritt's reagent, for detection of acetylmethylcarbinol
Solution A

Alpha-naphthol	5.0 g
Ethanol, absolute	95.0 ml

Note: Dissolve the alpha-naphthol in the ethanol with constant stirring.
Solution B

Potassium hydroxide	40.0 g
Creatine	0.3 g
Distilled water	100.0 ml

Note: Dissolve the potassium hydroxide in 75 ml of distilled water. The solution will become warm. Allow to cool to room temperature. Add the creatine and stir to dissolve. Add the remaining water. Store in a refrigerator.

Biotin-histidine solution, for Ames test

l-Histidine HCl	0.5 mM
Biotin	0.5 mM
Distilled water	10.0 ml

Buffered glycerol (pH 7.2), for immunofluorescence

Glycerin	90.0 ml
Phosphate buffered saline	10.0 ml

Diphenylamine reagent, for detection of nitrates
Dissolve 0.7 g diphenylamine in a mixture of 60 ml concentrated sulfuric acid and 28.8 ml of distilled water. Cool and slowly add 11.3 ml of concentrated hydrochloric acid. Allow to stand for 12 hours. Sedimentation indicates that the reagent is saturated.

Ferric chloride reagent

Ferric chloride	10.0 g
Distilled water	100.0 ml

Gram's iodine, for detection of starch
As in Gram's stain

Hydrogen peroxide, 3%, for detection of catalase activity
Note: Refrigerate when not in use.

Kovac's reagent, for detection of indole

p-Dimethylaminobenzaldehyde	5.0 g
Amyl alcohol	75.0 ml
Hydrochloric acid (concentrated)	25.0 ml

Note: Dissolve the p-dimethylaminobenzaldehyde in the amyl alcohol. Add the hydrochloric acid.

McFarland Barium Sulfate Standards, for API®
Staph-Ident procedure
Prepare 1% aqueous barium chloride and 1% aqueous sulfuric acid solutions. Using the following table, add the amounts of barium chloride and sulfuric acid to clean 15- × 150-mm screw-capped test tubes. Label the tubes 1 through 10.

Preparation of McFarland Standards

Tube	Barium Chloride 1% (ml)	Sulfuric Acid 1% (ml)	Corresponding Approximate Density of Bacteria (million/ml)
1	0.1	9.9	300
2	0.2	9.8	600
3	0.3	9.7	900
4	0.4	9.6	1,200
5	0.5	9.5	1,500
6	0.6	9.4	1,800
7	0.7	9.3	2,100
8	0.8	9.2	2,400
9	0.9	9.1	2,700
10	1.0	9.0	3,000

Methyl cellulose, for microscopic observation of protozoa

Methyl cellulose	10.0 g
Distilled water	90.0 ml

Methyl red solution, for detection of acid

Methyl red	0.1 g
Ethyl alcohol	300.0 ml
Distilled water	200.0 ml

Note: Dissolve the methyl red in the 95% ethyl alcohol. Dilute to 500 ml with distilled water.

Nessler's reagent, for detection of ammonia

Potassium iodide	50.0 g
Distilled water (ammonia-free)	35.0 ml

Add saturated aqueous solution of mercuric chloride until a slight precipitate persists.

Potassium hydroxide (50% aqueous)	400.0 ml

Note: Dilute to 1000 ml with ammonia-free distilled water. Let stand for 1 week, decant supernatant liquid, and store in a tightly capped amber bottle.

Nitrate test solution, for detection of nitrites

Solution A, Sulfanilic acid

Sulfanilic acid	8.0 g
Acetic acid, 5 N: 1 part glacial acetic acid to 2.5 parts distilled water	1000.0 ml

Solution B, Alpha-naphthylamine

Alpha-naphthylamine	5.0 g
Acetic acid, 5 N	1000.0 ml

Orthonitrophenyl-β-D-galactoside (ONPG), for enzyme induction

0.1 M sodium phosphate buffer (pH 7.0)	50.0 ml
ONPG (8×10^{-4} M)	12.5 mg

p-Aminodimethylaniline oxalate, for detection of oxidase activity

p-Aminodimethylaniline oxalate	0.5 g
Distilled water	50.0 ml

Note: To dissolve fully, gently warm the solution.

Phosphate-buffered saline, 1% (pH 7.2–7.4), for immunofluorescence

Solution A

Disodium phosphate	1.4 g
Distilled water	100.0 ml

Solution B

Sodium dihydrogen phosphate	1.4 g
Distilled water	100.0 ml

Note: Add 84.1 ml of Solution A to 15.9 ml of Solution B. Add 8.5 g of sodium chloride and q.s. to 1 liter.

Rabbit plasma, for detection of coagulase activity

Note: Store vials at 2°C to 8°C. Reconstitute by the addition of 7.5 ml of sterile water.

Sodium barbital buffer, for immunofluorescence

Sodium barbital	6.98 g
Sodium chloride	6.0 g
1 N hydrochloric acid	27.0 ml
Distilled water, q.s. to 1000 ml	

Toluidine blue solution, 0.1%, for detection of DNase activity

1% toluidine blue solution	0.1 ml
Distilled water	99.9 ml

Trommsdorf's reagent, for detection of nitrite

Slowly add 100 ml of 20% aqueous zinc chloride solution to a mixture of 4 g of starch in water. Heat until the starch is dissolved as much as possible and the solution is almost clear. Dilute with water and add 2 g of potassium iodide. Dilute to 1000 ml, filter, and store in an amber bottle.

Staining Reagents

Acid-Fast Stain

Carbol fuchsin (Ziehl's)
Solution A
Basic fuchsin (90% dye content)	0.3 g
Ethyl alcohol (95%)	10.0 ml

Solution B
Phenol	5.0 g
Distilled water	95.0 ml

Note: Mix Solutions A and B. Add 2 drops of Triton X per 100 ml of stain for use in heatless method.

Acid Alcohol
Ethyl alcohol (95%)	97.0 ml
Hydrochloric acid	3.0 ml

Methylene blue
Methylene blue	0.3 g
Distilled water	100.0 ml

Capsule Stain

Crystal violet (1%)
Crystal violet (85% dye content)	1.0 g
Distilled water	100.0 ml

Copper sulfate solution (20%)
Copper sulfate ($CuSO_4 \cdot 5H_2O$)	20.0 g
Distilled water	80.0 ml

Fungal Stains

Lactophenol–cotton-blue solution
Lactic acid	20.0 ml
Phenol	20.0 g
Glycerol	40.0 ml
Distilled water	20.0 ml
Aniline blue	0.05 g

Note: Heat gently in hot water (double boiler) to dissolve; then add aniline blue dye.

Water–iodine solution
Gram's iodine (as in Gram's stain)	10.0 ml
Distilled water	30.0 ml

Gram Stain

Crystal violet (Hucker's)
Solution A
Crystal violet (90% dye content)	2.0 g
Ethyl alcohol (95%)	20.0 ml

Solution B
Ammonium oxalate	0.8 g
Distilled water	80.0 ml

Note: Mix Solutions A and B.

Gram's iodine
Iodine	1.0 g
Potassium iodide	2.0 g
Distilled water	300.0 ml

Ethyl alcohol (95%)
Ethyl alcohol (100%)	95.0 ml
Distilled water	5.0 ml

Safranin
Safranin O	0.25 ml
Ethyl alcohol (95%)	10.0 ml
Distilled water	100.0 ml

Negative Stain

Nigrosin
Nigrosin, water-soluble	10.0 g
Distilled water	100.0 ml

Note: Immerse in boiling water bath for 30 minutes.
Formalin	0.5 ml

Note: Filter twice through double filter paper.

Spore Stain

Malachite green
Malachite green	5.0 g
Distilled water	100.0 ml

Safranin
Same as in Gram stain

Experimental Microorganisms

Cultures

Bacteria

Alcaligenes faecalis
Alcaligenes viscolactis
Aquaspirillum itersonii
Bacillus cereus
Bacillus megaterium
Bacillus stearothermophilus
Branhamella catarrhalis
Citrobacter freundii
Citrobacter intermedius
Clostridium butyricum
Clostridium sporogenes
Corynebacterium kutscheri
Corynebacterium xerosis
Enterobacter aerogenes
Enterococcus faecalis
Escherichia coli
Escherichia coli ATCC™ e 23724
Escherichia coli ATCC e 23725
Escherichia coli ATCC e 23735
Escherichia coli ATCC e 23740
Escherichia coli B
Halobacterium salinarium
Klebsiella ozaenae
Klebsiella pneumoniae ATCC e 15574
Lactobacillus casei
Lactobacillus delbrueckii
Lactobacillus fermenti
Lactococcus lactis
Leuconostoc mesenteroides
Micrococcus luteus
Micrococcus varians
Moraxella bovis
Mycobacterium smegmatis
Neisseria mucosa
Neisseria sicca
Proteus inconstans
Proteus mirabilis
Proteus rettgeri

Proteus vulgaris
Pseudomonas aeruginosa
Pseudomonas fluorescens
Pseudomonas mallei
Pseudomonas savastanoi
Salmonella typhimurium
Salmonella typhimurium ATCC e 29631
Serratia marcescens
Shigella dysenteriae
Staphylococcus aureus ATCC e 25923
Staphylococcus aureus ATCC e 27659
Staphylococcus aureus ATCC e 27660
Staphylococcus aureus ATCC e 27661
Staphylococcus aureus ATCC e 27691
Staphylococcus aureus ATCC e 27693
Staphylococcus aureus ATCC e 27697
Staphylococcus epidermidis
Staphylococcus saprophyticus ATCC e 15305
Streptococcus agalactiae
Streptococcus bovis
Streptococcus mitis
Streptococcus pneumoniae
Streptococcus pyogenes ATCC 12385
Streptococcus var. Lancefield group E
Streptomyces griseus

Fungi

Alternaria sp.
Aspergillus niger
Candida albicans
Cephalosporium sp.
Cladosporium sp.
Fusarium sp.
Mucor mucedo
Penicillium notatum
Rhizopus stolonifer
Rhodotorula rubra
Saccharomyces cerevisiae
Saccharomyces cerevisiae var. *ellipsoideus*
Schizosaccharomyces octosporus
Selenotila intestinalis

Viruses

T$_2$ coliphage

Prepared Slides

Bacteria

Aquaspirillum itersonii
Bacillus subtilis
Spirillum itersonii
Staphylococcus aureus

Fungi

Saccharomyces cerevisiae
Treponema denticola

Protozoa

Balantidium coli
Entamoeba histolytica
Giardia lamblia
Plasmodium vivax
Trypanosoma gambiense

Other

Blood smear

Page references followed by *fig* indicate an illustrated figure; those followed by *t* indicate a table.

Abbé condenser, 29
abscesses, 415
absorption (virus infection), 245
acid-alcohol, 77
acid-fast stain (Ziehl-Neelsen method), 77, 78, 79*fig*
acidic stains, 54*fig*–55*fig*
acne, 415
adaptive enzymes, 373
aerial mycelium, 225
aerobes, 117
aerotolerant anaerobes, 117
agar
  defining, 1
  differential media, 101–102
  enriched media, 102
  selective media, 101
agar deep tubes, 2
agar plates
  described, 2
  sensitivity method for evaluating antiseptic agents in, 297–298
agar slants, 2
agglutination, 467
agglutinins, 467
AIDS, 499
algae, 39*fig*
alkaline reaction, 180
alpha-hemolysis, 102, 423
Ames test, 389–391*fig*
amino acids
  casein hydrolysis and, 146
  gelatin hydrolysis and, 146
  proteolysis (peptonization) and, 180
  utilization of, 197–199*fig*
ammonification, 341, 342*fig*
amylase, 145
anaerobic microorganism cultivation
  GasPak system for, 125*fig*
  methods for, 124*fig*
  principles of, 123
  procedure for, 124–125
Anthony method capsule staining, 85–87, 86*fig*
antibiotic resistance microorganisms, 385–386

antibiotics
  formation of, 467
  prototype, 279*t*
  soil microorganisms
    isolation of, 359–361, 360*fig*
    role in producing, 340
  See also penicillin
antibodies
  Forssman heterophile, 495
  non-Forssman heterophile, 495
antibody tests
  for *Chlamydia* genus, 502
  differential adsorption test, 495
  ELISA, 491–493
  febrile test, 480
  fluorescent antibody technique, 483*fig*–484
  heterophile antibody test, 495
  latex agglutination test, 487–488
  Mono-Test for IM, 495–496
  RPR (rapid plasma reagin test) for syphilis, 499–500
  for STDs, 499–502
  VDRL (Venereal Disease Research Laboratory), 499–500
  See also pathogens
antigen (immunogen), 467
antimetabolite, 280
antimicrobial spectrum of isolates, 361
antiseptics
  agar plate–sensitivity method, 297–298
  defining, 261
  efficiency of, 296
  listed, 293*t*–295*t*
  phenol coefficient test, 296–297*t*
API (Analytical Profile Index) System, 416, 435*fig*–436, 437
Ascomycetes, 223, 224*t*
ascospores, 233
ascus, 233
atmospheric oxygen requirements
  principles of, 117
  procedure for determining, 118, 119*fig*
autoclave, 264*fig*
autotroph cultivation, 93

Azotobacter species
  characteristics of, 345*t*
  nitrogen fixation and, 344–345
  procedure for isolation of, 345–346

bacillary dysentery, 321
Bacitracin test, 424
Bactercult, 449*t*, 450
bacteremia, 453
bacteria
  classified by temperature requirements, 109
  found in pond water, 39*fig*
  identifying
    gram-negative, 462*fig*, 463*fig*
    gram-positive, 460*fig*–461*fig*
    unknown bacteria cultures, 203–204*t*, 205, 459–463*fig*
bacterial conjugation
  defining, 396
  principle of, 379
  procedure for, 380
bacterial count analysis, 309–310*fig*
bacterial genetics
  Ames test for chemical carcinogenicity, 389–391*fig*
  bacterial conjugation, 379*fig*–380
  enzyme induction, 373–375, 374*fig*
  introduction to, 371–372
  mutations
    isolation of streptomycin-resistant, 385–386
    types of, 371–372
  transformation, 379, 395–398*fig*, 399
bacterial growth curve
  graph on, 135*fig*
  indirect method of determining, 136*fig*
  method for mixing sample in dilution bottle, 137*fig*
  principles of, 135–136
  procedure for, 137
  spectrophotometric/dilution-plating procedure for, 138*fig*